Heavy Equipment Operations

Level 1

FOURTH EDITION

Pearson

NCCER National Center for Construction Education and Research

NCCER
President and Chief Executive Officer: Boyd Worsham
Vice President of Innovation and Advancement: Jennifer Wilkerson
Chief Learning Officer: Lisa Strite
Senior Curriculum Leader: Chris Wilson
Technical Writing Manager: Gary Ferguson
Technical Writer: Nick Goodnight, PhD
Art Manager: Bree Rodriguez
Technical Illustrators: Judd Ivines, Liza Wailes
Production Artist: Chris Kersten
Permissions Specialists: Adam Black, Sherry Davis
Project Coordinator: Colleen Duffy

Pearson
Commercial Product Manager – Associations: Andy Dunaway
Senior Digital Content Producer: Shannon Stanton
Associate Project Manager: Monica Perez-Kim
Content Producer: Alexandrina Wolf, Mary Roth
Executive Marketing Manager: Mark Marsden
Designer: Mary Siener
Rights and Permissions: Kaitlyn Coddington
Composition: Integra Software Services
Printer/Binder: Lakeside Book Company
Cover Printer: Lakeside Book Company
Text Typefaces: Palatino LT Pro and Helvetica Neue

Cover Image
Bobcat skid steer loaders are built to help operators accomplish more by tackling the most difficult jobs in construction, landscaping, agriculture, forestry, snow removal, and more. Paired with a huge family of nearly 200 genuine Bobcat attachments, each skid steer model offers operators more ways to conquer challenges on your jobsite. In the cover photo, the operator uses the Bobcat S650 skid steer loader with the Bobcat industrial grapple attachment to move material. As the inventor of the original skid steer loader, Bobcat remains the world's best-selling skid steer loader brand.

10 9 8 7 6 5 4 3 2 1

7 2025

Pearson

NCCER National Center for Construction Education and Research

ISBN-10: 0-13-830422-X
ISBN-13: 978-0-13-830422-5

PREFACE

To the Trainee

Welcome to the world of heavy equipment operations. Equipment operators work on a variety of projects, including the construction of buildings, roads, and bridges; mining and timber operations; and in many other sectors of the industry. New construction and infrastructure projects will continue to increase the demand for qualified equipment operators. The skills provided by equipment operators are crucial to clearing and grading sites, moving materials, and other earthmoving operations.

There are nearly a half million heavy equipment operators in the United States, but that number is expected to grow by nearly 40,000 by 2032. Operators work on a variety of equipment from small and nimble skid steers to big earthmoving motor graders and dozers. If you have the desire to move dirt, this program offers plenty of opportunity to get behind the controls and operate big machinery.

New with *Heavy Equipment Operations Level 1*

NCCER is proud to present *Heavy Equipment Operations Level 1*. In this new edition, we have added two new modules. *Introduction to GPS/GNSS* covers the use of GPS equipment controls for site layout and other earthmoving tasks. *Equipment Operation, Inspection, and Readiness* covers the walk-around inspection process and startup procedures for heavy equipment. The *Skid Steers* module was moved from Level 2 to Level 1 to offer trainees another opportunity to learn on real equipment. All of the art and images in the program have been updated to reflect modern equipment and safety components. This revision also includes less focus on equipment repair and maintenance, as that is typically the job of the diesel mechanic.

We wish you success as you progress through this training program. If you have any comments on how NCCER might improve upon this textbook, please complete the User Update form using the QR code on this page. NCCER appreciates and welcomes its customers' feedback. You may submit yours by emailing **support@nccer.org**. When doing so, please identify feedback on this title by listing *#Heavy Equipment Operations L1* in the subject line.

Our website, **www.nccer.org**, has information on the latest product releases and training.

SCAN ME

NCCER Standardized Curricula

NCCER is a not-for-profit 501(c)(3) education foundation established in 1996 by the world's largest and most progressive construction companies and national construction associations. It was founded to address the severe workforce shortage facing the industry and to develop standardized training processes and curricula. Today, NCCER is supported by hundreds of leading construction and maintenance companies, manufacturers, and national associations. The NCCER Standardized Curricula were developed by NCCER in partnership with Pearson, the world's largest educational publisher.

Some features of the NCCER Standardized Curricula are as follows:

- An industry-proven record of success
- Curricula developed by the industry, for the industry
- National standardization providing portability of learned job skills and educational credits
- Compliance with the Office of Apprenticeship requirements for related classroom training (*CFR 29:29*)
- Well-illustrated, up-to-date, and practical information

NCCER maintains a secure online database that provides certificates, digital credentials, transcripts, and wallet cards to individuals who successfully complete programs under an NCCER-accredited organization or through one of

NCCER's self-paced, online programs. This system also allows individuals and employers to track and verify industry-recognized credentials and certifications in real time.

For information on NCCER's credentials, contact NCCER Customer Service at 1-888-622-3720 or visit **https://www.nccer.org**.

Digital Credentials

Show off your industry-recognized credentials online with NCCER's digital credentials!

NCCER is now providing online credentials. Transform your knowledge, skills, and achievements into digital credentials that you can share across social media platforms, send to your network, and add to your resume. For more information, visit **www.nccer.org**.

Cover Image

Since 1958, Bobcat Company has empowered people to accomplish more. As a leading global manufacturer, Bobcat has a proud legacy of innovation, delivering smart solutions to customers' toughest challenges. Backed by the support of a global dealership network, Bobcat offers an extensive line of worksite solutions, including loaders, excavators, tractors, utility vehicles, telehandlers, mowers, turf renovation equipment, light compaction, portable power, industrial air, forklifts and industrial vehicles, attachments, implements, parts, and services.

With its North American headquarters in West Fargo, North Dakota, Bobcat leads the industry with its innovative offerings designed to transform how the world works, builds cities, and supports communities for a more sustainable future. The Bobcat brand is owned by Doosan Bobcat, Inc., a company within Doosan Group.

DESIGN FEATURES

Content is organized and presented in a functional structure that allows trainees to access the information where they need it.

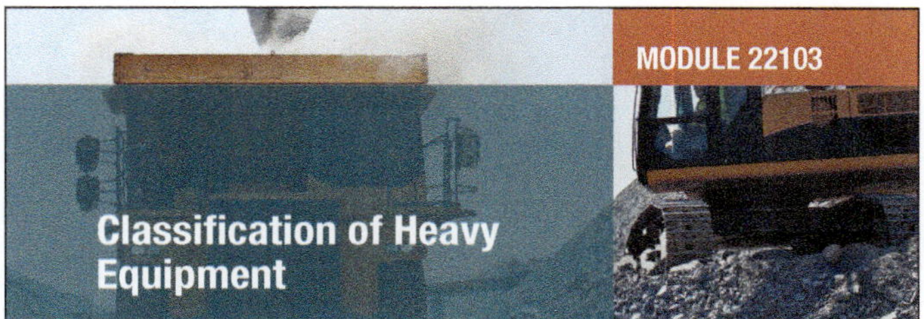

Source: Martin Barraud/iStock/Getty Images

Objectives

Successful completion of this module prepares you to do the following:

1. Identify and describe heavy equipment used in construction and mining.
   a. Identify and describe equipment used in excavating and grading construction sites.
   b. Identify and describe trenching equipment.
   c. Identify and describe heavy equipment used in mining.
2. Explain hydraulic systems and their components.
   a. Explain hydraulic system fundamentals.
   b. Identify and describe the equipment and components used in hydraulic systems.

Performance Tasks

Under supervision, you should be able to do the following:

1. Identify types of heavy equipment and explain their uses.
2. Identify the basic parts of each type of equipment and explain the differences in models among the types of equipment.

Overview

There are many different types of heavy equipment. Each machine is typically designed for a particular type of work, but many can be modified with attachments to perform a variety of functions. It is important for heavy equipment operators to know the capabilities and limitations of the machines they operate. They also must be familiar with the various attachments, and how to properly use them. When heavy equipment operators gain this type of knowledge, they will have additional opportunities for advancement.

NCCER Industry-Recognized Credentials

If you are training through an NCCER-accredited sponsor, you may be eligible for credentials from NCCER. The ID number for this module is 22103. Note that this module may have been used in other NCCER curricula and may apply to other level completions. Contact NCCER at 1.888.622.3720 or go to **www.nccer.org** for more information.

You can also show off your industry-recognized credentials online with NCCER's digital credentials. Transform your knowledge, skills, and achievements into credentials that you can share across social media platforms, send to your network, and add to your resume. For more information, visit **www.nccer.org**.

Digital Resources for Heavy Equipment Operations

SCAN ME

Scan this code using the camera on your phone or mobile device to view the digital resources related to this craft.

Trainees can navigate *Classification of Heavy Equipment* using color coded tabs on the upper right hand corner.

The Objectives list the skills and knowledge trainees need in order to complete the module successfully.

The Performance Tasks give you an opportunity to apply your knowledge to real-world tasks.

Section Openers provide a visual organizational structure for the information. Objectives and Performance Tasks are broken out for each section.

Performance Tasks	1.0.0 Work Zone Safety Objective
1. Place barricades and temporary traffic control devices for a highway construction zone. 2. Demonstrate how to use flags or paddles to control traffic.	Explain work zone safety requirements. a. Describe signs, barriers, and other devices and methods used for work zone safety. b. Describe traffic control safety methods.

2.1.0 Site Planning for GNSS Use

When surveying a site to set up a **universal total station (UTS)** for use with satellite-guided equipment, extensive staking and string lines are no longer needed. Using the UTS and GNSS navigation, the piece of equipment can easily be navigated to the proper locations and the blade can be angled exactly to create the proper grade specified by the site plans. The key to this type of guidance is setting up the site to be **geospatial information system (GIS)** compliant.

When surveying the site, the creation of X, Y, and Z coordinates (*Figure 14*) based on the GNSS location allows the surveyor to generate the needed information for the equipment to operate on a 3D axis. These coordinates are put into the surveying software to generate commands for the equipment that will operate on the site. As the equipment is operated on the construction site, antennas located on various components of the machine receive a position from the UTS setup on the site. The operator of the equipment downloads a 3D topographical map of the site into the machine that it uses to reference compared to the UTS on the site.

Universal total station (UTS): A combination of a digital theodolite and an electronic distance measuring device that can integrate with a GNSS module to provide a point of reference for 3D site modeling.

Geospatial information system (GIS): Computerized tools related to the physical mapping of data within a visual reference point.

Trade Terms appear on the page adjacent to the text where they are first presented.

Step-by-step presentations and math equations help make the concepts clear and easy to grasp.

Step 4 Rotate the laser transmitter 180 degrees so that the –X side is aimed at the laser detector. Mark and record the elevation, noting that it is for the –X axis.

Step 5 Compare the +X and –X elevation readings. If the difference between the readings is less than $^3/_{32}$" (±2.38 mm), the X axis is within calibration. If within calibration, go to Step 9. If the difference in readings is more than $^3/_{32}$" (±2.38 mm), X-axis calibration is required, as described in Steps 6 through 8.

Step 6 To correct for a calibration error, locate a new mark midway between the +X and –X marks. Move the receiver on the rod until its center-marking notch is aligned with the new midpoint mark.

Step 7 Adjust the X-axis calibration screw to obtain an on-grade laser beam reading at the midpoint line. Most detectors have move-up and move-down indicators that show the direction the beams need to be moved up or down relative to the on-grade point.

Step 8 Rotate the laser transmitter 180 degrees back to the original +X face. The on-grade reading should be on or within $^3/_{64}$" (±1.19 mm) of the midpoint line. If not, repeat Steps 3 through 8.

SCAN ME

QR codes link trainees directly to digital resources that highlight current content.

Important information is highlighted, illustrated, and presented to facilitate learning.

Placement of images near the text description and details such as callouts and labels help trainees absorb information.

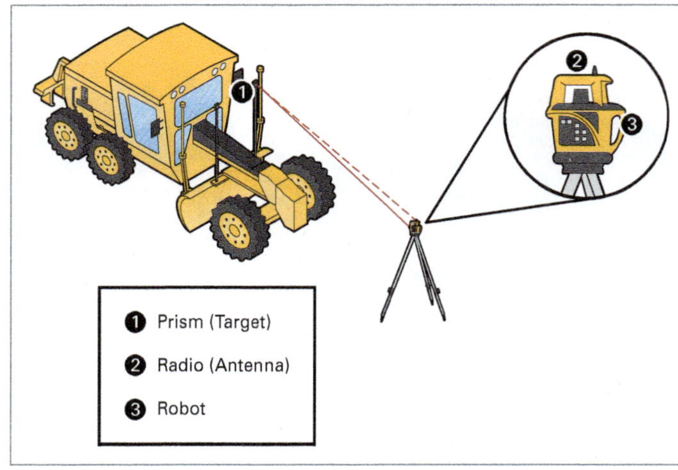

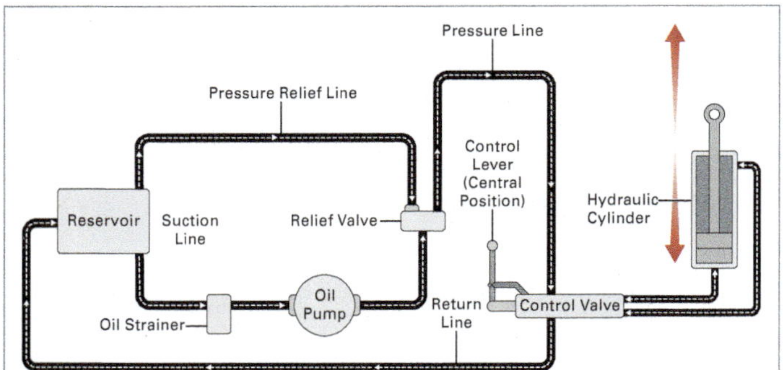

Hydraulic System

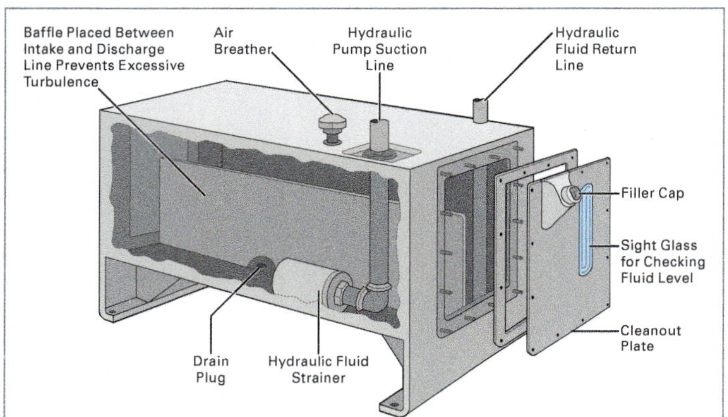

Reservoir Tank

(A) Motor Grader with One GPS Antenna

(B) GPS Base Station with Antenna and Radio Receiver

New boxes highlight safety and other important information for trainees. Warning boxes stress potentially dangerous situations, while Caution boxes alert trainees to dangers that may cause damage to equipment. Notes boxes provide additional information on a topic.

WARNING!

During the site development phase of a project, it is common to find many machines operating at the same time, as well as people who may be working near the machines. Operators must remain alert to their surroundings at all times in order to avoid accidents that could kill or injure personnel or damage expensive equipment.

CAUTION

Handle safety glasses and goggles with care. If they get scratched, replace them. The scratches will interfere with your vision. Clean the lenses regularly with lens tissues or a soft cloth.

NOTE

OSHA and MSHA require that approved seat belts and rollover protection structures (ROPS) be installed on virtually all heavy equipment. (Old equipment must be retrofitted.) Do not use any heavy equipment that is not equipped with these safety devices.

Going Green

Hydraulic Fluid

Hydraulic systems notoriously leak. The source of such leaks must be identified and stopped. Hydraulic spills must be prevented. Spilled hydraulic fluids harm the environment. If unable to prevent a hydraulic spill, follow SDS instructions and local laws for dealing with hydraulic fluid spills. Contaminated soil must be removed and disposed of in accordance with local guidelines.

Going Green looks at ways to preserve the environment, save energy, and make good choices regarding the health of the planet.

"Bull" Dozer

The original bulldozer was not a machine at all, and it had nothing to do with bulls. Rather, it was a blade pushed by mules. The first step in mechanizing this tool was accomplished when a tractor was converted to a continuous chain tread—or track crawler—format early in the twentieth century. The next step was to provide independent control for each track. The bulldozer came about when the blade was attached to a tractor a few years later. The final step occurred when a power control for the blade was added. Although the principle is still the same, dozers have come a long way since the early days.

These boxed features provide additional information that enhances the text.

Did You Know?

Latex

Latex is a common material for protective gloves, but latex gloves are not recommended for use around diesel fuel because diesel fuel dissolves latex.

Review questions at the end of each section and module allow trainees to measure their progress

3.0.0 Section Review

1. According to the definition of soil shear strength, soils with higher shear strengths _____.
   a. compact easier
   b. are less stable
   c. cannot form steep-sided slopes
   d. do not slide easily on an embankment

2. Which of the following statements about the logistics of soil compaction is true?
   a. Soil compaction works best with long, gentle precipitation.
   b. Soil compaction also shapes the grade to its specified profile and crown.
   c. Compaction occurs before soil stabilization to provide a firm base.
   d. Surface trimming shapes the compacted surface to accurate grade and crown specifications.

3. Which of the following is *not* a factor in determining the number of soil compaction passes required?
   a. Amount of compaction required
   b. Final layer thickness
   c. Initial soil density
   d. Moisture content

Module 22201 Review Questions

1. The solid earth layer that underlies the ground soil in the earth's crust is _____.
   a. the mantle
   b. common fill
   c. bedrock
   d. molten rock

2. The unique address given to a site that allows police, fire, and other emergency vehicles to find the location easily is called a(n) _____.
   a. fire road
   b. 911 address
   c. police locator
   d. emergency access

3. The soil type that aids cohesiveness in other soil types is _____.
   a. colloids
   b. silt
   c. sand
   d. gravel

NCCERconnect

This interactive online course is a unique web-based supplement in the form of an electronic book that provides a range of visual, auditory, and interactive elements to enhance training. Visit **www.nccerconnect.com** for more information!

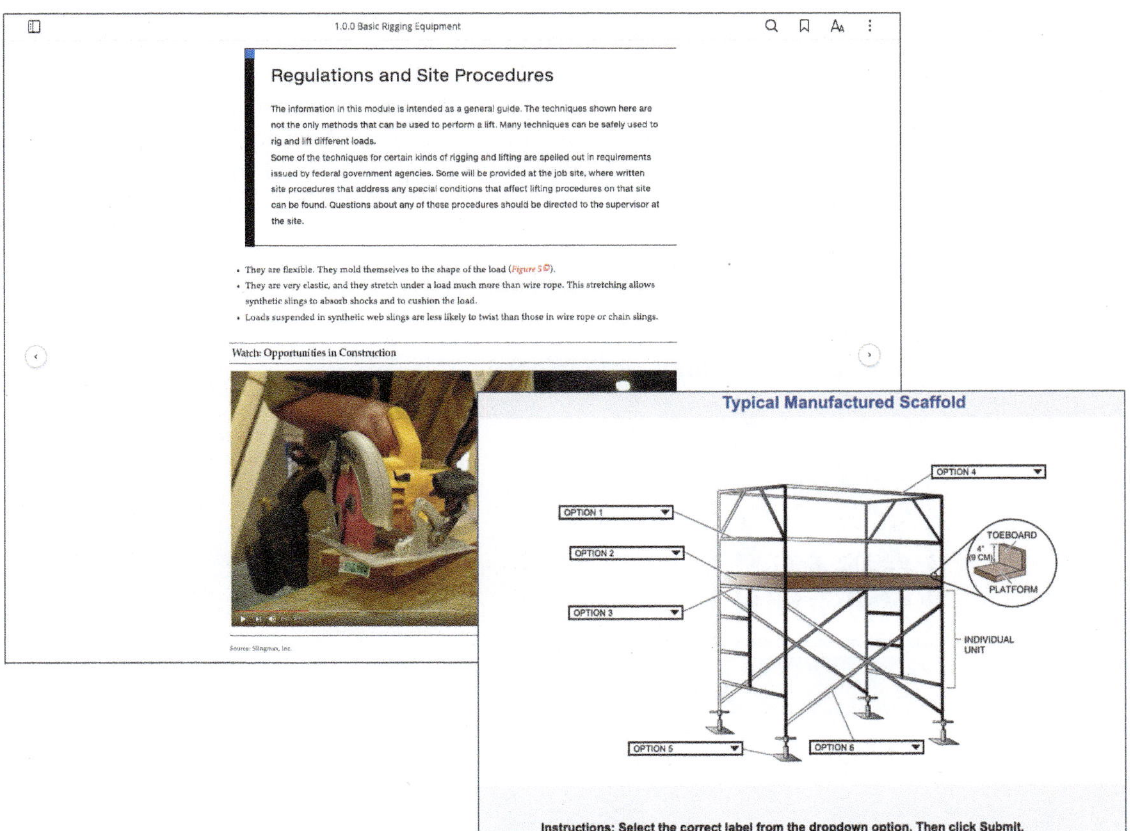

ACKNOWLEDGMENTS

This curriculum was revised as a result of the vision and leadership of the following sponsors:

AGC Inland Northwest Chapter
AGC New Mexico
Carolina Safety and Training
Fluor
Laurel Oaks Career Campus
Miami Valley Career Technology Center
North Arkansas College
T/J Inspection Training Center

This curriculum would not exist without the dedication and unselfish energy of those volunteers who served on the Authoring Team. A sincere thanks is extended to the following:

James Busch
Rusty Frommling
Chad Hathaway
Ron Kauffman
Joseph Montoya
Richard Oliver
Bert Rohrback
Rodney Scott
Tad Timmons
Dan Tucker

A final note: This book is the result of a collaborative effort involving the production, editorial, and development staff at Pearson Education, Inc., and NCCER. Thanks to all of the dedicated people involved in the many stages of this project.

NCCER PARTNERS

To see a full list of NCCER Partners, please visit **www.nccer.org/about-us/partners**.

NCCER

SCAN ME

HEO INTERIM CREDENTIALS

Credential Name	Module ID	Module Title	Level	Hours
Backhoe	22101	Introduction to Heavy Equipment	1	5
	22102	Heavy Equipment Operation Safety	1	15
	22108	Equipment Operation, Inspection, and Readiness	1	25
	22201	Introduction to Earthmoving	1	12.5
	22308	Soils	2	10
	22210	Site Work	2	25
	22303	Backhoes	2	30
			Total Hours	**122.5**

Credential Name	Module ID	Module Title	Level	Hours
Compaction Equipment	22101	Introduction to Heavy Equipment	1	5
	22102	Heavy Equipment Operation Safety	1	15
	22108	Equipment Operation, Inspection, and Readiness	1	25
	22308	Soils	2	10
	22210	Site Work	2	25
	22203	Compaction Equipment	3	25
			Total Hours	**105**

Credential Name	Module ID	Module Title	Level	Hours
Dozer	22101	Introduction to Heavy Equipment	1	5
	22102	Heavy Equipment Operation Safety	1	15
	22108	Equipment Operation, Inspection, and Readiness	1	25
	22201	Introduction to Earthmoving	1	12.5
	22109	Introduction to GPS/GNSS	1	5
	22207	Excavation Math	2	17.5
	22209	Interpreting Civil Drawings	2	20
	22308	Soils	2	10
	22210	Site Work	2	25
	22106	Grades and Grading	2	15
	22302	Dozers	2	30
			Total Hours	**180**

Credential Name	Module ID	Module Title	Level	Hours
Excavator	22101	Introduction to Heavy Equipment	1	5
	22102	Heavy Equipment Operation Safety	1	15
	22108	Equipment Operation, Inspection, and Readiness	1	25
	22201	Introduction to Earthmoving	1	12.5
	22109	Introduction to GPS/GNSS	1	5
	22207	Excavation Math	2	17.5
	22308	Soils	2	10
	22210	Site Work	2	25
	22304	Excavators	2	35
			Total Hours	**150**

(*continued*)

Credential Name	Module ID	Module Title	Level	Hours
Rough Terrain Forklift	22101	Introduction to Heavy Equipment	1	5
	22102	Heavy Equipment Operation Safety	1	15
	22108	Equipment Operation, Inspection, and Readiness	1	25
	22308	Soils	2	10
	22210	Site Work	2	25
	22206	Rough Terrain Forklifts	3	25
			Total Hours	**105**

Credential Name	Module ID	Module Title	Level	Hours
Conventional Forklift	22101	Introduction to Heavy Equipment	1	5
	22102	Heavy Equipment Operation Safety	1	15
	22108	Equipment Operation, Inspection, and Readiness	1	25
	22308	Soils	2	10
	22210	Site Work	2	25
	22107	Conventional Forklifts	3	10
			Total Hours	**90**

Credential Name	Module ID	Module Title	Level	Hours
Loader	22101	Introduction to Heavy Equipment	1	5
	22102	Heavy Equipment Operation Safety	1	15
	22108	Equipment Operation, Inspection, and Readiness	1	25
	22201	Introduction to Earthmoving	1	12.5
	22207	Excavation Math	2	17.5
	22209	Interpreting Civil Drawings	2	20
	22308	Soils	2	10
	22210	Site Work	2	25
	22106	Grades and Grading	2	15
	22205	Loaders	2	17.5
			Total Hours	**162.5**

Credential Name	Module ID	Module Title	Level	Hours
Motor Grader	22101	Introduction to Heavy Equipment	1	5
	22102	Heavy Equipment Operation Safety	1	15
	22108	Equipment Operation, Inspection, and Readiness	1	25
	22109	Introduction to GPS/GNSS	1	5
	22207	Excavation Math	2	17.5
	22209	Interpreting Civil Drawings	2	20
	22308	Soils	2	10
	22210	Site Work	2	25
	22106	Grades and Grading	2	15
	22305	Motor Graders	3	40
			Total Hours	**177.5**

Credential Name	Module ID	Module Title	Level	Hours
Skid Steer	22101	Introduction to Heavy Equipment	1	5
	22102	Heavy Equipment Operation Safety	1	15
	22108	Equipment Operation, Inspection, and Readiness	1	25
	22201	Introduction to Earthmoving	1	12.5
	22210	Site Work	2	25
	22212	Skid Steers	1	22.5
			Total Hours	**105**

(*continued*)

Credential Name	Module ID	Module Title	Level	Hours
Scrapers	22101	Introduction to Heavy Equipment	1	5
	22102	Heavy Equipment Operation Safety	1	15
	22108	Equipment Operation, Inspection, and Readiness	1	25
	22207	Excavation Math	2	17.5
	22209	Interpreting Civil Drawings	2	20
	22308	Soils	2	10
	22210	Site Work	2	25
	22106	Grades and Grading	2	15
	22204	Scrapers	3	20
			Total Hours	**152.5**

Credential Name	Module ID	Module Title	Level	Hours
HEO Utility Tractor	22101	Introduction to Heavy Equipment	1	5
	22102	Heavy Equipment Operation Safety	1	15
	22108	Equipment Operation, Inspection, and Readiness	1	25
	22105	Utility Tractors	1	17.5
			Total Hours	**62.5**

Credential Name	Module ID	Module Title	Level	Hours
HEO Dump Trucks	22101	Introduction to Heavy Equipment	1	5
	22102	Heavy Equipment Operation Safety	1	15
	22108	Equipment Operation, Inspection, and Readiness	1	25
	22308	Soils	2	10
	22210	Site Work	2	25
	22311	Dump Trucks	3	25
			Total Hours	**105**

CONTENTS

Module 22101 Introduction to Heavy Equipment

Module 22103 Classification of Heavy Equipment

Module 22102 Heavy Equipment Operations Safety

Module 22108 Equipment Operation, Inspection, and Readiness

Module 22201 Introduction to Earthmoving

Module 22109 Introduction to GPS/GNSS

Module 22105 Utility Tractors

Module 22212 Skid Steers

MODULE 22101

Introduction to Heavy Equipment

Source: Alexandr Baranov/iStock/Getty Images

Objectives

Successful completion of this module prepares you to do the following:

1. Explain the role of heavy equipment operators in the construction industry.
   a. Explain the uses of heavy equipment.
   b. Identify and describe the different types of heavy equipment.
   c. Describe the career opportunities associated with heavy equipment operation.
   d. Describe the training available to heavy equipment operators.
2. Explain the responsibilities and characteristics required of heavy equipment operators.
   a. Describe employee responsibilities.
   b. Explain the importance of human relations.
   c. Explain the importance of safety in the operation of heavy equipment.

Performance Tasks

This is a knowledge-based module. There are no Performance Tasks.

Overview

This module provides an overview of heavy equipment used in construction and mining work, including a description of each type of equipment and its uses. The module also explains the structure of an apprentice training program and addresses the values that help a person succeed in the industry. Because safety is such an important concern for anyone in the construction industry, the module also introduces safety issues that are expanded upon in later modules.

NCCER Industry-Recognized Credentials

If you are training through an NCCER-accredited sponsor, you may be eligible for credentials from NCCER. The ID number for this module is 22101. Note that this module may have been used in other NCCER curricula and may apply to other level completions. Contact NCCER at 888.622.3720 or go to **www.nccer.org** for more information.

You can also show off your industry-recognized credentials online with NCCER's digital credentials. Transform your knowledge, skills, and achievements into digital credentials that you can share across social media platforms, send to your network, and add to your resume. For more information, visit **www.nccer.org**.

Digital Resources for Heavy Equipment Operations

SCAN ME

Scan this code using the camera on your phone or mobile device to view the digital resources related to this craft.

1.0.0 Heavy Equipment Operator Training

Performance Tasks

This is a knowledge-based module. There are no Performance Tasks.

Objective

Explain the role of heavy equipment operators in the construction industry.

a. Explain the uses of heavy equipment.
b. Identify and describe the different types of heavy equipment.
c. Describe the career opportunities associated with heavy equipment operation.
d. Describe the training available to heavy equipment operators.

This module introduces the various types of heavy equipment used on construction sites. *Figure 1* shows examples of this equipment, which includes the following:

- Bulldozer
- Dump truck
- Excavator
- Backhoe loader
- Loader
- Motor grader
- Roller/compactor
- Forklift
- Scraper
- Construction tractor
- Heavy Construction Truck

(A) Bulldozer

(B) Articulated Off-Road Dump Truck

(C) Excavator

(D) Backhoe Loader

Figure 1 Heavy construction equipment.

Sources: (A) Maksim Safaniuk/iStock/Getty Images; (B) Image courtesy of Komatsu America Corp., (C, D) Courtesy of CASE Construction Equipment

(E) Skid Steer

(F) Motor Grader

(G) Vibratory Asphalt Compactor

(H) Straight Mast Rough Terrain Forklift

(I) Tandem Powered Scraper

(J) Utility Tractor

(K) Off-Highway Heavy Construction Truck

Figure 1 Heavy construction equipment (*continued*).

Sources: (E) Courtesy of CASE Construction Equipment, (F) Image courtesy of Komatsu America Corp., (H, I) Courtesy of JCB Inc.; (K) Image courtesy of Komatsu America Corp.

Heavy equipment operators are highly skilled workers who are needed on every construction site and in every mining operation. A skilled operator has many opportunities for advancement with an employer or as an entrepreneur. Many of the large excavating and site-development companies operating today were founded by equipment operators who started with a single machine.

1.1.0 Heavy Equipment Applications

Heavy construction equipment is used primarily in preparing a jobsite for the construction of buildings and other structures. Ten types of heavy equipment are covered in this module. A heavy equipment operator can specialize in one machine or can learn to operate several.

At the start of a project, bulldozers, excavators, loaders, and dump trucks are used either to demolish and remove existing structures or to clear a new site of trees, brush, and boulders. At the next phase of the project, this same equipment, along with scrapers, is used to add, remove, and relocate earth in order to bring the site to the grade established by the site design engineers. Preparing a site for construction often requires moving thousands of tons of material. Excavators and backhoes are used to dig trenches for utilities, such as water, sewer, and gas piping. At the final stage of site development, soil compactors and motor graders are used to prepare roads and parking lots for paving, and to prepare the building site for landscaping.

Dealing with Water Runoff

In most areas of the country, rainwater is absorbed by the soil. When land is developed, however, the soil is covered with buildings, roads, and parking lots, so the water can no longer be absorbed in those areas. If the site is not properly engineered, the runoff will have no place to go except onto neighboring lower property, which may cause flooding. For this reason, many construction sites are required to have detention ponds to receive the excess water. Although storm drainage systems are installed on most properties, a heavy rain can easily overwhelm these systems. The detention pond holds the water and lets it slowly drain.

A retention pond, on the other hand, always contains water. If the property includes natural water sources such as springs, for example, these sources may be diverted to a retention pond. Retention ponds are often used to create attractive water features for a residential or commercial development.

1.2.0 Types of Heavy Equipment

Heavy equipment is used in the construction of buildings, bridges, dams, and roads. It is also used extensively in mining operations. In fact, some of the largest equipment is designed for mining work (*Figure 2*).

Demolition: The destruction and removal of a structure.

Grading: Establishing the proper level or slope of the ground.

The primary uses of heavy equipment in construction are in **demolition** work, earth moving, trench digging, road construction, and site **grading**. An office building complex being developed on wooded land is a good example of an occasion when heavy equipment is needed. The first step would be to remove brush, trees, and rocks. This is known as clearing and grubbing and is typically done with a dozer (*Figure 3*). Then the site must be brought to subgrade, also known as rough grade, by removing or relocating soil and filling

(A) Underground Mining Loader **(B) Underground Mining Hauler**

Figure 2 Examples of mining equipment.
Sources: (A) Elena Bionysheva-Abramova/iStock/Getty Images; (B) Image courtesy of Komatsu America Corp.

Figure 3 Opening the site.
Source: Image courtesy of Komatsu America Corp.

Figure 4 Bringing the site to rough grade.
Source: Image courtesy of Komatsu America Corp.

depressions (*Figure 4*). Quite often, it is necessary to truck in material from outside sources, or to move excess material from one part of the site to another, in order to achieve the required grades (*Figure 5*).

Excavators are earth-moving machines used to dig foundations or move large amounts of soil and other material. The soil and rocks dug out by an excavator are usually removed from the site by dump trucks. The type of dump truck being used depends on the location of the worksite. On-road dump trucks are designed to travel on highways, and off-road dump trucks are used only on jobsites. The trucks may be loaded using excavators (*Figure 6*), loaders, or backhoes.

Figure 5 Transferring fill from the site.
Source: Philipp Berezhnoy/iStock/Getty Images

Figure 6 Excavator loading a truck.
Source: Courtesy of CASE Construction Equipment

Excavators and backhoes are used to dig trenches for utilities such as water, natural gas, sewer lines, and storm drains, as well as cables for electricity, telephone systems, and data lines. *Figure 7* shows a backhoe digging a trench.

After the site has been cleared and otherwise prepared, most of the heavy equipment work is finished and the operators go on to the next project. The exception is the equipment used to bring the site and its roads to finish grade (*Figure 8*).

When an existing building or other structure must be removed from the site to make way for new construction, heavy equipment operators are involved in the demolition (*Figure 9*). Those operators must be extremely careful in their actions because demolition activities on one area of a structure can cause other parts of the structure to fall. Excavators, loaders, and dump trucks are used to remove the debris from the site.

Figure 7 Trenching with a backhoe.
Source: Dmitrii Bykanov/iStock/Getty Images

Figure 8 Bringing the site to finish grade.
Source: LoveSilhouette/iStock/Getty Images

Figure 9 Excavator used in demolition work.
Source: Gabort71/iStock/Getty Images

Mechanical T-Rex

This demolition excavator has a special concrete pulverizing attachment.

Source: Nordroden/iStock/Getty Images

Much of the work performed with heavy equipment is associated with road building. The machines that scrape up and move dirt on those road-building projects are generally called scrapers. *Figure 10* shows an example of a scraper. Topsoil, when removed, is often stockpiled for reuse.

After the roadways have been roughly graded to the desired levels, other equipment is brought in to shape, pack, and smooth the dirt and gravel that form the roadbed. *Figure 11* shows a soil compactor being used to form and pack roadway gravel.

Figure 10 Earth-moving scraper.
Source: Iron_Man_/iStock/Getty Images

Figure 11 Soil compactor.
Source: No_limit_pictures/iStock/Getty Images

After the soil and gravel have been properly shaped and packed for a new road, the paving equipment is brought in. The new road may be poured concrete or asphalt. If it is asphalt, heavy asphalt compactors (*Figure 12*) are used to pack and smooth the freshly laid asphalt.

Figure 12 Asphalt compactor.
Source: Courtesy of CASE Construction Equipment

Did You Know?

Highway Construction Is a Need

Between new road construction and the widening of existing roads, highway construction in the United States averages more than 30,000 lane-miles of road per year. Historically, more than 60 percent of the new roads in the United States have been created for new residential development. Federal funding for maintenance of the US transportation infrastructure, including roads, bridges, tunnels, and dams, has been below requested amounts for several years, which means that work that needs to be done is not getting done. Some experts have attributed the collapse of Bridge 9340, which carried Interstate 35 West across the Mississippi River near Minneapolis, to a lack of inspection and maintenance.

1.3.0 Career Opportunities in Heavy Equipment

It is hard to imagine a construction project that does not require heavy equipment operators. Large commercial and industrial projects often need hundreds of them. According to the US Department of Labor's 2021 Occupational Outlook Handbook, the job opportunities for heavy equipment operators are expected to be good, with a projected growth of 5 percent, or about 22,000 new jobs between 2021 and 2031. According to the Department of Labor's statistics, construction equipment operators held almost 466,900 jobs in 2021.

Approximately three out of five heavy equipment operators work in building construction. Others work on roads, bridges, tunnels, and mining projects. Another area of heavy equipment use is power-generating facilities, including wind farms, solar farms, and biomass facilities. Due to increasing populations and the crumbling of existing **infrastructure**, jobs for heavy equipment operators are expected to be plentiful for the foreseeable future as those structures are repaired or replaced.

Infrastructure: A system of public works such as highways, bridges, tunnels, and dams.

There are more than 600,000 bridges in the United States. More than 12 percent of these bridges—as many as 75,000—are rated structurally deficient. A bridge is considered structurally deficient when it needs significant repair or replacement because of component deterioration. Based on Federal Highway Administration estimates, the country would need to spend $20.5 billion a year until 2028 in order to eliminate these deficiencies. This spending represents a huge increase in the demand for skilled craft workers in this sector of the construction industry.

Did You Know?
US Infrastructure

The nation's infrastructure consists of highways, bridges, tunnels, and dams. The US Interstate highway system alone comprises more than 47,000 miles of highway. In addition, there are about 160,000 miles of US-designated highways (US 1, US 66, etc.), along with hundreds of thousands of miles of state highways.

1.3.1 Getting Started

Heavy equipment operators have many work opportunities. Operators typically start out on lighter, simpler equipment, such as forklifts, trucks, skid steer loaders, and roller/compactors. To standardize a company's training programs and to document the training each trainee receives, many companies start their trainees out on interactive simulators run on a computer. Such programs teach the trainee about the machine to be operated and ensure that the trainee understands both the company's general safety **policies** and those associated with each machine. After a trainee has a proven understanding of the machinery to be operated, along with the related safety guidelines, the company will move the trainee on to the hands-on portion of the training.

Policies: General statements establishing guidelines for a specific activity.

With experience and further training, the operator can advance to more complex equipment, such as dozers, excavators, and graders. As an operator, your pay is generally linked to the type of equipment you are operating. The more types of machines you are qualified to operate, the more likely you are to find steady employment and earn a top income.

1.3.2 Other Job-Related Opportunities

Many companies run their own training programs for equipment operators, so qualified operators may have opportunities to serve as instructors in their company's training program. In addition, many independent schools that offer equipment operator training programs need qualified instructors. Construction companies often rent or lease additional equipment as needed. They get that extra equipment from companies that specialize in renting construction equipment. Another employment opportunity for equipment operators is to move into a sales or management role with such a rental company.

1.3.3 Supervisory and Management Positions

Advancement into administrative and management positions is also possible for equipment operators. A craftworker can advance to the position of crew leader, who gives work direction to a small group of workers and inspects their work. Beyond that, a person can be promoted to foreman, who may be responsible for all the members of a particular craft on a jobsite.

Many construction companies employ equipment coordinators to make sure that the required types and amounts of equipment, along with qualified operators, are available at each jobsite when they are needed. The equipment coordinator is typically someone with experience as an equipment operator. Most companies own some heavy equipment and rent or lease the rest as needed. Part of the equipment coordinator's responsibility is to arrange for timely rental of equipment to meet project needs.

Availability of equipment is only one problem a construction company faces. Equipment also needs to be serviced and repaired. Large construction companies often establish maintenance facilities to maintain their fleet of equipment. Such a company is likely to have equipment superintendents who are responsible for both the acquisition and maintenance of the company's equipment. Because experienced heavy equipment operators already have some experience maintaining heavy equipment, they may also work their way into equipment maintenance positions.

Many projects have both a project manager and a project superintendent. A project manager is generally headquartered at the contractor's facility but spends some time at the jobsite. Depending on project size, a project manager may be assigned to more than one project. A project superintendent is responsible for supervising the day-to-day activities of the project, and for that reason is generally located full time at the jobsite. In addition to supervising the construction, the superintendent coordinates subcontractor activities and manages work schedules. A large project may have more than one superintendent, all reporting to the project manager or general superintendent. Project superintendents are generally people who started as trade workers and worked their way up the ladder.

1.3.4 Self-Employed Contractor

Being your own boss is part of the American dream. If you look at the history of many excavation and site development companies you will find that many of them began with an equipment operator who bought a backhoe/loader or some other piece of equipment, along with a trailer and truck, to become an independent contractor. Many construction companies hire independent operators, either as a matter of course, or simply to help with peak loads. The way to be successful as an independent operator is to develop a reputation for being on the job when you are needed and for doing quality work.

1.4.0 Standardized Training By NCCER

NCCER is a not-for-profit education foundation established by the nation's leading construction companies. NCCER was created to provide the industry with standardized construction education materials and a system for tracking and recognizing students' training accomplishments—NCCER's National Registry. *Figure 13* shows examples of NCCER credentials.

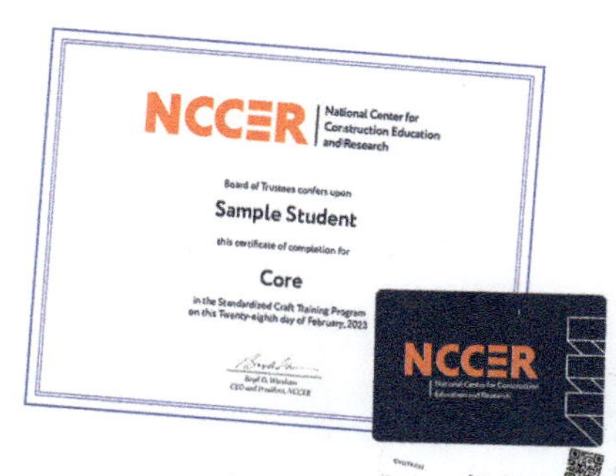

Figure 13 NCCER credentials.

NCCER also offers accreditation, instructor certification, and skills assessments. NCCER is committed to developing and maintaining a training process that is internationally recognized, standardized, portable, and competency-based.

Working in partnership with industry and academia, NCCER has developed a system for program accreditation similar to those found in institutions of higher learning. NCCER's accreditation process ensures that students receive quality training based on uniform standards and criteria. These standards are outlined in NCCER's Accreditation Guidelines and must be adhered to by NCCER Accredited Training Sponsors.

Millions of craft professionals and construction managers have received quality construction education through NCCER's network of Accredited Training Sponsors and the thousands of Training Units associated with the Sponsors. Every year the number of NCCER Accredited Training Sponsors increases significantly.

A craft instructor is a journeyman craft professional or career and technical educator trained and certified to teach the NCCER curriculum. This network of certified instructors ensures that NCCER training programs will meet the standards of instruction set by the industry. More information about NCCER is available at *www.nccer.org*.

Equipment Simulators

Equipment simulators, like the one shown here, can be used to train new operators, assess the skills of experienced operators, and screen potential new employees. The simulators duplicate the cab environment and controls. The simulation software that works behind the scenes causes the screen image to respond to the controls in exactly the same way as a real machine would. Trainees using the simulator learn to perform tasks in the safest and most efficient way. Using simulators creates significant savings in terms of machine ownership or rental, cost of fuel, and reduced wear and tear on machines. Simulators are available for most types of heavy equipment.

Source: ®2023 Simformotion™ LLC. Reprinted by permission. Cat® Simulators is a licensed brand. Simformotion™ LLC is a licensee of Caterpillar Inc.

On-the-job learning (OJL): Job-related learning acquired while working.

1.4.1 Apprenticeship Training

Registered apprenticeship programs must conform to federal and state requirements. All approved apprenticeship programs provide **on-the-job learning (OJL)** as well as classroom instruction. The related training need is fulfilled by all NCCER craft training programs. The main difference between NCCER training and registered apprenticeship programs is that apprenticeship has specific time limits in which the training must be completed. Apprenticeship standards set guidelines for recruiting and outreach, and a specific time limit for each of a variety of OJL tasks. Additionally, reporting requirements and audits ensure adherence to the apprenticeship standards. Companies and employer associations register their individual apprenticeship programs with the Office of Apprenticeship within the US Department of Labor, and in some instances, with a state apprenticeship council (SAC). The typical heavy equipment operator apprenticeship program is three years.

The traditional apprenticeship program required OJL of 2,000 hours per year and a minimum of 144 hours of classroom-related training. Apprenticeship programs therefore varied in length from 2,000 hours to 10,000 hours. However, on October 29, 2008, the US Department of Labor published new regulations to modernize the National Apprenticeship System. These regulations provide for more flexibility in how related technical instruction (RTI) can be delivered. They also provide registration agencies with the option of issuing interim credentials to offer active apprentices official recognition of their accomplishments and equip them with a portfolio of skills and incentives to complete their programs and continue their career preparation. Finally, these regulations allow program sponsors to offer three different ways for apprentices to complete a registered apprenticeship program:

- The traditional, time-based approach, which requires the apprentice to complete a specific number of OJL and RTI hours.
- A competency-based approach, requiring successful demonstration of acquired skills and knowledge by an apprentice, as verified by the program sponsor, with an on-the-job learning component and RTI.
- A hybrid approach, which requires the apprentice to complete a minimum number of OJL and RTI hours and demonstrate competency in the defined subject areas.

Did You Know?
Apprenticeship History

George Washington (surveyor), Benjamin Franklin (printer), and Paul Revere (silversmith) were apprentices. According to the US Department of Labor, more than 500,000 apprentices received registered apprenticeship training in the United States in 2021.

1.4.2 Youth Training and Apprenticeship Programs

Youth apprenticeship programs are available that allow students to begin their apprenticeship or craft training while still in high school. A student entering the program in the eleventh grade may complete as much as one year of the

NCCER training program by high school graduation. In addition, programs (in cooperation with local construction industry employers) allow students to work in the craft and earn money while still in school. Upon graduation, students can enter the industry at a higher level and with more pay than someone just starting in a training program.

Students participating in the NCCER or youth apprenticeship training are recognized through official transcripts and can enter the second level or year of the program wherever it is offered. They may also have the option of applying credits at two-year or four-year colleges that offer degree or certificate programs in their selected field of study.

Did You Know?

Labor Laws

Although youth apprentices may work in the field while they are in sponsored programs, labor laws prohibit employers from using workers under the age of 18 in hazardous duties.

1.0.0 Section Review

1. Which of these machines is used primarily in the final stage of site preparation?
   a. Motor grader
   b. Excavator
   c. Backhoe loader
   d. Dozer

2. Which of these machines is likely to be used for trenching?
   a. Dozer
   b. Compactor
   c. Excavator
   d. Scraper

3. About how many heavy equipment operators are there in the US?
   a. Less than 100,000
   b. Roughly 200,000
   c. More than 400,000
   d. More than 1 million

4. In addition to on-the-job learning, what is the minimum number of annual classroom hours a registered apprentice must complete in a traditional apprenticeship program?
   a. 50
   b. 84
   c. 100
   d. 144

2.0.0 Responsibilities and Characteristics of Operators

Objective

Explain the responsibilities and characteristics required of heavy equipment operators.

a. Describe employee responsibilities.
b. Explain the importance of human relations.
c. Explain the importance of safety in the operation of heavy equipment.

Performance Tasks

This is a knowledge-based module. There are no Performance Tasks.

To be successful, professionals must be able to use current trade materials, tools, and equipment to perform tasks quickly and efficiently. Equipment operators need to adapt to changing situations. The successful equipment operator must keep abreast of technical advancements and continually gain the skills to use them. A professional never takes chances with regard to personal safety or the safety of others.

2.1.0 Employee Responsibilities

Procedures: Instructions on how to carry out a policy.

Being an employee is more than just showing up and doing the assigned work. Employees must also follow company policies and **procedures**, accept the direction of assigned supervisors, work cooperatively with other employees and contractors, and accept responsibility for their actions.

2.1.1 Professionalism

The word *professionalism* is a broad term that describes the desired overall behavior and attitude expected in the workplace. Professionalism is too often absent from the construction site and the various trades. Most people would argue that professionalism must start at the top in order to be successful. It is true that management support of professionalism is important to its success in the workplace, but it is more important that individuals recognize their own responsibility for professionalism.

Productivity: The rate at which work is accomplished.

Professionalism includes honesty, **productivity**, safety, civility, cooperation, teamwork, clear and concise communication, punctuality, preparation for work, and regard for one's impact on one's coworkers. It can be demonstrated in a variety of ways every minute you are in the workplace. Most importantly, do not tolerate unprofessional behavior from coworkers. This is not to say that you shun the unprofessional worker; instead, you should demonstrate the benefits of professional behavior.

Professionalism is a benefit to both the employer and the employee. It is a personal responsibility. The industry is what each individual chooses to make of it; choose professionalism and the industry image will follow.

2.1.2 Honesty

Integrity: Soundness of character or commitment to moral and ethical principles.

Honesty and personal **integrity** are important traits of the successful professional. Professionals pride themselves in performing a job well, and in being punctual and dependable. Each job is completed in a professional way, never by cutting corners or skimping on materials to cut cost. A valued professional maintains work attitudes and ethics that protect property such as tools and materials belonging to employers, customers, and other trades from damage or theft at the shop or jobsite.

Honesty and success go hand-in-hand. It is not simply a choice between good and bad, but a choice between success and failure. Dishonesty always catches up with you. Whether you are stealing materials, tools, or equipment from the jobsite or simply lying about your work, it will not take long for your employer to find out. Of course, you can always go and find another employer, but this option will ultimately wear out.

If you plan to succeed and enjoy continuous employment, consistency of earnings, and being sought after by employers, then start out with the basic understanding of honesty in the workplace and you will reap the benefits.

Honesty means more, however, than just not taking things that do not belong to you. It means giving a fair day's work for a fair day's pay. It means that your words convey true meanings and actual happenings. Thoughts as well as actions should be honest. Employers place a high value on an employee who is strictly honest.

2.1.3 Loyalty

Employees expect employers to look out for their interests, to provide them with steady employment, and to promote them to better jobs as openings occur. Employers feel that they too have a right to expect their employees to be loyal to them, including doing the following:

- Keeping company interests in mind
- Speaking well of the company to others
- Keeping any minor troubles strictly within the plant or office
- Keeping absolutely confidential all matters that pertain to the business

Both employers and employees should keep in mind that loyalty is not something to be demanded; rather, it is something to be earned.

2.1.4 Willingness to Learn

Every office and plant has its own way of doing things. Employers expect their workers to be willing to learn these ways. Adapting to change and being willing to learn new methods and procedures as quickly as possible is key. Sometimes, a change in safety regulations or the purchase of new equipment makes it necessary for even experienced employees to learn new methods and operations. Employees often resent having to accept improvements because of the retraining that is involved. However, employers no doubt think they have a right to expect employees to put forth the necessary effort. Methods must be kept up to date in order to meet competition and show a profit. It is this profit that enables the owner to continue in business and provide jobs for the employees.

2.1.5 Willingness to Take Responsibility

Most employers expect their employees to see what needs to be done, then go ahead and do it. It is very tiresome to have to ask again and again that a certain job be done. It is obvious that having been asked once, an employee should assume the responsibility from then on. Once the responsibility has been delegated, the employee should continue to perform the duties without further direction. Every employee has the responsibility for working safely.

2.1.6 Willingness to Cooperate

To cooperate means to work together. In the modern business world, cooperation is the key to getting things done. Learn to work as a member of a team with your employer, supervisor, and fellow workers in a common effort to get the work done efficiently, safely, and on time.

2.1.7 Rules and Regulations

People can work well together only if there is some understanding about what work is to be done, when and how it will be done, and who will do it. Rules and regulations are necessary in any work situation and should be respected by all employees. Most companies have formal policies and procedures to help employees carry out their duties. A policy is a general statement establishing guidelines for a specific activity. Examples include policies on vacations, breaks, workplace safety, and checking out tools. Procedures are the instructions through which policies are carried out. For example, a procedure written to implement a policy on workplace safety would include guidelines for reporting accidents and general safety procedures that all employees are expected to follow.

An employee must be familiar with the company policies and procedures, especially with regard to safety practices. When OSHA inspectors visit a jobsite, they often question employees and crew leaders about the company policies related to safety. If they are investigating an accident, they will want to verify that the workers knew the applicable company policy and followed it.

> **Going Green**
> **Environmental Sensitivity**
>
> Some construction occurs around bodies of water, as well as land that may be environmentally sensitive. Everyone working on these projects needs to be aware of environmental concerns and the methods used to protect the environment during construction.

2.1.8 Tardiness and Absenteeism

Tardiness means being late for work, and absenteeism means being off the job for one reason or another. Consistent tardiness and frequent absences indicate poor work habits, unprofessional conduct, and a lack of commitment.

Humans are all creatures of habit. The habit of always being late may have begun during school days when it was hard to get up in the morning. This habit can get us into trouble at school, and it can continue getting us into trouble when we are through with school and go to work.

Work life is governed by the clock. People are required to be at work at a definite time. Failure to get to work on time results in confusion, lost time, and resentment on the part of those who do come on time. In addition, it may lead to penalties, including dismissal. Although it may be true that a few minutes out of a day are not very important, you must remember that a principle is involved. It is your obligation to be at work at the time indicated.

The terms of work are agreed upon when you accept a job. Perhaps it will help you to see things more clearly if you try to look at the matter from the point of view of the boss. Supervisors cannot keep track of people if they come in any time they please. It is not fair to others to ignore tardiness. Failure to be on time may hold up the work of fellow workers and affect a project schedule. Better planning of your morning routine will often keep you from being delayed and so prevent a breathless, late arrival. In fact, arriving a little early indicates your interest and enthusiasm for your work, which is appreciated by employers. The habit of being late is another one of those things that can stand in the way of promotion.

It is sometimes necessary to take time off from work. No one should be expected to work when sick or when there is serious trouble at home. However, it is possible to get into the habit of letting unimportant and unnecessary matters keep us from the job. This results in lost production and hardship on those who try to carry on the work with less help. Again, there is a principle involved. Employers have a right to expect you to be on the job unless there is a very good reason for staying away. Some trivial reason should not keep you at home. You should not stay up nights until you are too tired to go to work the next day. If you are ill, you should use the time at home to do all you can to recover quickly. This, after all, is no more than most employers would expect of a person they depend on to do a certain job.

If it is necessary to stay home, then at least phone the office early in the morning so that the boss can find another worker for the day. Time and again, employees have remained home without sending any word to the employer. This is the worst possible way to handle the matter. It leaves those at work uncertain about what to expect. They have no way of knowing whether you have merely been held up and will be in later, or whether immediate steps should be taken to assign your work to someone else. Courtesy alone demands that you let the boss know if you cannot come to work.

The most frequent causes of absenteeism are illness, death in the family, accidents, personal business, and dissatisfaction with the job. Some of the causes are legitimate and unavoidable, while others can be controlled. One can usually plan to carry on most personal business affairs after working hours. Frequent absences reflect unfavorably on a worker when promotions are being considered.

Employers sometimes resort to docking pay, demotion, and even dismissal in an effort to control tardiness and absenteeism. No employer likes to impose restrictions of this kind. However, in fairness to those workers who do come on time and who do not stay away from the job, an employer is sometimes forced to discipline those who do not follow the rules.

Ethical Principles for Members of the Construction Trades

- *Honesty* — Be honest and truthful in all dealings. Conduct business according to the highest professional standards. Faithfully fulfill all contracts and commitments. Do not deliberately mislead or deceive others.
- *Commitment to safety* — Focus on working safely to protect yourself and your coworkers from harm. Make sure the work environment is as safe as you can make it.
- *Integrity* — Demonstrate personal integrity and the courage of your convictions by doing what is right even where there is pressure to do otherwise. Do not sacrifice your principles because it seems easier.
- *Loyalty* — Be worthy of trust. Demonstrate fidelity and loyalty to companies, employers and sponsors, coworkers, and trade institutions and organizations.
- *Fairness* — Be fair and just in all dealings. Do not take undue advantage of another's mistakes or difficulties. Fair people are open-minded and committed to justice, equal treatment of individuals, and tolerance for and acceptance of diversity.
- *Respect for others* — Be courteous and treat all people with equal respect and dignity.
- *Obedience* — Abide by laws, rules, and regulations relating to all personal and business activities.
- *Commitment to excellence* — Pursue excellence in performing your duties, be well-informed and prepared, and constantly try to increase your proficiency by gaining new skills and knowledge.
- *Leadership* — By your own conduct, seek to be a positive role model for others.

2.2.0 The Importance of Human Relations

Most people underestimate the importance of working well with others. They tend to pass off human relations as nothing more than common sense. What exactly is involved in human relations? Part of human relations is being friendly, pleasant, courteous, cooperative, adaptable, and sociable.

2.2.1 Making Human Relations Work

Human relations are much more than just getting people to like you. You must also know how to handle difficult situations. Understanding human relations involves knowing how to work with supervisors who are often demanding and sometimes seem unfair. It also involves learning your own personality traits and those of your coworkers. Building sound working relationships with others is very important. If working relationships have deteriorated for one reason or another, restoring them is essential. Learning how to handle frustration without hurting others is important to human relations.

2.2.2 Human Relations and Productivity

Effective human relations are directly related to productivity, which is an important factor in achieving business success. Every employee is expected to produce at a certain level. Employers quickly lose interest in an employee who has a great attitude but is unable to complete work within budget.

All employees, both new and experienced, are measured by the amount of quality work they can safely turn out. The employer expects every employee to do their share of the workload.

However, doing one's share in itself is not enough. If you are to be productive, you must do your share (or more than your share) without antagonizing your fellow workers. You must perform your duties in a manner that encourages others to follow your example. It makes little difference how ambitious you are or how well you perform. You cannot become the kind of employee you want to be or the type of worker management wants you to be without learning how to work with your peers.

Employees must do everything they can to build strong, professional working relations with fellow employees, supervisors, and clients.

2.2.3 The Importance of a Positive Attitude

A positive attitude is essential to a successful career for the following reasons:

1. Being positive means being energetic, highly motivated, attentive, and alert. A positive attitude is essential to safety on the job.
2. A positive employee contributes to the productivity of others. Both negative and positive attitudes are transmitted to others on the job. A persistent negative attitude can spoil the positive attitudes of others. It is very difficult to maintain a high level of productivity while working next to a person with a negative attitude.
3. People favor a person who is positive. Being positive makes a person's job more interesting and exciting.
4. The kind of attitude transmitted to management has a great deal to do with an employee's future success in the company. Supervisors can determine workers' attitudes by observing their approach to the job, reactions to directives, and the way they handle problems.

A positive attitude is far more than a smile, which is only one example of an inner positive attitude. As a matter of fact, some people transmit a positive attitude even though they seldom smile. They do this by the way they treat others, the way they look at their responsibilities, and the approach they take when faced with problems. The following suggestions can help you maintain a positive attitude:

- Remember that your attitude follows you wherever you go. If someone makes a greater effort to be a more positive person in their social and personal lives, it automatically helps them on the job. The reverse is also true. One effort will complement the other.

- Negative comments are seldom welcomed by fellow workers on the job. Neither are they welcome on the social scene. The solution is to talk about positive things and be complimentary. Constant complainers do not build healthy and fulfilling relationships.
- Look for the good things in people on the job, especially your supervisor. Nobody is perfect, and almost everyone has a few worthwhile qualities. If you dwell on people's good features, it will be easier to work with them.
- Look for the good things where you work. What are the factors that make it a good place to work? Is it the hours, the physical environment, the people, or the actual work being done? Or is it the atmosphere? Keep in mind that you cannot expect to like everything. No work assignment is perfect. But if you concentrate on the good things, the negative factors will seem less important and bothersome.
- Look for the good things in the company. Just as there are no perfect assignments, there are no perfect companies. Nevertheless, almost all organizations have good features. Is the company progressive? What about promotional opportunities? Are there chances for self-improvement? What about the wage and benefit package? Is there a good training program? You cannot expect to have everything you would like, but there should be enough to keep you positive. In fact, if you decide to stick with a company for a long time, it is wise to focus on the good features. If you think positively, you will act the same way.
- You may not be able to change the negative attitude of another employee, but you can protect your own attitude from becoming negative.

Tips for a Positive Attitude

Here is a short checklist of things that you should keep in mind to help you develop and maintain a positive attitude:

- Remember that your attitude follows you wherever you go.
- Keep in mind that helpful suggestions and compliments are much more effective than negative comments.
- Look for the positive characteristics of your teammates and supervisors.

2.3.0 The Importance of Safety

An obligation is like a promise or a contract. In exchange for the benefits of your employment and your own well-being, you agree to work safely. In other words, you are obligated to work safely. You are also obligated to make sure anyone you happen to supervise or work with is working safely. Your employer is obligated to maintain a safe workplace for all employees. Safety is everyone's responsibility.

Some employers have safety committees. If you work for such an employer, you are then obligated to that committee to maintain a safe working environment. This means that you should follow the safety committee's rules for proper working procedures and practices and report any unsafe equipment and conditions directly to the committee or your supervisor.

On the job, if you see something that is not safe, report it! Do not ignore it. It will not correct itself. You have an obligation to report it.

In the long run, even if you do not think an unsafe condition affects you, it does. Do not mess around; always report unsafe conditions. Do not think your employer will be angry because your productivity suffers while the condition is corrected. On the contrary, your employer will be more likely to criticize you for not reporting a problem.

Your employer knows that the short time lost in making conditions safe again is nothing compared with the whole job being shut down because of a major disaster. If a disaster happens, you are out of work anyway. So do not ignore an unsafe condition.

2.3.1 OSHA

Occupational Safety and Health Administration (OSHA): The federal government agency established to ensure a safe and healthy environment in the workplace.

In 1970, the US Congress passed the Occupational Safety and Health Act. This act also created the **Occupational Safety and Health Administration (OSHA)**. OSHA is part of the US Department of Labor. The job of OSHA is to set occupational safety and health standards for all places of employment; enforce these standards; ensure that employers provide and maintain a safe workplace for all employees; and provide research and educational programs to support safe working practices. Safety standards apply to every part of the construction industry. Whether you work for a large contractor or a small subcontractor, you are obliged to report unsafe conditions. The easiest way to do this is to tell your supervisor. If that person ignores the unsafe condition, report it to the next highest supervisor. If it is the owner who is being unsafe, let that person know your concerns. If nothing is done about it, report it to OSHA. If you are worried about your job being on the line, think about it in terms of your life, or someone else's, being at risk.

OSHA was adopted with the stated purpose "to assure as far as possible every working man and woman in the nation safe and healthful working conditions and to preserve our human resources." OSHA requires every employer to provide a safe and hazard-free working environment. OSHA also requires that employees comply with OSHA rules and regulations that relate to their conduct on the job. To gain compliance, OSHA can perform spot inspections of jobsites, impose fines for violations, and even stop any more work from proceeding until the jobsite is safe.

According to OSHA standards, you are entitled to on-the-job safety training. As a new employee, you are entitled to the following:

- Being shown how to do your job safely
- Being provided with the required personal protective equipment (PPE)
- Being warned about specific hazards
- Being supervised for safety while performing the work

The enforcement of this act of Congress is provided by the federal and state safety inspectors, who have the legal authority to make employers pay fines for safety violations. The law allows states to have their own safety regulations and agencies to enforce them, but they must first be approved by the US Secretary of Labor. For states that do not develop such regulations and agencies, federal OSHA standards must be obeyed.

These standards are listed in *OSHA Safety and Health Standards for the Construction Industry (29 CFR, Part 1926)*, sometimes called *OSHA Standards 1926*. Other safety standards that apply to construction are published in *OSHA Safety and Health Standards for General Industry (29 CFR, Parts 1900 to 1910)*.

The most important general requirements that OSHA places on employers in the construction industry are as follows:

- The employer must perform frequent and regular jobsite safety inspections of equipment.
- The employer must instruct all employees to recognize and avoid unsafe conditions, and to know the regulations that pertain to the job so they may control or eliminate any hazards.
- No one may use any tools, equipment, machines, or materials that do not comply with *OSHA Standards 1926*.
- The employer must ensure that only qualified individuals operate tools, equipment, and machines.

2.3.2 MSHA

Heavy equipment operators may be operating machinery in mining environments, both above ground and under the earth's surface. In addition to the OSHA safety guidelines, there are safety guidelines set forth by the **Mine Safety and Health Administration (MSHA)**, which also falls under the US Department of Labor. In 1977, a new Mine Safety and Health Act was created. The following year (1978), MSHA was put in place. During 1977, 242 miners died in mining accidents. As a result of the stricter MSHA safety guidelines, mining accidents resulting in death dropped by more than 50 percent over the next 10 years and continued to fall thereafter.

Mine Safety and Health Administration (MSHA): A federal government agency established to ensure the safety of mining operations.

MSHA breaks mining into two major groups: coal mining and mining of metal and nonmetal resources. Anyone expecting to operate heavy equipment in mining operations can expect to receive additional training related only to mining safety. This training must be site-specific and task-oriented and must be delivered by a certified mine instructor. The MSHA web site (*www.msha.gov*) has links related to mining safety. This information should be reviewed by heavy equipment operators going into any mining operations.

2.3.3 General Safety Requirements

The equipment you will be operating is expensive, so one of your major responsibilities as an operator will be to take good care of the equipment. Never overload it or perform tasks for which it was not intended. Most heavy equipment manuals have load charts that explain how much the machine can carry, and how it must be operated when loaded. Always perform the required checks and inspections prior to, during, and after operations. Above all, keep in mind that the equipment, because of its size, represents a potential safety hazard to you and other workers on the site. Always operate the equipment in a safe manner and stay within the operating guidelines provided by the manufacturer.

2.0.0 Section Review

1. The term that defines a person achieving the desired overall behavior and attitude expected in the workplace is _____.
   a. loyalty
   b. honesty
   c. professionalism
   d. attitude

2. Which of the following characteristics is an aspect of human relations?
   a. Wearing the required PPE
   b. Willingness to learn
   c. Willingness to take responsibility
   d. Maintaining a positive attitude

3. Which of the following statements describes the mission of OSHA?
   a. Inspect jobsites and punish offenders.
   b. Set and enforce occupational safety and health standards for all places of employment.
   c. Ensure that all workers are wearing the required PPE.
   d. Advise Congress on construction health and safety.

Module 22101 Review Questions

Figure RQ01
Source: Courtesy of CASE Construction Equipment

1. The equipment shown in *Figure RQ01* is a(n) _____.
   a. backhoe loader
   b. skid steer
   c. motor grader
   d. excavator

Figure RQ02
Source: Maksim Safaniuk/iStock/Getty Images

2. The heavy equipment item shown in *Figure RQ02* is a(n) _____.
   a. excavator
   b. motor grader
   c. dozer
   d. underground mining hauler

3. Which of the following items is likely to be used for clearing land?
   a. Bulldozer
   b. Forklift
   c. Motor grader
   d. Compactor

4. Demolition of a building is commonly done with a(n) _____.
   a. motor grader
   b. compactor
   c. excavator
   d. forklift

Figure RQ03
Source: SweetyMommy/iStock/Getty Images

5. The heavy equipment item shown in *Figure RQ03* is a(n) _____.
   a. off-road truck
   b. motor grader
   c. excavator
   d. scraper

6. Which of the following machines is one that would be used in the initial stages of operator training?
   a. Scraper
   b. Motor grader
   c. Forklift
   d. Excavator

7. A person who is responsible for supervising the work of a small number of craftworkers is a _____.
   a. foreman
   b. crew leader
   c. project manager
   d. superintendent

8. The three types of apprentice programs are time-based, competency-based, and _____.
   a. traditional
   b. hybrid
   c. instructor-led
   d. self-paced

9. A document that is a general statement establishing guidelines for a specific activity is a _____.
   a. procedure
   b. manual
   c. policy
   d. work order

10. Frequent and regular jobsite safety inspections of equipment must be performed by _____.
   a. OSHA
   b. MHSA
   c. the employer
   d. the operator

Answers to odd-numbered questions are found in the Module Review Answer Key at the back of this book.

Answers to Section Review Questions

Answer	Section	Objective
Section One		
1. a	1.1.0	1a
2. c	1.2.0	1b
3. c	1.3.0	1c
4. d	1.4.1	1d
Section Two		
1. c	2.1.1	2a
2. d	2.2.3	2b
3. b	2.3.1	2c

MODULE 22103

Classification of Heavy Equipment

Source: Martin Barraud/iStock/Getty Images

Objectives

Successful completion of this module prepares you to do the following:

1. Identify and describe heavy equipment used in construction and mining.
   a. Identify and describe equipment used in excavating and grading construction sites.
   b. Identify and describe trenching equipment.
   c. Identify and describe heavy equipment used in mining.
2. Explain hydraulic systems and their components.
   a. Explain hydraulic system fundamentals.
   b. Identify and describe the equipment and components used in hydraulic systems.

Performance Tasks

Under supervision, you should be able to do the following:

1. Identify types of heavy equipment and explain their uses.
2. Identify the basic parts of each type of equipment and explain the differences in models among the types of equipment.

Overview

There are many different types of heavy equipment. Each machine is typically designed for a particular type of work, but many can be modified with attachments to perform a variety of functions. It is important for heavy equipment operators to know the capabilities and limitations of the machines they operate. They also must be familiar with the various attachments, and how to properly use them. When heavy equipment operators gain this type of knowledge, they will have additional opportunities for advancement.

1.0.0 Construction and Mining Equipment

Performance Tasks

Under supervision, you should be able to do the following:

1. Identify types of heavy equipment and explain their uses.
2. Identify the basic parts of each type of equipment and explain the differences in models among the types of equipment.

Objective

Identify and describe heavy equipment used in construction and mining.

a. Identify and describe equipment used in excavating and grading construction sites.
b. Identify and describe trenching equipment.
c. Identify and describe heavy equipment used in mining.

Ripper: A towed attachment with teeth used on dozers, motor graders, and other machines to loosen heavily compacted soil and soft rock.

Today's construction equipment is designed for versatility. In fact, many of the machines are designed to perform two or more tasks. For example, the backhoe loader (*Figure 1*) is used to dig trenches and to move or load soil, gravel, stone, and other materials. Available accessories, such as a tool carrier attachment, **ripper**, broom, and hydraulic hammer, can turn it into the workhorse of small construction projects.

Figure 1 Backhoe loader.
Source: Image Provided Courtesy of John Deere

1.1.0 Excavating and Grading Equipment

All the machines described in this module are available in a variety of sizes and types to fit all kinds of construction environments. Within each equipment classification, there are often models designed for specific applications. Certain types of excavators are made for areas with a tight turning radius. Other models are designed specifically to reach heights or to stretch out a great distance to dig in areas where the ground in front of the excavator is wet or unstable.

The following types of heavy equipment are used in excavating and grading work:

- Construction tractors
- Dump trucks

- Compactors
- Scrapers
- Dozers
- Loaders
- Forklifts
- Motor graders
- Backhoe loaders
- Excavators
- Trenchers

1.1.1 Construction Tractors

The type of tractors normally used in agricultural work are used as utility vehicles on some jobsites. The smaller tractors typically fall into the 60- to 100-horsepower range, but some of the larger ones have engines that range between 300 and 600 horsepower. They are usually equipped with a hitch, which allows them to pull trailers and other attachments that do not require any kind of drive power supplied by the tractor. These tractors are also equipped with a power takeoff (PTO) and hydraulic systems that allow them to power a variety of construction-related implements that may be used for grading, sweeping, or even post-hole digging. *Figure 2* shows an example of two tractors used on construction sites.

(A) Utility Tractor

(B) Tractor Pulling Scrapers

Figure 2 Tractors used in construction.
Sources: (A) Courtesy of New Holland; (B) photosbyjim/iStock/Getty Images

Large, **articulated** tractors are used for heavy-duty tasks such as pulling scrapers, oversized rollers, and water tanks. These tractors are usually 375 horsepower and higher. Some companies may prefer to use rubber-track vehicles for heavy work. Although they tend to cost more than rubber-tire machines, rubber-track vehicles deliver greater pull in difficult terrain.

Articulated: Two parts connected by a joint so as to move independently.

WARNING!

During the site development phase of a project, it is common to find many machines operating at the same time, as well as people who may be working near the machines. Operators must remain alert to their surroundings at all times in order to avoid accidents that could kill or injure personnel or damage expensive equipment.

1.1.2 Dump Trucks

Construction dump trucks are used to haul soil or debris from one location to another. The two basic types of dump trucks are on-road and off-road. The on-road dump trucks (*Figure 3*) are designed for use on highways and must be registered for that purpose. Anyone operating these trucks on the highway must have a commercial driver's license (CDL). The truck shown in *Figure 3 (A)* is a tandem-axle truck, which means it has two rear axles under the load, along with the front axle under the cab. A single-axle dump truck has only one rear axle and carries a smaller load.

(A) Tandem-Axle Dump Truck **(B) Auxiliary-Axle Dump Truck**

Figure 3 On-road dump trucks.
Source: (B) Navistar Inc.

The truck shown in *Figure 3 (B)* has two additional axles located in front of its two rear axles. These two auxiliary axles are carried in the raised position to save fuel and tire costs when the truck is empty or lightly loaded. When the truck is traveling with a load, the two extra axles and wheels are lowered to distribute the load and provide extra braking. A cab-operated air bag or spring-retraction system is used to raise and lower the axle assembly. Single-wheel tag axle versions without braking capability are also available.

Off-road dump trucks (*Figure 4*) are intended for hauling material on the site. They are usually larger than the on-road trucks. Off-road trucks are designed for traveling over the rough terrain normally found on a construction site. The articulated truck in *Figure 4 (A)* is the most popular of the off-road dump trucks. It consists of a trailer bed hitched to a large tractor. The tractor does not operate independently of the cargo bed, however. Rather, the truck is articulated so that it can make sharp turns. Articulated trucks have capacities of up to about 40 tons.

Rigid-frame trucks, as shown in *Figure 4 (B)*, are used for high-volume operations. The capacities of these trucks start about where the articulated trucks

(A) Articulated Truck **(B) Rigid-Frame Dump Truck** **(C) Crawler**

Figure 4 Off-road dump trucks.
Sources: (A, B) Image courtesy of Komatsu America Corp.; (C) Yoshi0511/Shutterstock

leave off and go up to more than 300 tons. In recent years, crawler dump trucks on rubber tracks, as shown in *Figure 4 (C)*, have become popular for use on wet construction sites.

A dump truck is usually loaded using an excavator or loader (*Figure 5*). Once loaded, the truck hauls the material to another location and dumps it using its hydraulic bed-dumping mechanism. While most dump trucks use a hydraulic lifting mechanism to raise the front of the truck bed, some trucks are designed with a bottom dump feature. Still other dump trucks are equipped with a side-dump capability.

Figure 5 An excavator loading an off-road ridged frame dump truck.
Source: Alexandr Baranov/iStock/Getty Images

1.1.3 Roller/Compactors

The foundations of buildings and roads are placed on soil, which ultimately carries the weight of the entire structure and its contents. It is important, therefore, that the soil base be as smooth and solid as possible. To create this base, the soil must be compacted. Compaction is the process in which the soil is rolled, tamped, vibrated, and/or pressed to make it denser. The denser the soil, the more likely it is to support the weight of the structure without sinking. If the soil gives way under the weight, the structure is likely to crack.

Roller/compactors are used to smooth and compact earth, gravel, and asphalt. They compress the material to a desired density so it can support specified loads without excessive settling. There are four basic types of roller/compactors, as shown in *Figure 6*:

- Sheep's foot soil compactor
- Two-drum, steel wheel compactor
- Vibratory steel wheel soil compactor
- **Pneumatic** tire compactor

Pneumatic: Inflated with compressed air.

Sheep's foot compactors, shown in *Figure 6 (A)*, have pads (feet) attached to the roller. Sheep's foot compactors are especially good for soils containing clay, but are not very effective in sandy soils. Steel wheel compactors, shown in *Figure 6 (B)*, are designed primarily for rolling pavement such as asphalt, but are also used for rolling gravel roads, road bases, and some subgrades. Vibratory compactors, shown in *Figure 6 (C)*, were developed to provide high densities in granular soils. The vibrating action moves the particles, and the static pressure forces them into a compact structure. Pneumatic tire compactors, shown

(A) Sheep's Foot Soil Compactor

(B) Two-Drum, Steel Wheel Roller (Compactor)

(C) Vibratory Soil Compactor

(D) Pneumatic Tire Roller (Compactor)

Figure 6 Roller/compactors.

Sources: (A) fjdelvalle/iStock/Getty Images; (B) vovashevchuk/iStock/Getty Images; (C) Nilsz/iStock/Getty Images; (D) Vadym Terelyuk/iStock/Getty Images

in *Figure 6 (D)*, provide uniform coverage and equal wheel loading. The rear wheels are staggered in relation to the front wheels to roll over the spaces left by the front wheels. This helps achieve uniform coverage. This offset adds a kneading action as the material is pressed toward the spaces between tires.

1.1.4 Scrapers

A scraper is used to remove dirt and other material from a site by scraping it from the surface into the bowl as it moves along. Scrapers are also used to spread dirt and other material, which can be placed in the bowl by a loader. The following four basic types of scrapers are shown in *Figure 7*:

- Standard or self-propelled scraper
- Elevating scraper
- Tandem-powered scraper
- Pull-type scraper

A standard scraper, shown in *Figure 7 (A)*, is a self-propelled unit consisting of a bowl pulled by a single tractor. The cutting edge at the bottom of the bowl can be adjusted to dig to two feet deep or more. The bowls of some units can hold more than 40 cubic yards of earth. The width of the cut can range up to 13', also depending on the model in use. The bowl contains a mechanism for ejecting the material. The forward wall of the bowl is a movable section called the apron. Standard scrapers are often pushed by a dozer in difficult terrain and on slopes.

(A) Self-Propelled Scraper **(B) Elevating Scraper**

(C) Tandem-Powered Scraper **(D) Pull-Type Scraper Towed by a Tractor**

Figure 7 Scrapers.

Sources: (A) Iron_Man_/iStock/Getty Images; (B) SweetyMommy/iStock/Getty Images; (C) Iron_Man_/iStock/Getty Images; (D) ghornephoto/iStock/Getty Images

The bowl of the elevating scraper, shown in *Figure 7 (B)* and also called a paddle wheel scraper, contains a conveyor system that moves the material toward the rear of the bowl as it is picked up from the ground. The capacity of these units is not as great as that of the standard scraper because of the space taken by the elevator system. This type of scraper is used for fine soils and sand and for fine grading work.

Tandem-powered (push-pull) scrapers, shown in *Figure 7 (C)*, have two engines, one in the tractor and a second at the rear of the scraper unit. The advantage offered by these units is that they can work on steeper grades and rougher terrain. Tandem-powered scrapers are also available in a push-pull model that is equipped with a cushioned plate on the front and a hook at the rear of the bowl. When operating in a cut, the second unit pushes the first unit. When the first unit is full, it then uses its engines to pull the second scraper through the cut. Push-pull scrapers are designed for heavier loads, typically 50 percent greater than a standard tandem scraper.

An auger attachment is available for self-propelled and tandem scrapers. These systems have an independently powered auger at the center of the bowl. The auger is designed to evenly distribute the material that flows over the blade. Auger units have the same capacity as standard units but are able to load and eject at a much greater rate.

The pulled, or towed, scraper, shown in *Figure 7 (D)*, is pulled along by a heavy-duty tractor. These types of scrapers have large capacities and can be towed in tandem to pick up twice the material in a single run.

Wheeled Dozers

This Caterpillar 854G wheel dozer has a blade that can push up to 58 cubic yards of material. It weighs over 200,000 pounds (lb) and has an 800-horsepower diesel engine.

Source: Schroptschop/iStock/Getty Images

Haul road: A compacted dirt road used to move material and equipment on and off the site.

1.1.5 Dozers

A dozer is a heavy-duty tractor with a pusher blade mounted on the front. Depending on the manufacturer, these machines may be called tractors, dozers, or bulldozers.

The blade of a dozer can be hydraulically raised, lowered, or positioned at an angle. Most dozers are tracked vehicles, although there are some wheel versions, including heavy-duty four-wheel-drive dozers such as in *Figure 8 (A)*.

Dozers are used for a variety of purposes, including the following:

- Clearing land, including pulling out tree stumps
- Moving soil, rock, and other material
- Spreading material dumped on the site
- Rough and finish grading
- Maintaining a **haul road**

Dozers are configured in different ways to serve different purposes. For example, there are two types of crawler dozers. The low-track machine, as shown in *Figure 8 (B)*, is used primarily as a grading machine. The high-track type, as shown in *Figure 8 (C)*, is a high-powered machine used primarily as a pusher. Compact versions of these machines can be used for lighter grading work.

(A) Wheeled Dozer **(B) Low-Track Dozer** **(C) High-Track Dozer**

Figure 8 Dozers.

Source: (A) By High Contrast – Own work, CC BY 3.0 de, https://commons.wikimedia.org/w/index.php?curid=6398546

Many dozers are equipped with a powered winch such as in *Figure 9 (A)* so they can pull stumps, drag logs, pull other machines out of mud, and even pull themselves out of trouble. Another common attachment is the ripper, which is used to break up hard soil, rock, and other materials, such as concrete and asphalt (*Figure 9 (B)*).

(A) Winch **(B) Ripper**

Figure 9 Dozer attachments.

Sources: (A) Cturtletrax/iStock/Getty Images; (B) Zorandimzr/iStock/Getty Images

"Bull" Dozer

The original bulldozer was not a machine at all, and it had nothing to do with bulls. Rather, it was a blade pushed by mules. The first step in mechanizing this tool was accomplished when a tractor was converted to a continuous chain tread—or track crawler—format early in the twentieth century. The next step was to provide independent control for each track. The bulldozer came about when the blade was attached to a tractor a few years later. The final step occurred when a power control for the blade was added. Although the principle is still the same, dozers have come a long way since the early days.

1.1.6 Loaders

Loaders are used for a variety of purposes, including the following:

- Loading trucks
- Stockpiling soil, gravel, and other material
- Rough grading
- Digging
- Hauling material and equipment

Like many of the other types of equipment described, loaders come in crawler and wheel configurations. However, unlike excavators and dozers, the wheel version of the loader, shown in *Figure 10 (A)*, is more common and has a wider range of models and sizes than the track version *Figure 10 (B)*.

(A) Wheel Loader

(B) Crawler Loader

(C) Wheeled Skid Steer with Bucket

(D) Tracked Skid Steer with Auger

Figure 10 Loaders.

Sources: (A) Aguus/iStock/Getty Images; (B) Rob Byron/Shutterstock; (C) Maksim Safaniuk/iStock/Getty Images (D) Courtesy of CASE Construction Equipment

Attachments Used with a Skid Steer

A skid steer can be used to perform the functions of many other machines when equipped with attachments available from the manufacturer. This skid steer is equipped with a pallet forks attachment, which allows it to be used as a forklift. Many different attachments are made by the manufacturer for use with this machine, including backhoe, auger, and trenching attachments.

Source: Courtesy of CASE Construction Equipment

Compact loaders are equipped with either wheels (*Figure 10 (C)*) or tracks (*Figure 10 (D)*). Wheeled loaders are known as skid-steer loaders, while tracked loaders are called compact track loaders. Tracked loaders have better traction. Compact loaders are steered by pushing or pulling on two levers in the cab. If the levers are moved in opposite directions, the machine will counter rotate. This ability to turn in small spaces is where the term *skid steer* came from. Like the backhoe loader, compact loaders are available to use with a range of attachments, such as pallet forks, augers, brooms and grapple buckets.

1.1.7 Forklifts

Forklifts are used to move equipment and materials and to unload materials from trucks and trailers. The forklifts used in construction are generally diesel-powered, rough-terrain forklifts. These forklifts have a higher ground clearance and larger tires than warehouse forklifts. The two basic types of forklifts are fixed-mast and telescoping-boom.

Fixed-Mast Forklifts — The upright member along which the forks travel is called the mast. On a fixed-mast forklift (*Figure 11*), the forks can be raised as high as the limits of the mast allow. The mast may be able to be tilted forward 15 to 20 degrees and backward by a few degrees. These forklifts must be able to get very close to the pickup or landing point because of the limited movement range of the mast.

Telescoping-Boom Forklifts — A telescoping-boom forklift (*Figure 12*) is able to lift material much higher than fixed-mast forklifts and has much more flexibility in positioning a load. Some telescoping-boom forklifts have a level-reach fork carriage, which is often called a squirt boom. A squirt boom allows the fork carriage to be moved in the horizontal plane while the boom remains stationary.

(A) **(B)**

Figure 11 Fixed-mast forklifts.
Source: (A, B) Courtesy of CASE Construction Equipment

Articulating Forklifts — Articulating forklifts (*Figure 13*) are built with a fork carriage that is hinged so that it pivots left or right of centerline. The operator uses hydraulic controls and cylinders to move the fork carriage. Most articulating forklifts are designed for operations within narrow, confined spaces such as the narrow aisles inside a warehouse. They are also used inside truck trailers.

1.1.8 Motor Graders

A motor grader is a machine with a centrally located blade, also called a moldboard, which can be angled to push material to the side. On most motor graders, the moldboard can be positioned several feet outside the rear wheels and angled to work on slopes. The grader is used to establish a finish grade but can also be used for cut-and-fill work.

At one time, motor graders were made with a rigid frame. Modern motor graders have a hydraulically controlled articulating frame that pivots at a point

Figure 12 Telescoping-boom forklift.
Source: JLG Industries, Inc.

Figure 13 Articulating forklift.
Source: Landoll

just forward of the engine. The articulated frame provides greater maneuverability and a tighter turn radius. For example, articulation gives the machine the ability to work on a slope with the front wheels and moldboard on the slope and the rear wheels on the roadway for greater stability and power. The moldboard can be raised and lowered, turned in the horizontal plane, and have one end raised to grade slopes (*Figure 14*).

Figure 14 Grading a slope.
Source: Image courtesy of Komatsu America Corp.

Regardless of how the grader is being used, getting its blade or moldboard set to the proper angle is most critical. Graders are now equipped with a global positioning system (GPS) that allows the position of the moldboard to be set by a computer system. One or two GPS antennas are mounted near the outer tips of the moldboard (*Figure 15 (A)*). A GPS base station (surveying system) is set up near where the grader is working. The base station works with the GPS devices on the grader to control the positioning of the moldboard. Motor grader operators must learn how to work with these GPS devices. The GPS base station shown in *Figure 15 (B)* is an Ashtech ProFlex 500 GNSS receiver (with antenna) mated with a radio receiver.

Laser-Guided Machines

Laser-controlled graders are being used more and more because they provide very precise grading and significantly reduce grading errors and rework. The laser transmitter is set up to establish the desired grade. It then transmits a laser beam to sensors attached to a mast mounted on the grader blade. If the blade is off grade, the sensor detects the difference. In a fully automatic system, the sensor sends a signal to the hydraulic controls to adjust the blade. In a semiautomatic system, the sensor signal is sent to a cab display unit that tells the operator how to adjust the blade position. In addition to graders, this technology is being used on dozers to control cut and fill, as well as on excavating equipment to control the depth and slope of trenches.

Source: Grigorenko/iStock/Getty Images

(A) Motor Grader with One GPS Antenna

(B) GPS Base Station with Antenna and Radio Receiver

Figure 15 Motor grader with GPS.

Sources: (A) Courtesy of CASE Construction Equipment; (B) Image Courtesy of Trimble Inc.

Scarifier: An attachment with teeth used on motor graders to loosen soil.

The attachments commonly used on motor graders are a **scarifier** and a ripper (*Figure 16*). Scarifiers are attached to either the front of the grader or just in front of the moldboard (blade). Scarifiers rip up the ground so that the moldboard has an easier job of breaking up and moving the dirt.

A ripper mounted on the rear of a grader breaks up the ground as the grader passes over a given section of ground. The grader's moldboard may or may not be down and cutting the dirt at the same time the ripper is being used.

(A) Scarifier Shanks on a Motor Grader

(B) Ripper on the Rear of a Motor Grader

Figure 16 Common motor grader attachments.

Sources: (A) Larry Dallaire/iStock/Getty Images; (B) Getty_dumy67/iStock/Getty Images

1.2.0 Trenching and Excavating Equipment

Excavators and backhoe loaders are commonly used in preparing sites for construction. For example, either of these machines may be used to load soil onto a dump truck. Another important use for these machines is in digging trenches for the installation of pipelines and utilities.

1.2.1 Backhoe Loaders

A backhoe loader, shown in *Figure 17 (A)*, is often just called a *backhoe* and consists of three separate parts: a tractor, a loader attachment, and a backhoe attachment. The loader attachment is mounted on the front of the tractor while the backhoe attachment is mounted at the rear of the tractor. Other attachments are available to replace the loader bucket, such as the tool/material carrier shown in *Figure 17 (B)*.

The backhoe attachment consists of three components known as the boom, the stick, and the bucket, shown in *Figure 17 (C)*. The backhoe is used to dig and clean ditches, excavate hard-to-reach areas, and hoist light loads. One common use is digging trenches for pipes and other utilities. Many backhoes, including the one shown, have a boom that can slide left and right, which adds versatility to the machine. The backhoe buckets on larger backhoe loaders are able to lift over 400 lb, or 14 cubic feet of material, and can be extended more than 20'.

Figure 17 Backhoe loaders.
Sources: (A) Alder7/iStock/Getty Images; (B) Roman023/iStock/Getty Images (C) Courtesy of CASE Construction Equipment

The loader attachment is used primarily to move soil, gravel, stone, and similar material. It is also used to fill in ditches after the pipes or other items have been installed. Like its larger cousin the loader, the loader attachment on a backhoe loader can lift material from the ground and dump it into the bed of a dump truck. The largest loader buckets can lift more than a yard and a half of soil.

Backhoes are equipped with stabilizer legs (outriggers) that carry the weight when the backhoe is digging. Stabilizers distribute the lateral loads created by the bucket as it digs. They keep the backhoe from tipping or rolling, as well as widen the stance of the machine and increase its capacity.

The operator faces the loader end when moving the machine and when operating the loader. The backhoe has separate controls, and the operator simply turns the seat around to operate that attachment.

1.2.2 Excavators

The excavator is the workhorse on most construction sites. The excavator is used for the following types of work:

- Digging trenches and foundations
- Loading soil, stone, gravel, and other materials into dump trucks and scrapers
- Demolishing structures
- Clearing land
- Lifting and placing tools and materials

Most excavators are track vehicles, such as the medium construction excavator shown in *Figure 18 (A)*. Some manufacturers also make a few wheeled excavators. The one shown in *Figure 18 (B)* is a wheeled version of the excavator shown in *Figure 18 (A)*. The compact excavator, shown in *Figure 18 (C)*, has become popular, especially for working on sites with very little working clearance. Telescoping excavators are preferred for some applications. There are both track and wheel versions of the telescoping excavator.

(A) Hydraulic Crawler Excavator **(B) Hydraulic Wheel Excavator** **(C) Compact Excavator**

Figure 18 Excavators.

Sources: (A) Faiz Dila/iStock/Getty Images; (B) Sergii Petruk/iStock/Getty Images; (C) Photovs/iStock/Getty Images

There are many types of excavator attachments and many variations within the types. The bucket (*Figure 19*) is the most common attachment, and each excavator comes equipped with a bucket. One excavator manufacturer lists five different types of buckets for tasks such as trenching, excavation, rock ripping, utility work, and ditch cleaning. Each type of bucket comes in a range of sizes to match the excavator's capacity. Most buckets have teeth on the lower edge. The number and type of teeth is a function of the bucket width and purpose. Typically, excavators have from three to seven teeth. Buckets made for ditch cleaning do not have teeth.

(A) Excavator Bucket with Teeth **(B) Ditch-Cleaning Excavator Bucket without Teeth** **(C) Excavator with Replaceable Bucket Teeth**

Figure 19 Excavator buckets.

Excavators can be used for many tasks, including demolition. In some cases, the bucket attachment can be used to knock over or break up a structure. In other cases, specialized attachments such as powerful hydraulic shears and hammers are needed (*Figure 20*). Excavators can also be used to knock over and remove trees in clearing a site for construction. To collect and hold demolition debris, roots and stumps, or other such materials, excavator buckets can be equipped with an extra part often called a thumb.

(A) Rock Hammer **(B) Bucket with Thumb** **(C) Shears**

Figure 20 Demolition attachments.
Sources: (A) Dalomo84/iStock/Getty Images; (B) Jason Finn/iStock/Getty Images (C) Stanley LaBounty

Like the backhoe, an excavating assembly consists of a boom and stick (*Figure 21*). Many combinations of these two components are used to satisfy various needs. The boom and stick are matched to the application. Some have a short reach of less than 10'; others can reach 50' or more. Some are designed for an extended reach, while others are designed for work in areas where the swing radius is short. Different amounts of counterweight at the back of the excavator are used to balance the machine. The longer the reach and the greater the load capacity, the more counterweight is needed.

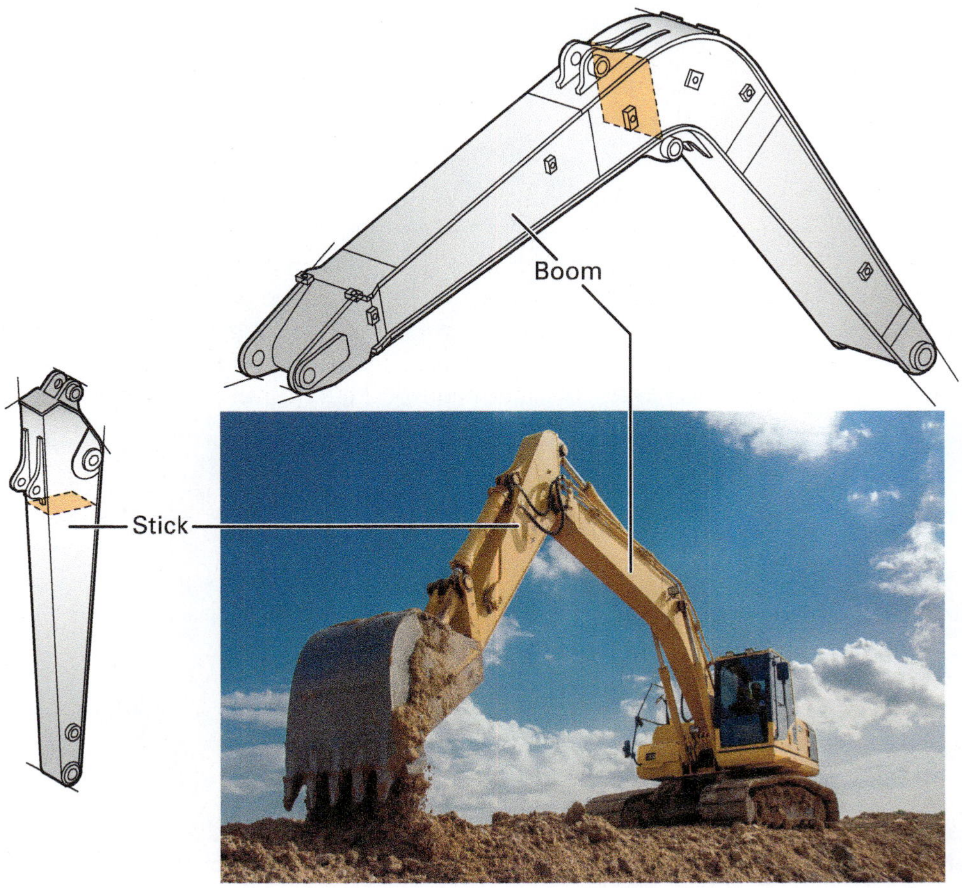

Figure 21 Stick and boom.
Source: (Photo) Chaiyaporn1144/iStock/Getty Images

1.2.3 Trenchers

Trenchers are machines used to dig trenches for the installation of underground services. Different styles of trenchers are available to match the scale of the work to be performed (*Figure* 22).

Pedestrian Trenchers — Operating pedestrian trenchers requires the operator to manually control the machine with handles. The operator must have good upper body strength to control the machine. The operator must also be very careful to maintain stable footing while operating the machine. Most pedestrian trenchers have a kill handle that must be held down to keep the machine running. If the operator loses control of the machine and releases the kill handle, the machine stops.

Compact Trenchers — The operator rides on the rear of a compact trencher. Most compact trenchers allow for trenching forward or backward. Like any other trencher, the compact trencher kicks the dirt out the side as it progresses along the way. Operators of these compact trenchers must be very careful about the grades over which they travel with these trenchers. The operator standing on the rear end of the trencher creates additional weight on that end of the machine. If the compact trencher is digging backward and is allowed to go over an incline, it may flip backward with the additional weight of the operator on it. If it is digging forward and moves across an incline, the weight of the digging bar and chain can make the machine flip forward, even with the operator's weight on the rear. Operators of compact trenchers need to be constantly aware of the trencher's balance.

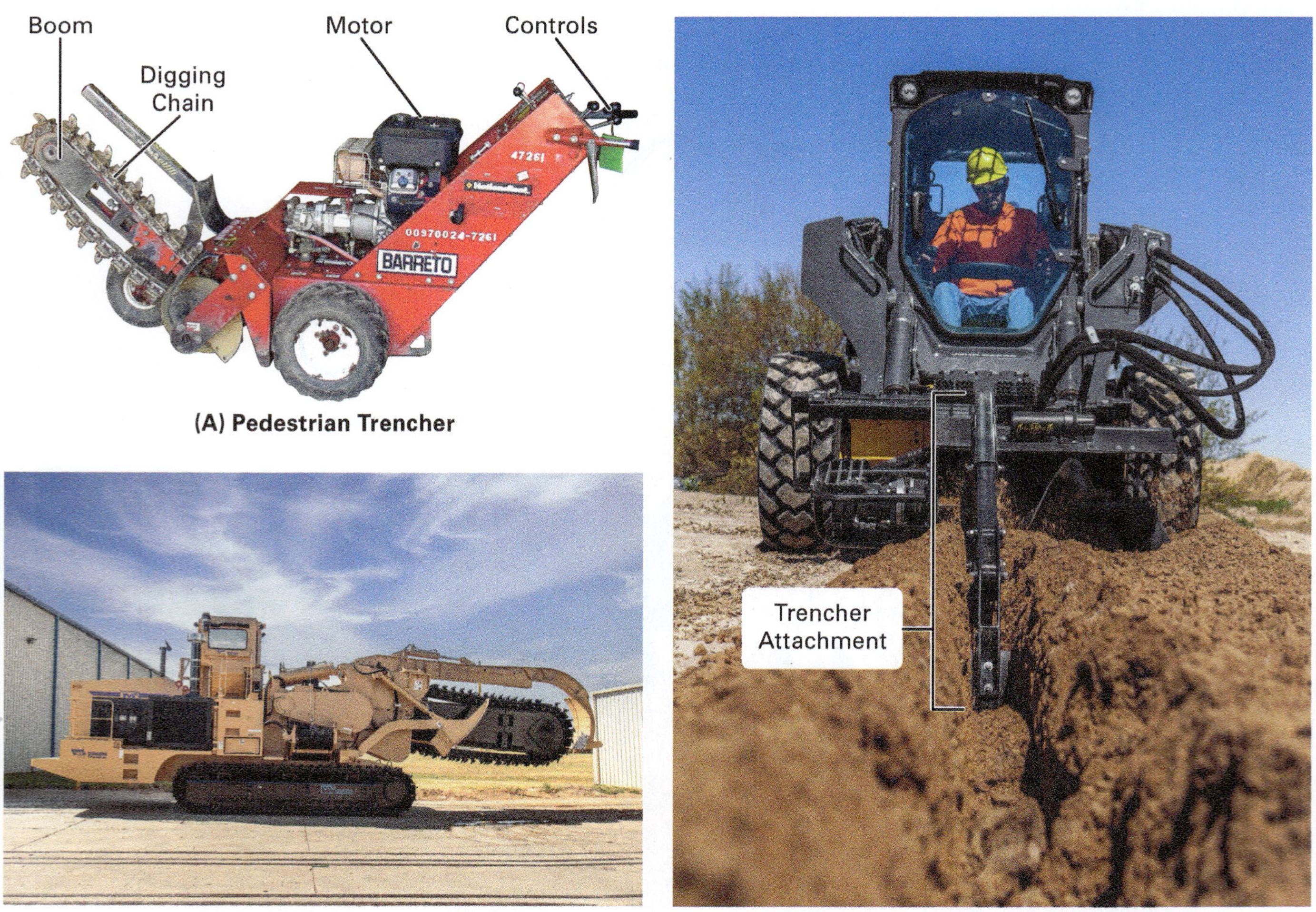

(A) Pedestrian Trencher

(B) Large Construction Trencher

(C) Skid Steer with Trencher Attachment

Figure 22 Examples of trenchers.

Sources: (B) TESMEC USA M5 Trencher; (C) Image Provided Courtesy of John Deere

Large Trenchers — Large trenchers are simply large construction vehicles. They are operated much like an excavator in that controlling their tracks controls the direction in which they move. Their digging boom and chain assembly is a lot like the boom, stick, and bucket of an excavator. It is raised for travel and lowered for trenching. The operator inside the cab must stay alert to his or her surroundings. The operator's concentration is on the trenching equipment, but he or she still must stay alert to anything going on around the trenching process.

Trencher Attachments – These attachments are used frequently with skid steers and utility tractors. *Figure 10 (C)* is an example of a trencher attached to a skid steer. Skid steers operate in reverse when using a trenching attachment. When used with utility tractors, trencher attachments are pulled behind the tractor after being connected to the three-point hitch and PTO shaft.

Trenchers use either a gasoline or diesel engine to drive a hydraulic system that provides power to the drive wheels and digging chain mounted on the digging boom. The digging chain and boom are very similar to the blade and chain of a chainsaw. When activated, the chain moves forward along the top of the boom, around the end of the boom, and returns to the mounting end of the boom. When the boom is lowered onto the ground, the chain begins to dig into the ground. As the digging progresses, the dirt and rock being removed are kicked out the sides. Also, as the digging progresses, the operator makes the digging boom go farther and farther into the ground until the desired depth of the new trench is reached. At that point, the trenching machine is moved on to dig another section of the trench. On some trenchers, the trenching is done as the trencher machine is backing up. If that is the case, the dirt is kicked back toward the machine. Most large trenchers move forward and kick the dirt out and away from the machine.

Any kind of trenching requires that the trenching machinery dig into the ground. Trenching operations must therefore be well planned prior to any trenching work. Underground utility lines must be well marked by the utility companies owning them. Trenching operators need to walk the area to be trenched and verify that all underground utility lines are well marked and easily seen from the operator's position behind, on, or in the trenching machine being used.

1.3.0 Mining Equipment

Heavy equipment operators may find themselves working in mining. Mining is a form of earthmoving work that is carried out both above ground and below ground. In addition to the usual safety guidelines required on any construction site, personnel doing mining work must also comply with the guidelines set forth by the Mine Safety and Health Administration (MSHA). MSHA is charged with guarding the safety of miners to prevent death, disease, and injury resulting from mining work.

1.3.1 Aboveground Mining Equipment

Aboveground mining involves digging down through the surface of the earth to reach particular minerals, such as coal, mineral ore, or stone. Construction materials, such as sand, fill dirt, and gravel, are taken from open-pit quarries located all over the country. The equipment used for aboveground mining is usually large dozers, excavators, loaders, and rigid-frame trucks. *Figure 23* shows a few of the heavy equipment machines often found in aboveground mining operations.

(A) Large Excavator with Shovel Attachment **(B) Shovel Excavator Loading Rigid Frame Truck**

Figure 23 Aboveground mining equipment.
Sources: (A) Adhi Warsito/iStock/Getty Images; (B) nikitos77/iStock/Getty Images

Giant Excavator

This monster of an excavator is used in mine and quarry operations. It can lift 50 cubic yards of material in one scoop.

Source: Image courtesy of Komatsu America Corp.

1.3.2 Belowground Mining Equipment

Belowground mining involves digging a tunnel into Earth and then working underground to extract the desired materials. Any time a tunnel is dug into the ground, the ground above the work area must be either strong enough to support any weight above the open area or be supported by fabricated bracing that prevents a collapse of the ground above the opening. To help reduce the risk of a collapse, the ceiling or overhead surfaces above the underground work areas are kept low, which means that the machines used in underground mining are built low to the ground. The machines that are primarily used in underground mining are underground mining loaders and underground mining trucks. Construction excavators and loaders may be used for higher-ceiling areas. *Figure 24* shows machines often used for underground mining.

Because underground mining is done in a confined space, additional attention must be paid to the air in the area. Workers and the machines they are using must be able to breathe the air without problems. The exhaust that the machines produce must not pollute the air to the point where it hurts the workers.

(A) Wheeled Loader and Hydraulic Excavator Digging a Tunnel **(B) Underground Mining Loader** **(C) Underground Truck**

Figure 24 Underground mining equipment.
Sources: (A) agnormark/iStock/Getty Images; (B) Nordroden/iStock/Getty Images; (C) Image courtesy of Komatsu America Corp.

Underground mining means digging through the dirt, rock, and minerals that make up the earth's crust, which may at times include gas pockets or veins of water. Such potential hazards should be identified and eliminated by mining safety engineers before the heavy equipment operators start their work in the mines. Heavy equipment operators must review all MSHA guidelines and all company safe-operating procedures before doing any underground mining work.

1.0.0 Section Review

1. The type of compactor used primarily for rolling paving material such as asphalt is the _____.
   a. steel wheel compactor
   b. pneumatic tire compactor
   c. vibratory compactor
   d. sheepsfoot compactor

2. The machine that uses stabilizer legs is the _____.
   a. excavator
   b. backhoe loader
   c. scraper
   d. motor grader

3. The machines commonly used in underground mining are the loader and the _____.
   a. backhoe
   b. excavator
   c. mining truck
   d. scraper

2.0.0 Hydraulic Systems

Objective

Explain hydraulic systems and their components.

a. Explain hydraulic system fundamentals.

b. Identify and describe the equipment and components used in hydraulic systems.

Performance Tasks

There are no Perfomance Tasks in this section.

Most modern heavy equipment uses the engine to power a hydraulic pump (or multiple hydraulic pumps). These pumps force high-pressure hydraulic oil to do the work of moving the drive train, blades, buckets, and other tools used on most heavy construction equipment. A hydraulic system uses high-pressure hydraulic pumps and hydraulic fluids to do its work. Be prepared to follow all safety precautions when working on hydraulic systems. Because hydraulic pressure must be released during system maintenance or repair, it may be necessary to block cylinders from retracting and arms from falling. Rotating equipment may need to be **dogged**.

Dogged: Held or secured from moving with a mechanical device.

Going Green

Hydraulic Fluid

Hydraulic systems notoriously leak. The source of such leaks must be identified and stopped. Hydraulic spills must be prevented. Spilled hydraulic fluids harm the environment. If unable to prevent a hydraulic spill, follow SDS instructions and local laws for dealing with hydraulic fluid spills. Contaminated soil must be removed and disposed of in accordance with local guidelines.

2.1.0 Hydraulic Fundamentals

Pascal's law: A scientific law stating that pressure at any one point in a closed static fluid system will be applied to all points in the system in every direction and will exert equal forces on equal areas. It is the basis of hydraulic theory and was developed by Blaise Pascal (1623–1662), a French philosopher and mathematician.

The transmitting of power in a hydraulic system requires an understanding of **Pascal's law** of hydraulics. This law states that any pressure applied to a fluid in a closed system is applied to all points and surfaces in the system equally. Look at *Figure 25*. It shows a smaller piston with a force of 10 lb applied to it, transmitting a force to the larger piston, which has five times the surface area of the smaller piston. This results in a force of 50 lb generated by the larger piston. However, the larger piston was only moved one-fifth the distance of the smaller piston.

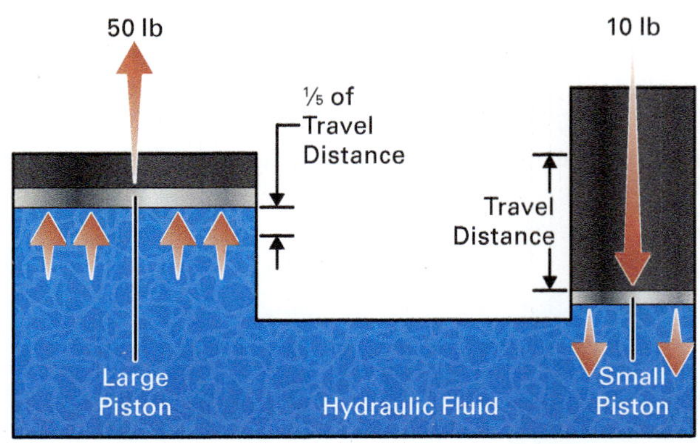

Formulas for Pascal's Law

$$\text{Pressure (psi)} = \frac{\text{Force (lb)}}{\text{Area (in}^2\text{)}}$$

$$\text{Force (lb)} = \text{Pressure (psi)} \times \text{Area (in}^2\text{)}$$

Figure 25 Applying Pascal's law.

Hydraulic actuator: A device that converts the fluid energy of the hydraulic system into mechanical energy using a hydraulic cylinder or motor.

A typical hydraulic system (*Figure 26*) consists of four basic components: a hydraulic pump, a reservoir with a strainer, a valve, and a **hydraulic actuator**, in this case a cylinder. The pump moves the hydraulic fluid through the system. The reservoir is the source of the hydraulic fluid to be pumped into the system. It also allows for the return of the fluid. The valve controls the fluid as it does its work by porting the pressurized fluid to the proper place. The actuator is the device that performs the work. The actuator can be a cylinder that provides for linear work or a hydraulic motor for rotating work.

Because hydraulics increase a system's mechanical advantage, they are frequently used to provide power assistance for many of the mechanical systems that an equipment operator must position manually, such as brakes and clutches.

WARNING!

When checking, filling, or working on a hydraulic system, wear proper eye, face, and hand protection. Hydraulic oil may be very hot if the machine has been operating. Also, ensure that any hydraulic cylinders are retracted as much as possible and that any hoist motors are dogged before releasing pressure on the system or any pressurized reservoir tank. Hydraulic system pressures can be very high if the system is under static load. Retracting any hydraulic cylinders will prevent overfilling of the reservoir tank.

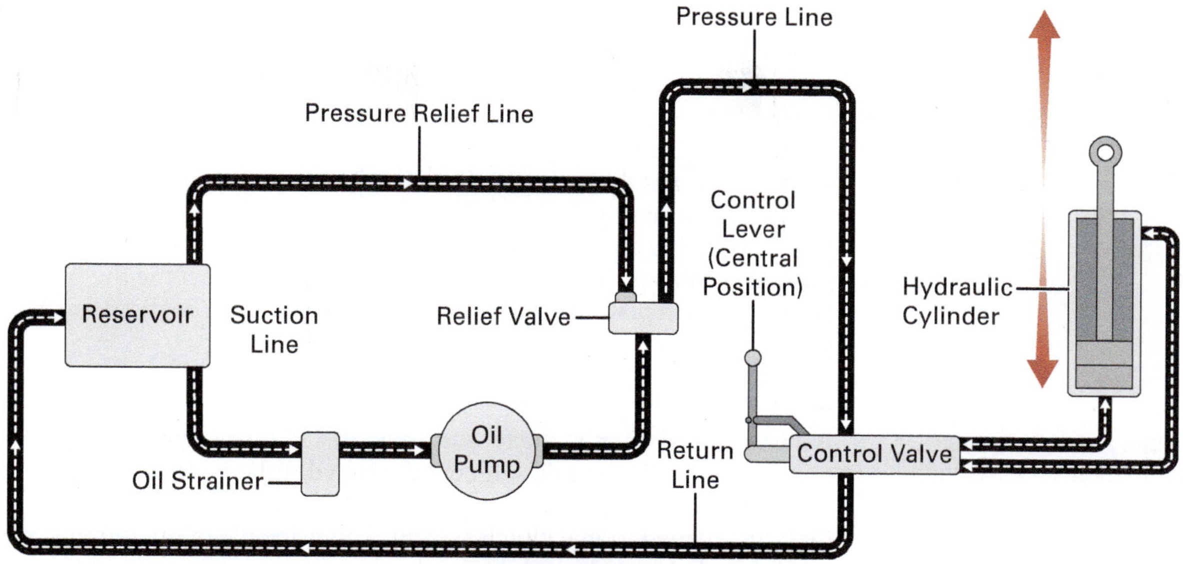

Hydraulic System

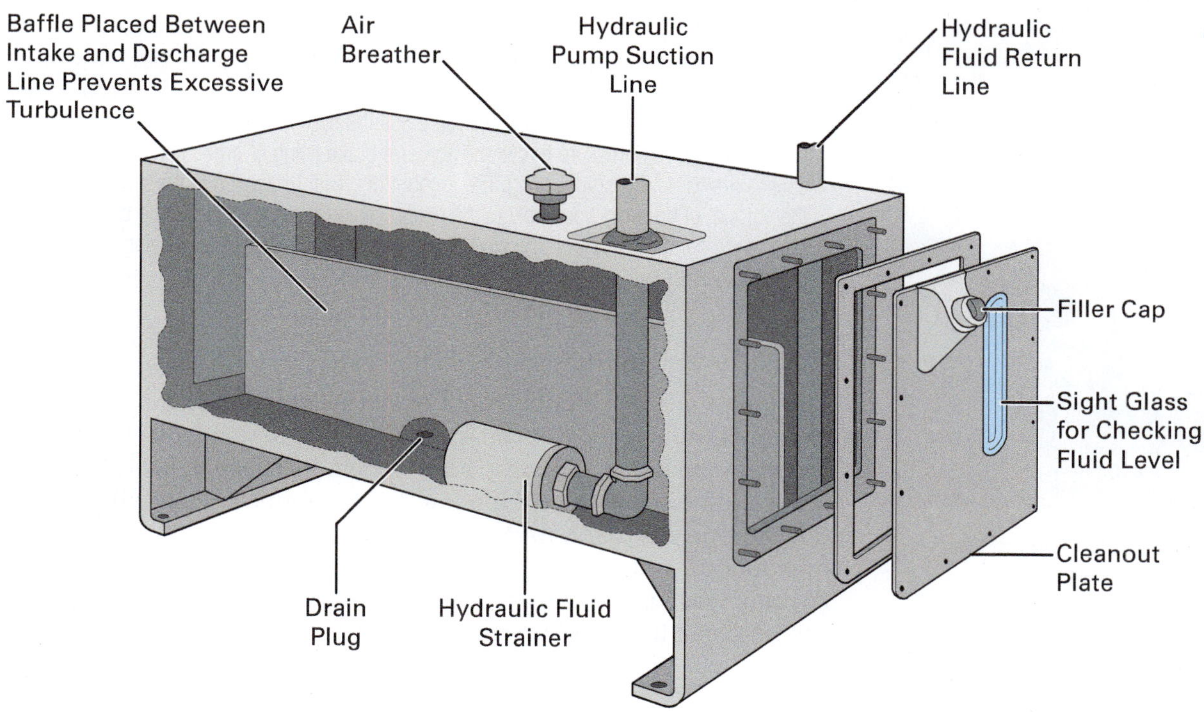

Reservoir Tank

Figure 26 Typical hydraulic system and reservoir tank.
Source: United States Navy

2.2.0 Hydraulic Equipment and Components

Hydraulic systems are made up of components and hydraulic hoses. The main components of hydraulic systems used in heavy equipment are brakes, clutches, couplers, and pumps.

2.2.1 Hydraulic Brakes and Clutches

Power-assist brakes are used on some machines. This system uses a small reservoir of brake fluid with a piston. As the operating pedal is pushed, it transmits the pressure to a brake cylinder, which in turn applies the brake pad to the drum or disc. This same type of system may be used for other braking systems, such as the swing brake and hoist brake on some machines.

The clutch system uses a similar layout. This system typically involves a pedal operating a piston in a reservoir. The pressure applied when the pedal is pushed is transmitted to a cylinder that moves to overcome the spring pressure of the pressure plate. This allows the engine power to be disconnected from the power distribution system.

2.2.2 Hydraulic Power Couplers

Torque converter: A device that uses fluid, usually oil, to transmit torque from one shaft to another. Also known as a *fluid coupler*.

A system used often is the **torque converter**. A torque converter, also known as a fluid coupler, is a small, self-contained hydraulic system. It is used to connect the main power source to the transmission or gearbox. The torque converter replaces a conventional clutch system. The principle of operation is simple. It is an oil-filled, donut-shaped device that has pump vanes connected to the engine surrounding turbine vanes. The vanes are connected to the transmission (*Figure 27*). As the pump is turned by the engine, it forces fluid into the turbine vanes, making them move. This movement causes the transmission gearing to transfer the torque created by the movement to the power distribution system.

Figure 27 Torque converter.
Source: Ellefson Off Highway, Inc.

Hydraulic Pressure and Systems

A pinpoint leak in a hydraulic system can be very dangerous. In some cases, the pressure from such a leak can be great enough to puncture the skin. A good example of a hydraulic system is the braking system on your car. When you press the brake pedal, hydraulic fluid in the brake lines is pressurized. The pressure forces the abrasive brake lining or brake pad against the rotor, which causes the wheels to stop turning. The amount of pressure applied to the brakes is proportional to the amount of pressure you apply to the pedal. That is why you can use the brake to simply slow down or to come to a screeching stop. The same principle is used to position buckets and blades on heavy equipment.

2.2.3 Hydraulic Pumps

The pump is a major component of the hydraulic system. The pump is driven from the output of the transmission or gearbox to supply the fluid flow necessary for operation of the hydraulic power plant. Its speed depends on engine speed. If the engine speed is kept too low, the hydraulic system will have insufficient flow to operate the system. This reduced flow can cause overheating of the hydraulic oil and prevent the operation of some or all of the hydraulic actuators. For this reason, it is important that the equipment operator maintain enough engine speed to allow for proper operation of the hydraulic system.

2.0.0 Section Review

1. The hazards associated with hydraulic systems are _____.
   a. heat and high pressure
   b. high pressure and toxic fumes
   c. heat and toxic fumes
   d. heat and corrosive acid

2. The device used to connect the engine to the transmission in place of a clutch is the _____.
   a. hydraulic actuator
   b. hydrostatic shifter
   c. pressure regulator
   d. torque converter

Module 22103 Review Questions

1. The purpose of a PTO on a tractor is to _____.
   a. improve the tractor's acceleration
   b. provide power for attachments
   c. augment the tractor's brakes
   d. provide power to the transmission

2. Articulated trucks normally have capacities of up to _____.
   a. 20 tons
   b. 40 tons
   c. 100 tons
   d. 300 tons

Figure RQ01
Source: Vadym Terelyuk/iStock/Getty Images

3. The machine shown in *Figure RQ01* is a _____.
   a. wheel loader
   b. wheel dozer
   c. tractor
   d. roller/compactor

Figure RQ02
Source: Iron_Man_/iStock/Getty Images

4. The machine shown in *Figure RQ02* is a(n) _____.
   a. scraper
   b. excavator
   c. skid steer
   d. motor grader

Figure RQ03

5. The machine shown in *Figure RQ03* is a(n) _____.
   a. dozer
   b. excavator
   c. wheel loader
   d. backhoe loader

Figure RQ04
Source: Maksim Safaniuk/iStock/Getty Images

6. The machine shown in *Figure RQ04* is a(n) _____.
   a. motor grader
   b. excavator
   c. skid steer
   d. scraper

Figure RQ05
Source: Courtesy of CASE Construction Equipment

7. The machine shown in *Figure RQ05* is a(n) _____.
   a. wheeled loader
   b. fixed-mast forklift
   c. wheeled excavator
   d. telescoping boom forklift

Figure RQ06
Source: Image courtesy of Komatsu America Corp.

8. The machine shown in *Figure RQ06* is used primarily as a(n) _____.
   a. pusher
   b. loader
   c. grading machine
   d. excavating machine

9. Which of these machines is equipped with stabilizers?
   a. Roller
   b. Dozer
   c. Motor grader
   d. Backhoe loader

10. The number of teeth on an excavator bucket is a function of the _____.
    a. bucket's width
    b. capacity of the bucket
    c. capacity of the excavator
    d. material being excavated

11. A small, self-contained hydraulic system that connects the main power to the gearbox is called a _____.
    a. torque converter
    b. transmission
    c. hydraulic actuator
    d. boom

12. In a heavy equipment hydraulic system, the pump is driven from the _____.
    a. battery
    b. transmission
    c. water pump
    d. alternator

Answers to odd-numbered review questions are found in the Module Review Answer Key at the back of this book.

Answers to Section Review Questions

Answer	Section	Objective
Section 1.0.0		
1. a	1.1.3	1a
2. b	1.2.1	1b
3. c	1.3.2	1c
Section 2.0.0		
1. a	2.1.0	2a
2. d	2.2.2	2b

MODULE 22102

Heavy Equipment Operations Safety

Source: Sam Edwards/iStock/Getty Images

Objectives

Successful completion of this module prepares you to do the following:

1. Explain work zone safety requirements.
   a. Describe signs, barriers, and other devices and methods used for work zone safety.
   b. Describe traffic control safety methods.
2. Explain hazard communication.
   a. Explain the purpose and content of safety data sheets (SDS).
   b. Describe the labeling systems used for hazardous materials.
   c. Describe noise control methods.
3. Identify and describe safety practices associated with heavy equipment operation.
   a. List jobsite and equipment safety requirements.
   b. Explain the safety hazards associated with weather.
4. Identify safety practices associated with trenches and excavations.
   a. Explain how soil composition relates to trench failures.
   b. Describe the methods used to protect workers in trenches and excavations.
   c. Identify personal safety hazards related to dust and soils.

Performance Tasks

Under supervision, you should be able to do the following:

1. Place barricades and temporary traffic control devices for a highway construction zone.
2. Demonstrate how to use flags or paddles to control traffic.
3. Demonstrate how to use various types of personal protective equipment (PPE):
   - Hard hat
   - Safety glasses
   - Ear protection
   - Gloves
   - Safety harness
   - Reflective safety vest

Digital Resources for Heavy Equipment Operations

SCAN ME

Scan this code using the camera on your phone or mobile device to view the digital resources related to this craft.

Overview

Working in and around heavy equipment is hazardous. Heavy equipment operators and other workers need to work together to keep the jobsite safe. The heavy equipment operator's job is to operate the equipment in a manner that protects both the operator, other workers, and the equipment. When working with and around heavy equipment, operators must act in a way that does not place them or their coworkers in danger.

This module covers dangers associated with working around heavy equipment. Because heavy equipment work is hazardous, specific rules and procedures must be in place to decrease the chances of having an accident. This module highlights many of those rules and practices that will help you stay safe while operating heavy equipment.

NCCER Industry-Recognized Credentials

If you are training through an NCCER-accredited sponsor, you may be eligible for credentials from NCCER. The ID number for this module is 22102. Note that this module may have been used in other NCCER curricula and may apply to other level completions. Contact NCCER at 1.888.622.3720 or go to **www.nccer.org** for more information.

You can also show off your industry-recognized credentials online with NCCER's digital credentials. Transform your knowledge, skills, and achievements into credentials that you can share across social media platforms, send to your network, and add to your resume. For more information, visit **www.nccer.org**.

1.0.0 Work Zone Safety

Performance Tasks

1. Place barricades and temporary traffic control devices for a highway construction zone.
2. Demonstrate how to use flags or paddles to control traffic.

Objective

Explain work zone safety requirements.

a. Describe signs, barriers, and other devices and methods used for work zone safety.
b. Describe traffic control safety methods.

The Occupational Safety and Health Administration (OSHA) and Mine Safety and Health Administration (MSHA) are the bodies of authority who provide general principles for what are safe and healthy working conditions at jobsites. MSHA has a few industry-specific requirements, but much of what OSHA and MSHA regulate are the same for worksites across multiple industries. Safety training is required for all activities to prevent as many worksite accidents as possible.

One of the ways accidents can be avoided is by using safety devices such as signs, safety tags, barricades, barriers, and other devices. When used properly, these devices can not only prevent accidents but can also assist in the smooth flow of work and permit the job to be completed on schedule.

Construction work zones are often located near public areas and sometimes even on public roads (*Figure 1*). This creates the additional danger of having pedestrian and motor traffic in the surrounding area. Work on public roads is so dangerous that every US state has enacted laws that allow added fines and penalties for motorists who speed and commit other traffic violations in roadway construction zones. Specific standard procedures must be used without fail in such work zones to ensure everyone's safety.

Safety devices, policies, and procedures are like tools: they cannot prevent accidents unless they are used correctly. Workers are the best tools for preventing accidents. They should stay alert for information and conditions that can affect everyone's safety (*Figure* 2) and they should take responsibility for their own safety. Always think a job through before starting it. Look around for potential hazards, and stay alert for changes in the work area. Remember the following:

- Accidents can kill and disable.
- Accidents cost you and your company money.
- Accidents can be prevented.

Figure 1 Traffic control.
Source: Kent Weakley/Shutterstock

Figure 2 Stay alert for important safety information.

Case History

A 23-year-old apprentice power line worker died of injuries he received after being run over by the tandem dual rear tires of a digger derrick truck. He was part of a five-person crew that had been working on the job for three days. At the time of the incident, the victim and two other crew members had just finished setting and back-filling around a utility pole. They proceeded to the next pole requiring framing and setting, walking 30 feet ahead of the digger derrick in one lane of the road. The digger derrick moved slowly in reverse to the same pole. At a point approximately midway between the two poles, the victim knelt with his back to the truck, apparently to inscribe a word or initials into some seal coating on the roadway. He was hit and run over by the backing truck's passenger-side tandem dual rear tires. The digger derrick truck did not have a back-up alarm system.

The Bottom Line: Keep your mind on the job. Communication with other workers can save your life. Check equipment for warning signals before you use it. If it does not have warning signals, establish hand or radio signals. Always check the surrounding area for potential hazards.

Source: The National Institute for Occupational Safety and Health (NIOSH)

1.1.0 Signs, Barriers, and Other Safety Devices

To ensure everyone's safety, unauthorized people and vehicles must be kept away from the work area. Creating a clear work zone is an important part of working safely (*Figure* 3). Signs, tags, lockout/tagout devices, and barricades and barriers are some of the ways to mark the jobsite and alert workers and the public of hazardous conditions. Operators should always be aware of state and local regulations before beginning a job.

(A) Barricades

Red	Indicates immediate and severe danger, or presence of fire protection equipment.
Orange	Marks the dangerous parts of machines or energized equipment which may cut, crush, shock or injure employees. Orange emphasizes these hazards when the guards or the enclosure around them is open.
Yellow	Designates caution and should be used for marking physical hazards such as: striking against, stumbling, falling, tripping, and 'caught inbetween'.
Blue	Used to mark out-of-service equipment which should not be used.
Green	Indicates either the location of safety equipment such as first aid materials or conveys safety information.
Purple	Used to mark radiation hazards. It may contain a combination of purple and yellow.

(B) Safety-Related Colors

Figure 3 Typical use of barricades at a jobsite.
Source: (A) J Cardamone/Shutterstock; (B) Shutterstock ID 1992703211

1.1.1 Signs

All worksites have specific markings and signs to identify hazards and provide important information. You must recognize all types of signs related to each jobsite and make sure they are properly placed and working correctly. Doing this can save a life.

One type of sign is the danger sign (*Figure 4*). Danger signs are usually red, black, and white. They are used to inform workers that an immediate hazard exists.

Another type of sign is the caution sign (*Figure 5*). Caution signs are used to inform workers of potential hazards or unsafe conditions. They are usually yellow with black letters. When you see a caution sign, take action to protect yourself.

Temporary traffic control (TTC): Device and plan used to safely divert traffic when the normal use of the road is disrupted due to construction.

Another type of sign seen at worksites located on or near roads is a **temporary traffic control (TTC)** sign (*Figure 6*). These signs are used to redirect traffic and provide information to motorists and pedestrians during construction and maintenance of roadways and at other worksites. These signs are orange with black letters.

Safety alert symbols are sometimes seen on signs. The safety alert symbol is an exclamation point inside of a triangle. It is an internationally recognized symbol that means, Attention! Become Alert! These signs are seen in manuals, on machines, and on safety signs. When you see a safety alert symbol, pay attention to the message. It could save your life. If unsure of the meaning, check with a supervisor.

Figure 4 Typical danger sign.

Figure 5 Typical caution sign.

Figure 6 Temporary traffic control signs.

Safety alert symbols are organized into the following three safety levels:

- *Danger* — Danger symbols indicate immediate hazards that, if not avoided, will result in death or serious injury (*Figure 7*).

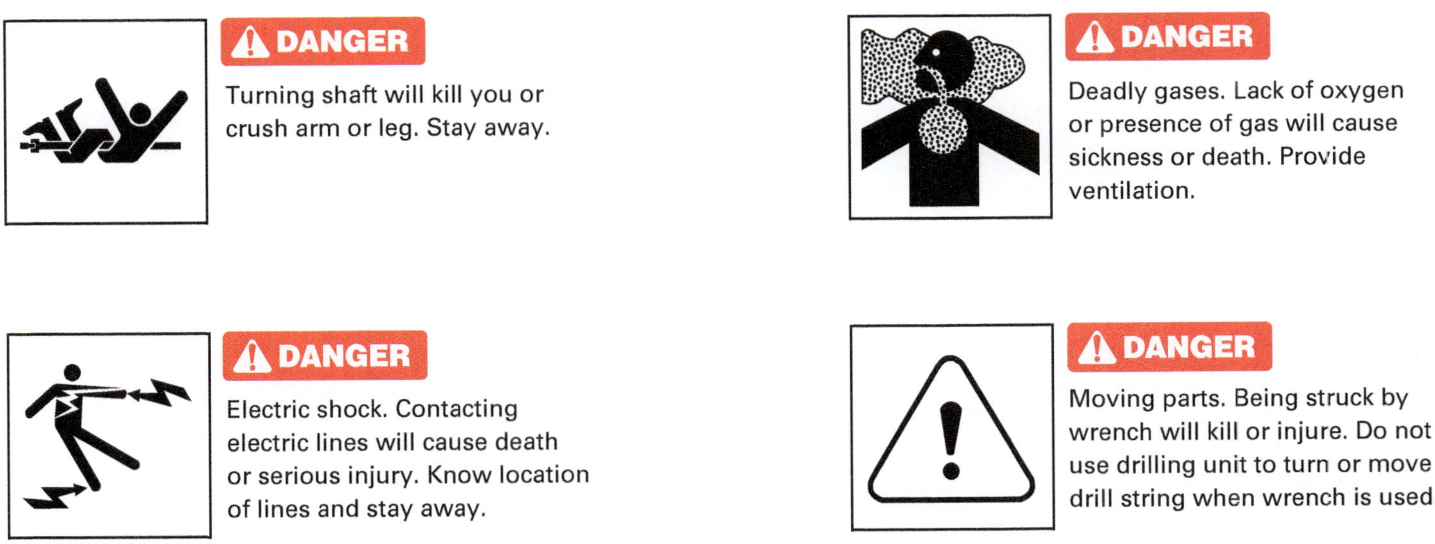

Figure 7 Danger symbols.

- *Warning* — Warning symbols (*Figure 8*) indicate potential hazards that, if not avoided, could result in death or serious injury.
- *Caution* — Caution symbols (*Figure 9*) indicate hazards that, if not avoided, may result in minor or moderate injury.

On-site traffic signs help with the safe movement of vehicles and pedestrians. Just as traffic signs must be obeyed on public highways and streets, they must also be observed in the workplace.

⚠WARNING

Explosion possible. Serious injury or equipment damage could occur. Follow directions carefully.

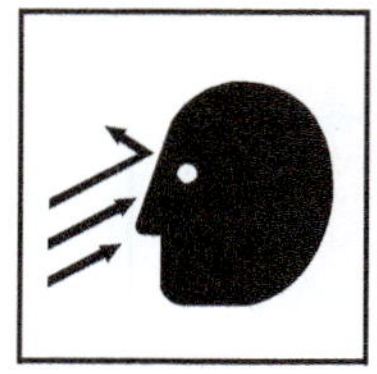

⚠WARNING

Looking into fiber-optic cable could result in permanent vision damage. Do not look into ends of fiber-optic or unidentified cable.

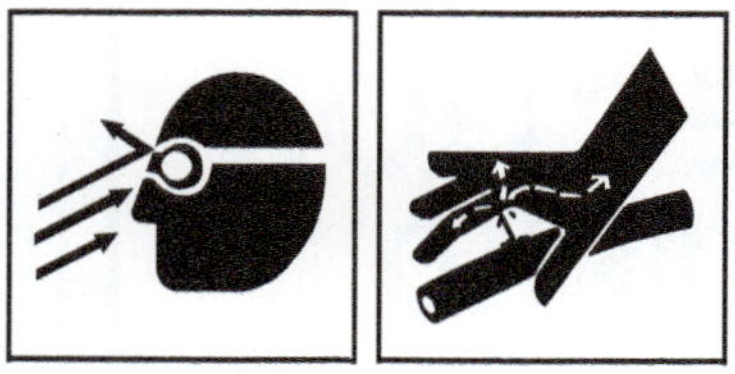

⚠WARNING

Fluid or air pressure could pierce skin and cause injury or death. Stay away.

⚠WARNING

Fire or explosion possible. Fumes could ignite and cause burns. No smoking, no flame, no spark.

⚠WARNING

Jobsite hazards could cause death or serious injury. Use correct equipment and work methods. Use and maintain proper safety equipment.

⚠WARNING

Moving traffic—hazardous situation. Death or serious injury could result. Avoid moving vehicles, wear high visibility clothing, post appropriate warning signs.

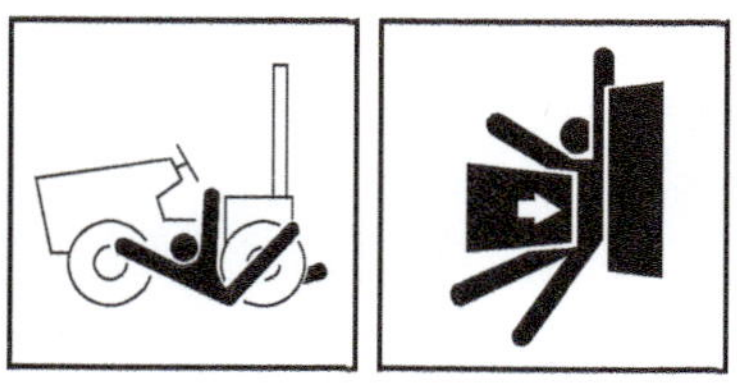

⚠WARNING

Crushing weight could cause death or serious injury. Use proper procedures and equipment or stay away.

⚠WARNING

Hot pressurized cooling system fluid could cause serious burns. Allow to cool before servicing.

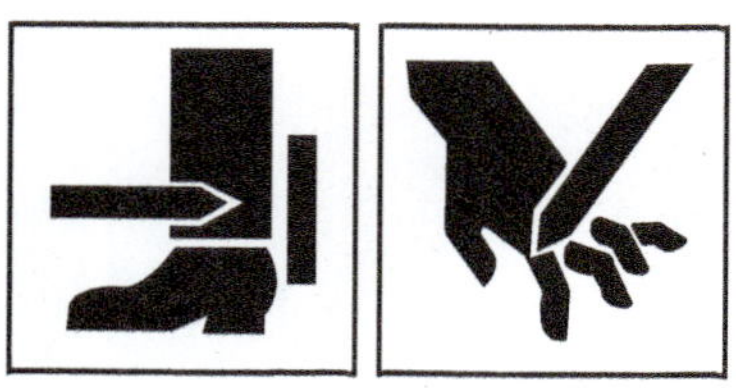

⚠WARNING

Moving parts could cut off hand or foot. Stay away.

⚠WARNING

Improper control function could cause death or serious injury. If control does not work as described in instructions, stop machine and have it serviced.

Figure 8 Warning symbols.

⚠CAUTION

Flying objects may cause injury. Wear hard hat and safety glasses.

⚠CAUTION

Fall possible. Slips or trips may result in injury. Keep area clean.

⚠CAUTION

Hot parts may cause burns. Do not touch until cool.

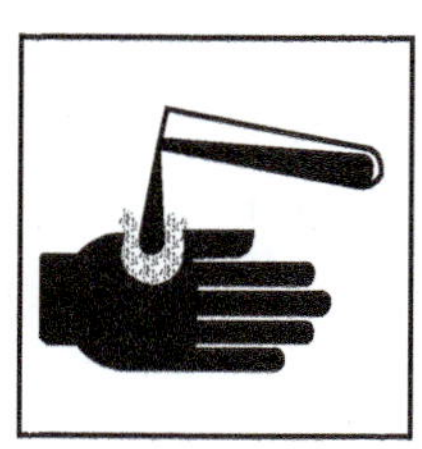

⚠CAUTION

Battery acid may cause burns. Avoid contact.

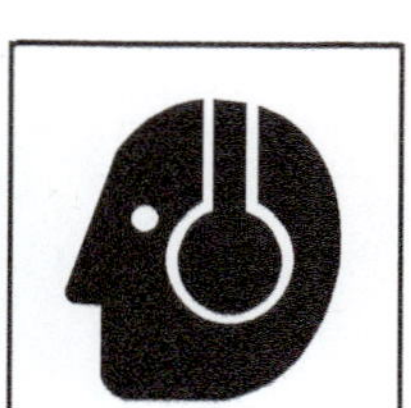

⚠CAUTION

Exposure to high noise levels may cause hearing loss. Wear hearing protection.

⚠CAUTION

Improper handling or use of chemicals may result in illness, injury, or equipment damage. Follow instructions on labels and in safety data sheets (SDSs).

Figure 9 Caution symbols.

A slow-moving-vehicle emblem is used on vehicles that move at speeds of 25 miles per hour (mph) or less. The emblem is a fluorescent yellow-orange triangle with dark red reflective border (*Figure 10*). It is frequently seen on construction and farm equipment.

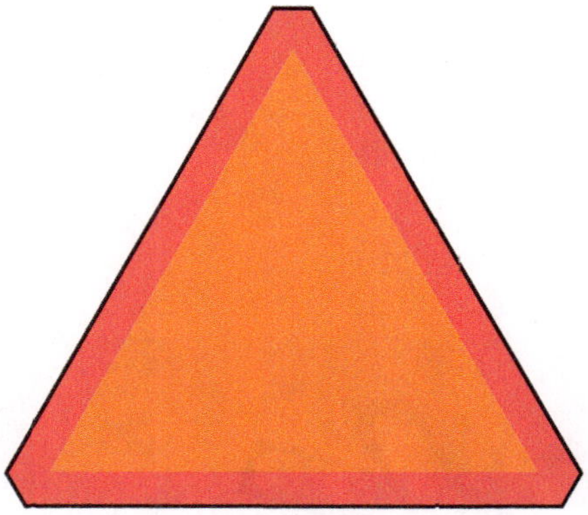

Figure 10 Slow-moving-vehicle sign.

1.1.2 Safety Tags

Accident prevention tags (*Figure 11*) are used as a temporary warning for workers of immediate and potential hazards. Tags are similar to signs, but they are not meant to be used in place of signs or as a permanent means of protection. For example, a tag that states DO NOT START may be used on a forklift while it is being serviced.

Figure 11 Examples of safety tags.

Tags can be an effective way of protecting workers and property when used correctly. Use tags on the equipment whenever it is out of order or being serviced, or when other workers need special information about the equipment. Only the person who placed the tag is authorized to remove it.

WARNING!

Never remove a safety tag that you did not place. You could put others in danger.

1.1.3 Lockout Devices

Another safety tool that you may be required to use is a lockout/tagout device, which ensures no one uses the equipment that is locked out. Some of these devices are very secure, because they permit a part or switch to be locked with a keyed padlock (*Figure 12*). Others are less secure than the padlock because they may simply be tags that have the words LOCKED OUT on them. Both devices mean the same thing and should be removed only by the person who attached the device.

Figure 12 Lockout/tagout devices.

1.1.4 Barricades and Barriers

Barricades and barriers are used to alert workers to potential hazards and to help prevent accidents. Each work zone may have different policies and procedures for how and when to use them, so learn and follow the rules at the jobsite. Sometimes barriers are placed in an area to prevent injury to workers, and other times they are placed to alert heavy equipment operators of dangers. For example, a barrier may be erected to keep a bulldozer off soft, graded ground. *Figure 13* shows the various types of barricades and barriers.

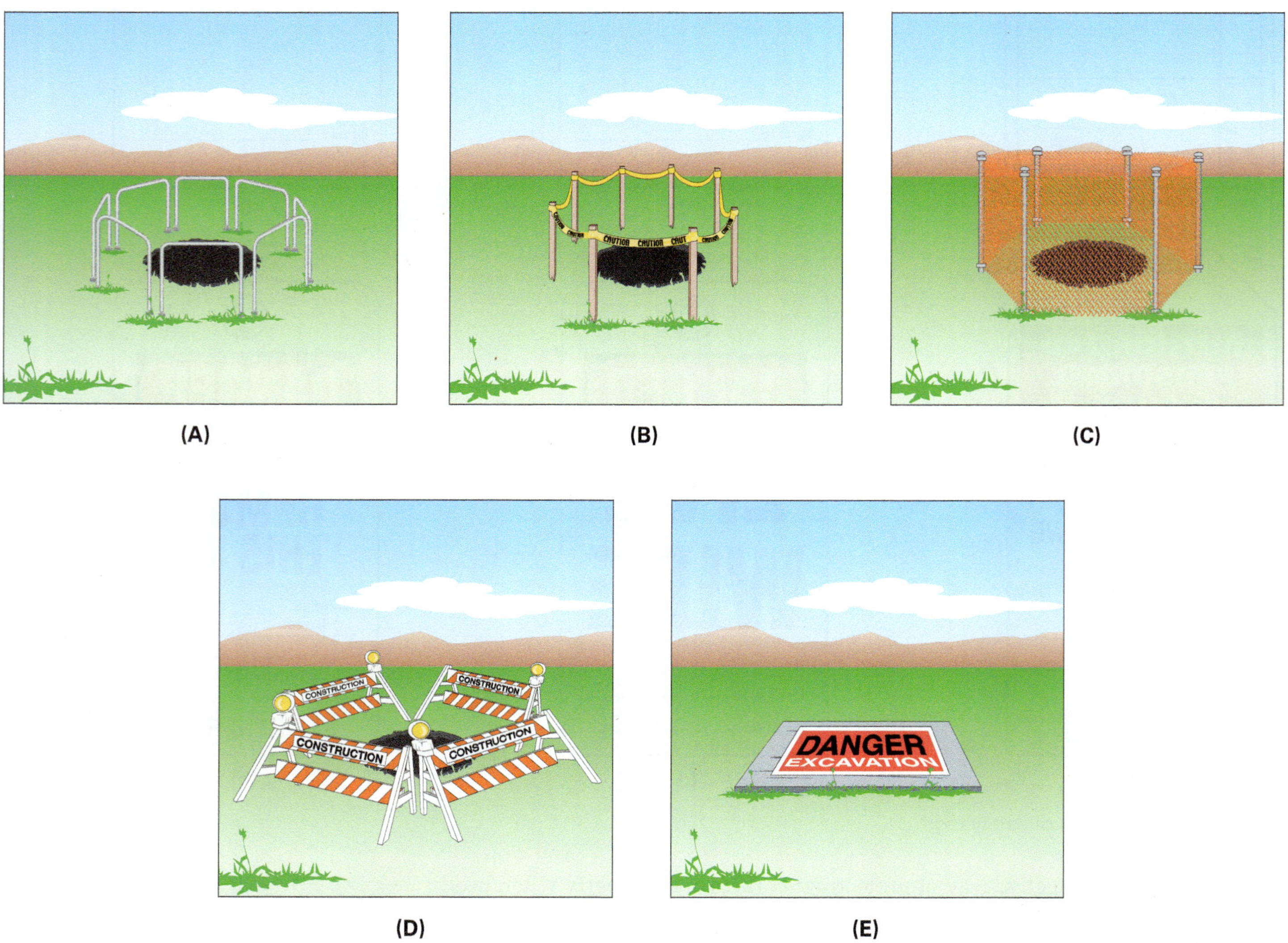

Figure 13 Common types of barricades and barriers.

Barricades and barriers help to prevent accidents and injuries, but they are not foolproof. Even with barricades or barriers in place, there can still be dangers, so when you are assigned to a jobsite, carefully examine the area where you are working. Before moving heavy equipment, walk the area to ensure that there are no hidden hazards. This is especially important when there are obstructions such as snow, mud, leaves, and other debris in the area. Never move a barricade, even in an emergency, unless you are certain it is safe to enter the barricaded area.

1.1.5 Audible and Hand Signals

Audible signals such as alarms, bells, buzzers, whistles, and horns can be used to communicate hazards to workers. For example, back-up alarms are used on forklifts, trucks, and other heavy equipment. Fire alarms are used to clear work areas. Conveyer belt lines have buzzers, bells, and/or whistles to let workers know they are about to start. However, at times it is useful to communicate through hand signals (*Figure 14*) because it is too loud to use your voice or another type of audible alarm.

There can be various meanings for the same hand signal. Always establish with your coworkers the hand signals that will be used and their meanings before starting the task. Some of the common hand signals and their meanings are listed in *Table 1*.

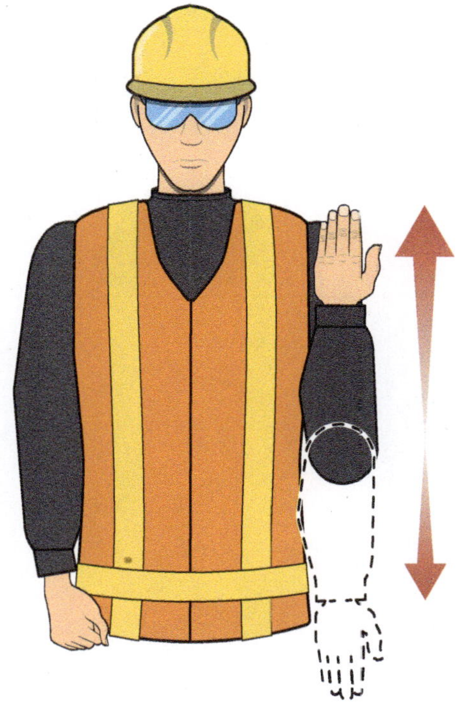

Figure 14 Example of a hand signal (come here).

TABLE 1 Common Hand Signals

Signal	Meaning
Point index finger toward self	I/me
Point index finger toward object	It/them
Point index finger toward person	You/them
Circle index finger at group	We/us/all of us
Beckon by moving arm toward self	Come here
Extend arm and point hand to the right	Move right
Extend arm and point hand to the left	Move left
Point with thumb in a particular direction	Move this way/go this way
Bring index finger across throat	Quit
Slowly ease palm face down	Relax/slow down
Put palm over brow	Scout it out/check it out
Hold index finger up near head	Wait
Hands on top of head	I'm OK
Thumb up	Good/OK
Thumb down	Bad/not OK
Slap forehead	Bad idea
Palm down and rotated from side to side	Unsure/can't decide
Wave goodbye	Goodbye
Form a circle with thumb and index finger	OK/I understand/agree
Military salute	I understand and will comply
Shake head from side to side	No/disagree

NOTE

Standards may vary from state to state. Be sure to use hand signals for your local jurisdiction.

Hand signals can be safely used over long distances, but those using hand signals must never be out of each other's sight. Always use slow and exaggerated gestures to be sure that you are understood. Heavy equipment operators who lose sight of their coworker must immediately bring their vehicle to a safe stop. Always make eye contact with nearby personnel to be sure they are aware of your presence.

1.2.0 Traffic Control Safety

When construction sites are near or in public roads, moving traffic presents an additional hazard to construction workers. Dangers to construction personnel working near moving traffic are so great that all states have laws that provide for additional fines for drivers who speed or commit other traffic violations in a traffic work zone. When the worksite is located near traffic, supervisors study the area and make plans to ensure that the job can be completed safely and efficiently. However, safety is your responsibility, so stay alert.

When the normal use of a road is disrupted, temporary traffic control (TTC) measures must be used. TTC measures help ensure that motorists, pedestrians, and bicyclists can use the road safely while construction is going on. Planning traffic flow is key to keeping traffic moving smoothly and managing traffic incidents. Specially trained workers usually create traffic flow plans. They also decide on the need for and position of TTC devices such as barricades, cones, and signs, but you may be called on to help place some of these devices. In addition, it may be necessary to temporarily change the TTC during an accident or other emergency.

The US Department of Transportation Federal Highway Administration publishes a manual that states the basic principles for changing the flow of traffic at a worksite. It also explains how to design and use traffic control devices. This manual is called the Manual on Uniform Traffic Control Devices (MUTCD) for Streets and Highways. It is used for streets and highways open to the public, regardless of type or class or the public agency having control of the road.

For a traffic control device to be useful, it must meet the following five basic requirements:

- Fulfill a need
- Attract attention
- Have a clear and simple meaning
- Command respect from road users
- Provide enough time for proper response

TTC measures and devices are used to provide for safe and orderly movement of traffic through or around work zones and to protect workers, responders to traffic accidents, and equipment. At the same time, the TTC zone must permit workers to complete the project quickly and well.

Flaggers: Workers who are specially trained to direct traffic through and around a work zone.

Conditions in TTC zones are always changing. These conditions include lane closures and merges; speed changes; **flaggers**, trucks and other equipment entering and leaving the roadway; and uneven pavement. Such conditions can be very confusing to motorists and dangerous for construction workers and heavy equipment operators.

Various methods and devices are used to mark the TTC zone (*Figure 15*). It is important to understand how the TTC zone is set up and how TTC devices are used to keep workers safe. It is also important to know the purpose and use of flaggers in the TTC zone, and to be aware of the signals that flaggers use, just in case you need to control traffic in an emergency.

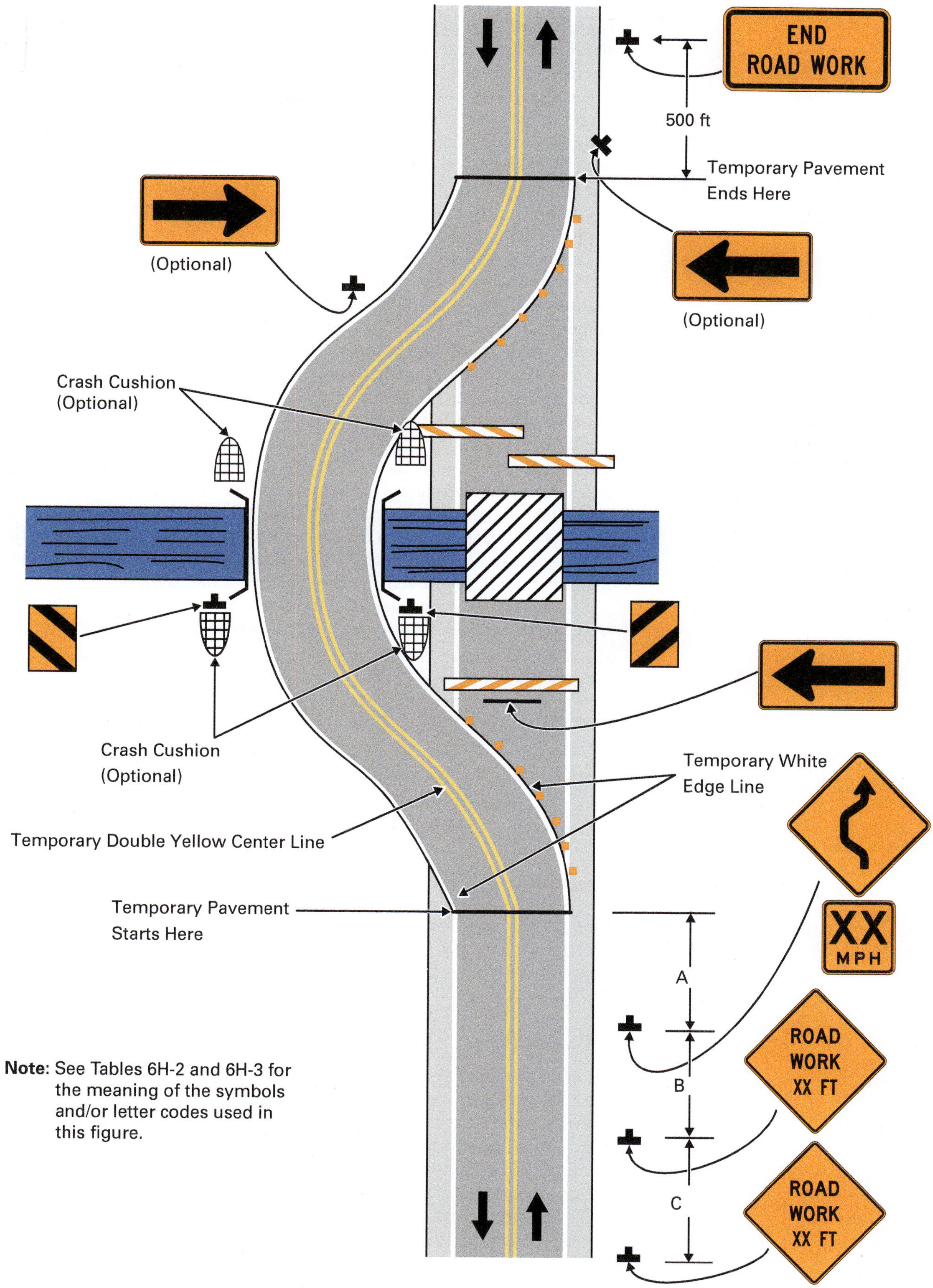

Figure 15 TTC zones can be confusing.
Source: MUTCD, 2009 edition, published by FHWA at http://mutcd.fhwa.dot.gov

1.2.1 Temporary Traffic Control Zones

Because the TTC zone is always changing, a single set of TTC devices cannot satisfy all conditions for a project. At the same time, making detailed plans that cover all situations is not practical. This section covers some common uses for TTC devices. The exact devices used on each jobsite depend on the type of road, road conditions, length of the job, physical environment, and how close the workspace is to vehicle and pedestrian traffic. Although you may not be planning TTC zones, you need to understand the layout and purpose of the zone (*Figure 16*) in order to work safely.

The first part of the TTC zone is the advance warning area. Warning signs (*Figure 17*) in this area let motorists know what to expect ahead. The actual signs depend on the type of work being done. The spacing of the signs depends on the normal posted speed of the road.

The second part of the TTC zone is the transition area. This area is used to gradually move traffic out of its normal flow and into the desired path. Sometimes this area has flashing, lighted signs (*Figure 18*). Orange and white barrels, cones, and other types of markers, which are called **channeling devices** (*Figure 19*), are used to guide traffic into the desired path. The length of the transition area depends on the amount of traffic that is normal for the road, as well as other road conditions.

Channeling devices: Cones, barrels, or markers that are used to move traffic from its normal lane to a desired lane.

The next area in the TTC zone is the activity area, consisting of a longitudinal buffer zone and the workspace. If there is no room for a buffer zone, truck-mounted attenuators are required. The buffer zone (*Figure 20*) and workspace have some type of barrier device on the traffic side to mark the border.

A lateral buffer is next to the border and is designed to protect workers from closely passing vehicles. Buffer zones must be clear areas. They are there to give reaction time in case a vehicle breaks through the barriers.

The workspace is for workers, equipment, and material. Stay in this area. Do not leave this area on foot or move the equipment from this area unless authorized to do so.

WARNING!

Be aware of the workspace borders at all times. Leaving the workspace can cause an accident.

The final area in the TTC zone is the termination area. This is the end of the construction area. It uses channeling devices and is marked with signs to let motorists know that they are leaving the construction zone and to merge back into their lane.

WARNING!

While working in the TTC zone, someone may be assigned to monitor the barriers, channeling devices, and signs to ensure that they stay in position. Heavy equipment operators need to be aware of the movements of this worker to prevent injuries or death.

Case History

A seven-person crew was paving two lanes of a four-lane highway. One worker was walking back and forth along the highway to ensure that the traffic cones in the center of the highway and the construction-zone warning signs remained in position. The foreman operated the paving machine, and a ride-on roller/compactor followed behind compacting the newly laid asphalt. The roller operator made a pass with the roller, stopped, and then reversed the machine. The machine traveled approximately 10 feet when the operator sensed that something was wrong and stopped the roller. The worker assigned to monitor the cones and signs was discovered lying face down with his head crushed. Emergency rescue personnel arrived at the site within 15 minutes, but the worker was pronounced dead at the site.

The Bottom Line: The TTC work zone is always changing. Heavy equipment operators and other workers need to be constantly aware of their surroundings.

Source: The National Institute for Occupational Safety and Health (NIOSH)

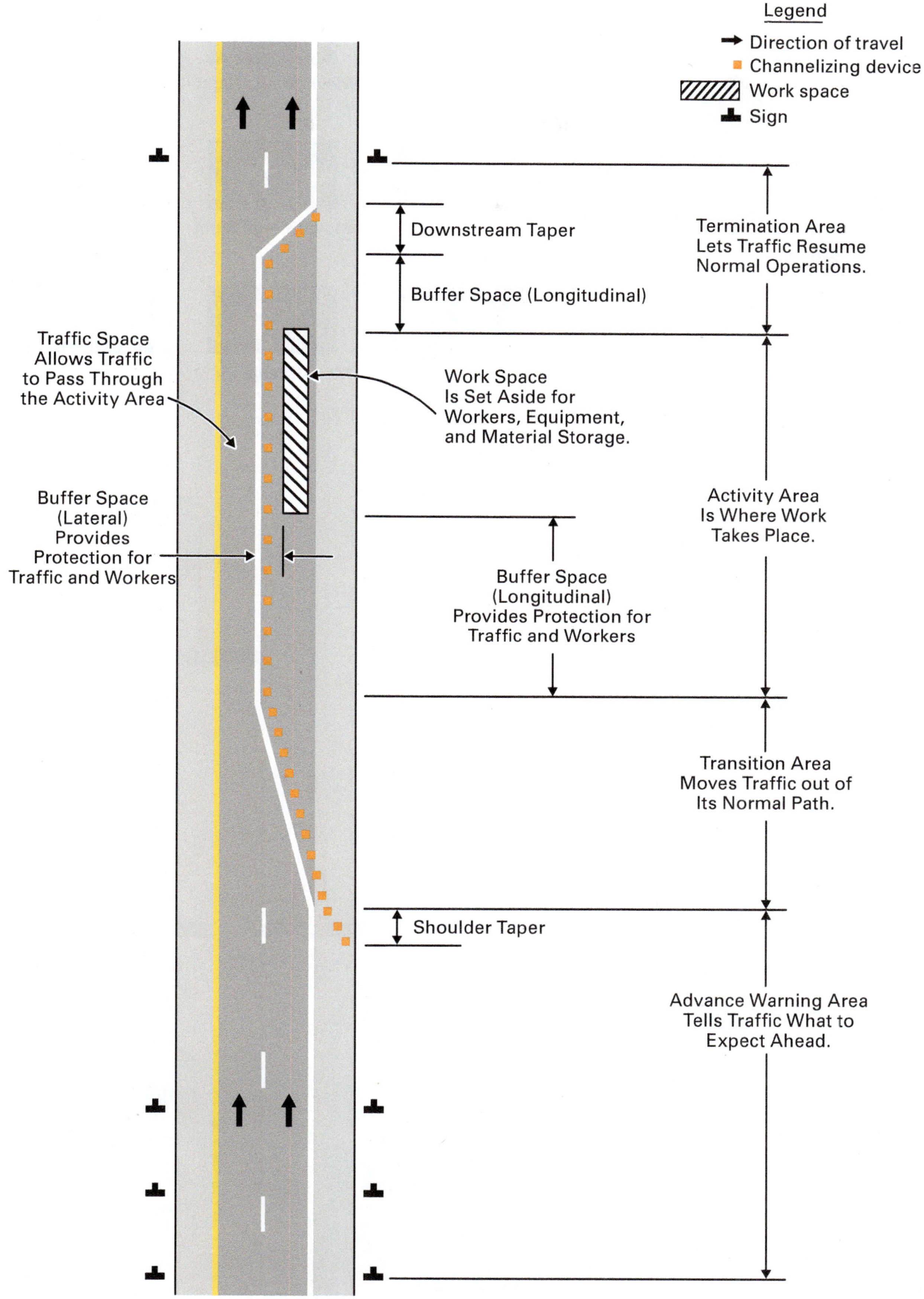

Figure 16 A typical TTC zone.
Source: MUTCD, 2009 edition, published by FHWA at http://mutcd.fhwa.dot.gov

Figure 17 Warning signs.

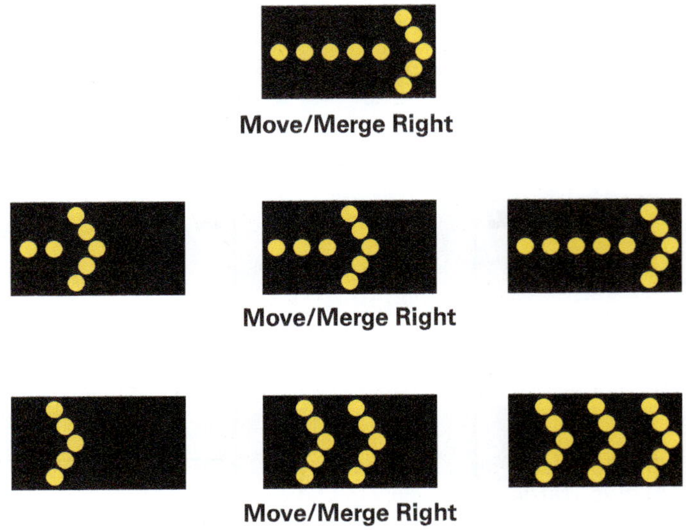

Figure 18 Merge right arrow signs.

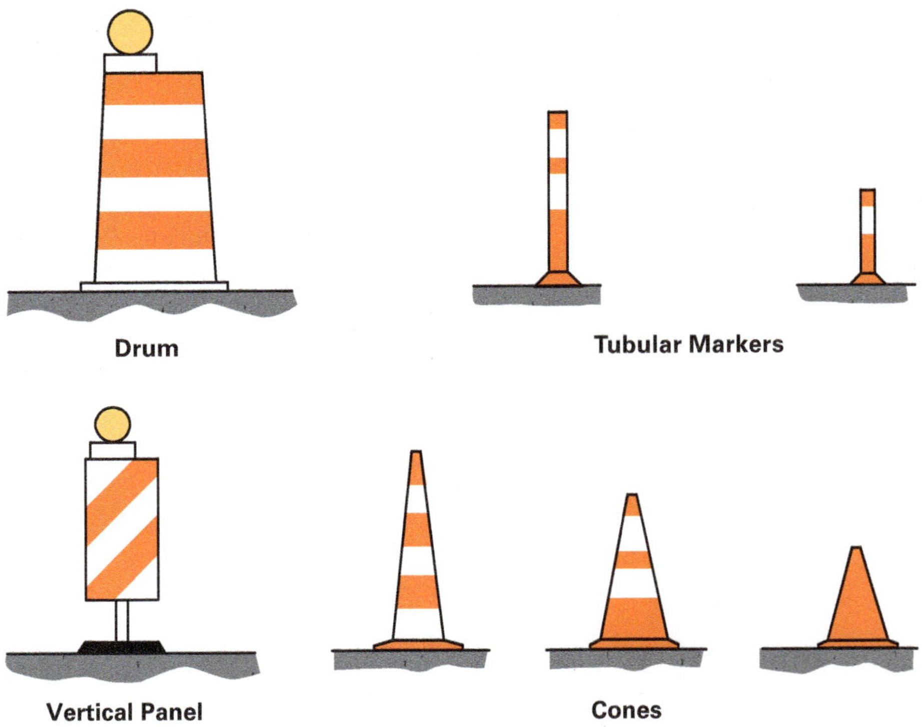

Figure 19 Channeling devices.

Figure 20 Typical buffer zone in road construction.
Source: Rainer Fuhrmann/Shutterstock

1.2.2 Flagger Control

Flaggers are part of the construction team. They are responsible for public safety, so they must be trained in safe traffic control practices. In some states, flaggers must be certified. Flaggers must keep themselves and the construction team safe, as well as direct traffic using signs, paddles, or flags. On most jobs, it is unlikely that an equipment operator will need to be a flagger. In case of an emergency, however, the operator may need to perform this task for a short period. A flagger must be able to do the following:

- Clearly, firmly, and politely give specific instructions.
- Move quickly to avoid danger.
- Use signaling devices, such as paddles and flags, to provide clear direction to drivers.
- Apply safe traffic control practices, sometimes in stressful or emergency conditions.
- Identify unsafe traffic situations and warn workers in time to avoid injury.

One of the ways that flaggers keep themselves safe is by wearing highly visible safety gear. Class 2 material is required for daytime work and Class 3 is required at night. The apparel background (outer) material color must be fluorescent orange-red, fluorescent yellow-green, or a combination of the two as defined in the American National Standards Institute (ANSI) standard. The **retroreflective** material must be orange, yellow, white, silver, yellow-green, or a fluorescent version of these colors. It must be visible at a minimum distance of 1,000 feet. The retroreflective safety apparel must clearly identify the wearer as a person.

Retroreflective: A device or surface that reflects light with very little scattering.

Hand-signaling devices, such as Stop/Slow paddles, and lights are used to direct motorists through TTC zones. The Stop/Slow paddle is the best device to use because it is easy for drivers to understand. Red flags may be used in emergencies.

The Stop/Slow paddle is octagonal in shape, and the Stop side looks like the Stop signs you see at intersections. The Slow side has an orange background with black letters and borders. When used at night, the paddle must be made of material that reflects light. The Stop/Slow paddle can also have either white or red flashing lights on the Stop side and either white or yellow flashing lights on the Slow side.

When flags are used, they must be at least 24 inches square and made of a good grade of red material. The free edge of a flag must be weighted so the flag hangs vertically, even in heavy winds. When used at night, flags must be made of red retroreflective material.

WARNING!

Never step in front of a vehicle in order to stop it.

The paddle must be held still, with the flagger's arm extended horizontally away from the body (*Figure 21*). The other arm is used to give motorists additional directions. To stop traffic, the flagger's free hand is held above shoulder level with the palm facing oncoming traffic. To direct stopped traffic to continue, the Slow side of the paddle must face oncoming traffic, and the flagger's free arm is used to motion drivers to continue. To slow moving traffic, the flagger can slowly motion up and down with the free hand, palm down.

When a red flag is used in an emergency situation, the flagger must stand facing oncoming traffic with the red flag in the hand closest to traffic. To stop traffic, the flagger extends the flag horizontally across the road users' lane so that the full area of the flag is visibly hanging below the staff. The free arm is used in the same way as described when using the Stop/Slow paddle. To release stopped traffic, the flag is hung down at the flagger's side and the free arm is used to wave traffic on. To slow traffic, the flagger must stand with the free hand down and slowly wave the flag up and down without raising the arm above shoulder level. Motorists often mistake this gesture as a stop signal.

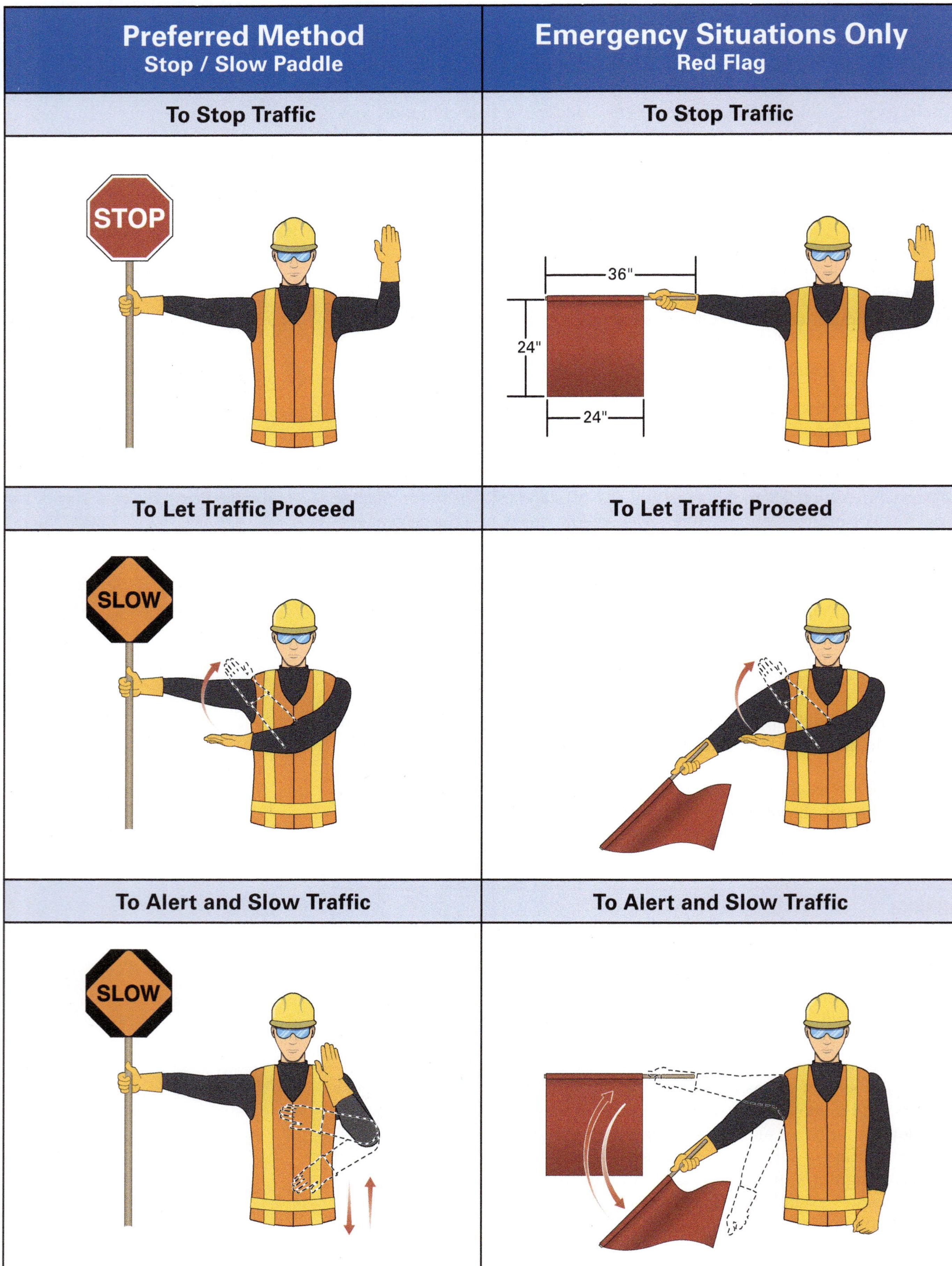

Figure 21 Flagger signals.

Source: Manual on Uniform Traffic Control Devices (MUTCD) US Department of Transportation Federal Highway Administration www.mutcd.fhwa.dot.gov

1.2.3 Personal Safety

Working in and around heavy equipment at a jobsite adds another hazardous element to any work. It is easy to become so engrossed in a task that you forget what is going on around you. When you are operating heavy equipment, you may find yourself lulled into a trance by the sound and movement of the equipment. If working in a vehicle with a temperature-controlled cab, you may forget about outside conditions. Responsible operators remain alert and stay aware of their surroundings at all times.

Heavy equipment operators can stay alert by taking every opportunity to get out of the operator's seat. Walking increases blood circulation, thereby decreasing drowsiness, and reduces muscle discomfort that comes from sitting in the same position for long periods. In newer models of heavy equipment, the cabs are designed to reduce vibration and noise and to increase driver visibility. But heavy equipment is built to last, and an older model of equipment may be used. You can still help to relieve stress and muscle fatigue even when unable to leave the operator's seat. First, sit up straight while working. Good posture goes a long way in not only reducing fatigue, but also reducing muscle strain. Next, every time the vehicle stops, set the brake and stretch your arms and legs while sitting in the operator's seat, then do a few neck and shoulder rolls. Once in the habit of stretching, it becomes second nature.

When heavy equipment is in use, it is best to minimize the number of workers on foot around the worksite. Because this is not practical, workers should be sure to maintain eye contact with the operator when working near the equipment. Drivers must make eye contact and acknowledge the presence of other workers and take care not to lose sight of them. Workers on foot need to stay in the driver's view and when leaving the area must do so in plain sight of the driver. If conditions make it impossible for the driver and other personnel to maintain eye contact, a third person should be positioned to maintain constant eye contact with both the driver and other personnel.

Case History

An 18-year-old flagger, outfitted in full reflective vest, pants, and hard hat, was directing traffic at one end of a bridge approach during a night milling operation. The work zone was correctly marked with cones and signs, and the entire bridge was lit with streetlights. The flagger was standing under portable floodlights in the opposing lane close to the center line, facing oncoming traffic. A pickup truck traveling in the wrong lane at an estimated speed of 55 to 60 miles per hour struck the flagger head on and carried him approximately 200 feet. He died of multiple traumatic injuries at the scene.

The Bottom Line: Highway construction is very dangerous. Even when following all of the safety rules, you are still at risk. Stay alert.

Source: The National Institute for Occupational Safety and Health (NIOSH)

1.0.0 Section Review

1. The type of safety sign that is usually yellow with black letters is the _____.
   a. warning sign
   b. temporary traffic control sign
   c. caution sign
   d. safety alert

2. Information on how to design and use traffic control devices is provided in the _____.
   a. MUTCD
   b. OSHA handbook
   c. SDS
   d. TTC handbook

2.0.0 Hazard Communication

Performance Tasks

There are no Performance Tasks in this section.

Objective

Explain hazard communication.

a. Explain the purpose and content of safety data sheets (SDS).
b. Describe the labeling systems used for hazardous materials.
c. Describe noise control methods.

Construction workers are exposed to a wide range of health hazards, especially when working around heavy equipment. In addition to the obvious hazards of rollover injuries and collisions, heavy equipment introduces less obvious hazards, some of which may not be apparent until working in construction for many years.

As a heavy equipment operator, you will be responsible for the operation, and perhaps the maintenance, of the equipment. While performing these tasks, you will be exposed to many types of chemicals used in products, such as diesel fuel, hydraulic fluid, and cleaning solvents. In addition, you will be exposed to diesel exhaust and noise from the equipment. Take steps to protect yourself from these hazards.

2.1.0 Safety Data Sheets

Hazard communication (HazCom): An OSHA standard that requires contractors to educate employees about working with or near hazardous chemicals.

OSHA and MSHA require that all employers inform and educate their employees about hazardous chemicals at the jobsite as part of a **hazard communication (HazCom)** program. As part of this program, OSHA and MSHA require that all hazardous chemicals in a workplace be clearly identified. Further, they require that suppliers of hazardous chemicals provide a safety data sheet (SDS) for each hazardous substance in a shipment. This information must be available at the worksite, but it is up to you to use the information. A file containing the SDSs for any potentially harmful materials on the site, or associated with the job, should be located near the work area. For construction jobs, such files may be kept in the supervisor's truck or in the project manager's office.

SDSs are available so that workers can protect themselves. Employers are required to supply you with safety equipment that you may need when working around hazardous chemicals. Although the employer provides the SDSs and the safety equipment, it is still up to employees to use the SDS information and the protective equipment to protect themselves. It is part of your job to check the SDS and evaluate the work environment to determine whether protective equipment is needed.

The content of an SDS is specified by the OSHA Hazard Communication Standard. Although SDSs do not all look alike, they should all have 16 sections containing specific types of information. The 16 sections are as follows:

- Identification
- Hazard(s) identification
- Composition/information on ingredients
- First-aid measures
- Firefighting measures
- Accidental release measures
- Handling and storage
- Exposure controls/personal protection
- Physical and chemical properties
- Stability and reactivity

- Toxicological information
- Ecological information (nonmandatory)
- Disposal considerations (nonmandatory)
- Transport information (nonmandatory)
- Regulatory information (nonmandatory)
- Other information

2.1.1 Section 1—Product and Company Identification

The product and company identification section of an SDS provides the proper name(s) of the product; the chemical family that the product is associated with; the name of the manufacturing company; and the phone numbers for contacting the manufacturer.

2.1.2 Section 2—Hazards Identification

The hazards identification section gives an emergency overview of the product. The product's appearance is described, and its combustibility is addressed. Warnings are given about actions to avoid, such as inadvertent suction of the liquid into the lungs. This section also explains what reactions you may experience if the product is inhaled or ingested, or if it comes in contact with skin or eyes. If the product is a carcinogen, which can cause some form of cancer, that information is also listed in this section.

WARNING!

The SDS lists typical immediate effects of the hazardous chemical on the body. A reaction that is not listed may be an allergic reaction. Allergic reactions can be very mild, like redness on the skin, or they can be so severe that they are life threatening. Severe reactions need to be treated by a doctor as soon as possible. Symptoms of a severe allergic reaction include shortness of breath.

2.1.3 Section 3—Composition

This section identifies the hazardous ingredients that make up the product and lists the maximum exposure allowed in order to prevent skin damage.

2.1.4 Section 4—First Aid

Anyone working with heavy equipment will sooner or later be involved in refueling the machines. Even when spills do not occur, vapors from the diesel fuel will still fill the fueling area. Section 4 of an SDS lists the first-aid actions to take if the product is inhaled, ingested, or splattered onto skin or into eyes. Such splashes may aggravate existing skin conditions.

Understand that these measures are the minimum actions to be taken after exposure to diesel fuel. Each worksite may have stricter policies that must be followed. Know where to find first-aid equipment before you need it. *OSHA CFR 1926.50* requires that employers provide first-aid kits at every jobsite. These kits typically contain gauze pads, rolled gauze bandage, wound cleaning agent, scissors, adhesive tape, bandages, adhesive bandages, a splint, tweezers, resuscitation equipment, and a blanket. Because of the potential for workers to be exposed to blood-borne pathogens, there should also be a bodily fluid pickup kit or personal protection kit. These kits usually contain eye and hand protection, fluid absorbents, sanitizing and disposal products, and a CPR mask.

2.1.5 Section 5—Firefighting Measures

When heavy equipment is being used to clear land, the potential for fire exists as the equipment moves through underbrush that may be dry enough to easily catch fire. A diesel fuel leak could lead to a fire. Heavy equipment being used in mining operations also are operating in a potentially dangerous environment where fires can create explosions. Section 5 lists specific hazards associated with the product, and the products recommended for extinguishing any fire involving the product. This section also lists any special PPE for fighting a fire fueled by this product.

Going Green

Containing Runoff

If water is used to cool exposed surfaces close to a fire, make sure to contain any runoff water. Keep it out of sewers and streams.

WARNING!

In some situations, there are certain substances that should not be used on fires. Make sure to review such warnings before attempting to extinguish any fire.

2.1.6 Section 6—Accidental Release Measures

Some products simply need to be isolated from public access. The accidental release section of the SDS addresses what to do if a product is accidentally released by way of spills, leaks, or unintentional releases. If applicable, this section may list specific products recommended for absorbing the spilled or leaked product.

2.1.7 Section 7—Handling and Storage

Section 7 explains how a product should be handled and stored. With products that may ignite, special grounding procedures may be needed to keep any static electricity from igniting the product. Products such as fuels often need to be kept cool and ventilated. Never cut, drill, grind, or weld on an empty container that has held potentially flammable materials.

Diesel fuel must be stored properly to ensure it does not represent a hazard to the workplace. There are ways of handling and storing flammable and combustible materials such as diesel fuel that decrease explosion and fire risks. Only approved containers and approved portable tanks may be used for flammable and combustible liquid storage and handling. Approved storage and handling containers, such as safety cans and safety cabinets, have been inspected by UL (formerly Underwriters Laboratories) and have gained a certification by UL. It is your responsibility to know what the handling and storage procedures are on the jobsite.

Safety cans are special vessels used for safe transportation and storage of flammable materials (*Figure 22*). These essential pieces of equipment are available in a variety of styles and colors. Many jobsites use color-coded safety cans to designate different types of liquids, such as gasoline, kerosene, diesel fuel, and oils. Color coding not only helps segregate liquids that should not be used together, but it also helps ensure the right liquids are used for each job. Safety cans have many built-in safety features such as vapor control, flame arresters, and pressure-relief valves. The features and usage requirements vary by manufacturer, so make sure you are familiar with the specifics for the safety cans on your site.

Figure 22 Safety can.
Source: Justrite Safety Group

WARNING!

If you do not know how to use safety cans properly, ask your supervisor for instructions. Improper use is dangerous!

Safety cabinets (*Figure 23*) are important for the proper storage of flammables and combustibles. These cabinets are common on jobsites, including construction, maintenance, and renovation locations. Just like with safety cans, it is important that these cabinets be used properly. Appropriate safety cabinets are designed specifically to reduce the risk of explosion and fire. Features include vents with flame arresters, special lips on the shelves to catch spills, and grounding connectors to prevent buildup of static electricity. Safety cabinets are made of steel designed to withstand intense heat and are tailored to meet OSHA, MSHA, and National Fire Protection Association (NFPA) guidelines. Some manufacturers produce self-closing models that automatically close in response to fire-level heat. Because characteristics of these cabinets vary by model and manufacturer, familiarize yourself with the specific cabinets and procedures used on your site.

Figure 23 Safety cabinet.
Source: Justrite Safety Group

WARNING!

If you do not know how to use safety cabinets properly, ask your supervisor for instructions. Improper use is dangerous!

NOTE

OSHA 29 CFR 1910.106 regulates the amount of flammable or combustible liquids that may be located near or inside a gas storage cabinet or in any one fire area of a building. The standard prohibits more than 60 gallons of Class I or Class II liquids or more than 120 gallons of Class III liquids in one storage cabinet. No more than 25 gallons may be stored outside a protected storage area or cabinet.

Gas storage cabinets must be designed to limit the internal temperature not to exceed 325°F. Metal cabinets should be constructed of 18-gauge sheet metal and should be double-walled. For added safety, gas storage cabinets should be self-closing. Cabinets must be labeled in obvious lettering with the statement "Flammable—Keep Fire Away." No more than 25 gallons of flammable or combustible liquids may be stored in a room outside of an approved storage cabinet.

Source: Occupational Safety and Health Administration website. Regulations (*Standards – 29 CFR*), *Flammable and Combustible Liquids – 1926.152*.

Static Electricity Can Kill

In recent incidents reported to the National Institute for Occupational Safety and Health (NIOSH), fires spontaneously ignited when workers or others attempted to fill portable gasoline containers (gas cans) in the backs of pickup trucks equipped with plastic bed liners or in cars with carpeted surfaces. Serious skin burns and other injuries resulted. Similar incidents in the last few years have resulted in warning bulletins from several private and government organizations.

These fires result from the buildup of static electricity. The insulating effect of the bed liner or carpet prevents the static charge generated by gasoline flowing into the container or other sources from grounding. The discharge of this buildup to the grounded gasoline dispenser nozzle may cause a spark and ignite the gasoline. Both ungrounded metal (most hazardous) and plastic gas containers have been involved in these incidents.

Source: The National Institute for Occupational Safety and Health (NIOSH)

2.1.8 Section 8—Exposure Controls/Personal Protection

Section 8 addresses both the personal protection equipment (PPE) needed to prevent injuries that could occur if exposed to the product and any engineering measures that should be taken to exhaust any vapors or fumes generated by the product. This section lists specific PPE, such as specific respiratory filters (*Figure 24*) or specific types of gloves, that should be used for protection. If splashing is possible, this section may indicate that safety glasses/goggles (*Figure 25*) or even face shields should be used.

(A) Self-Contained Breathing Apparatus (SCBA)

(B) Supplied Air Mask

(C) Full Facepiece Mask

(D) Half Mask

Figure 24 Respirators.
Sources: (A, B, C) Used with permission from MSA; (D) Becki Swinehart

Figure 25 Safety glasses and goggles.
Source: Used with permission from MSA

CAUTION

Handle safety glasses and goggles with care. If they get scratched, replace them. The scratches will interfere with your vision. Clean the lenses regularly with lens tissues or a soft cloth.

You should refer to section 8 often for all chemicals that are handled. Section 8 of the SDS for a solvent-cement (shown in *Appendix 22102A*) recommends wearing safety glasses with brow guards and side shields when there is a danger of splashing, wearing protective gloves when there is a danger of skin contact, and using air-purifying respirators when there is danger from fumes. When using an air-purifying respirator, be aware that this type of respirator needs a filter cartridge to work correctly. *Table 2* lists the color codes for cartridges. Always use the correct filter. For example, a black or yellow filter (organic vapor) is the correct filter to select for diesel fumes.

TABLE 2 Respirator Cartridge Color Codes

Respirator Filter Cartridge Color Codes	
Organic Vapor	Black
Acid Gas	White
Organic Vapor/Acid Gas	Yellow
Ammonia	Green
Particle (HEPA)	Magenta
Multi-Gas/Vapor	Olive

Diesel exhaust, also called diesel fumes, is a mixture of gases and particles. One of these gases is carbon monoxide, which is a deadly gas that is produced when any fuel is burned. Soot (the black smoke seen coming out of the exhaust pipe) is a major part of diesel exhaust (60 to 80 percent). In the short term, breathing diesel exhaust can cause coughing, itchy or burning eyes, wheezing, and difficulty breathing. These effects disappear when the worker is away from the source of the exhaust.

Longer exposure to diesel exhaust can cause carbon monoxide poisoning. Carbon monoxide gas cannot be seen or smelled, and it can build up quickly, even in areas where there appears to be adequate air circulation. It can overcome workers without warning, producing weakness and confusion.

Recognizing carbon monoxide poisoning can be difficult because early symptoms, such as headache, dizziness, and nausea, can be easily mistaken for a minor illness. Advanced symptoms of carbon monoxide poisoning are confusion and weakness. Victims of carbon monoxide poisoning must be immediately removed from the site and given 100-percent oxygen. Some cases are so bad that the victim needs to be placed in a special device called a hyperbaric chamber, which provides oxygen under pressure.

The soot in diesel exhaust can get into a worker's lungs and cause a number of lung problems, maybe even cancer. There is also evidence that the fine particles in soot can worsen heart problems and respiratory illnesses such as asthma.

Several safety measures can reduce exposure to diesel exhaust. The first is to use low-sulfur diesel fuel, which burns cleaner than high-sulfur fuel. Another is to direct the equipment exhaust pipe away from the driver and nearby workers. This can make a significant difference in the amount of exhaust to which workers are exposed.

Another form of protection is to wear a respirator, but workers must have a medical evaluation, be fit-tested by a competent person, and undergo training before wearing them. Beards and even stubble may prevent a proper seal. The respirator must be approved by the National Institute of Occupational Safety and Health (NIOSH) and carry an approval number. Respirators are only a temporary measure. The primary goal is to eliminate or control diesel emissions.

Finally, turn off equipment when it is not needed. Idling engines burn fuel and put out poisonous gases. Taking some of these steps and using protective equipment can mean a healthier life for everyone.

Did You Know?
Latex
Latex is a common material for protective gloves, but latex gloves are not recommended for use around diesel fuel because diesel fuel dissolves latex.

2.1.9 Section 9—Physical and Chemical Properties

Section 9 lists the physical and chemical properties of the product. Those properties include things like its appearance and whether it is a solid, liquid, or gas. The section also lists other bits of information that are related to how the product reacts to different temperatures and pressures, such as the product's boiling point, melting point, and flash point.

2.1.10 Section 10—Stability and Reactivity

Section 10 lists information about how stable the product is at given temperatures and pressures. It also lists materials to keep separated from the product, and conditions to avoid that could otherwise cause the product to ignite or explode.

2.1.11 Section 11—Toxicological Information

Toxic: The degree to which a substance can harm humans or animals.

The word **toxic** is usually a red flag indicating a potential health hazard. Toxicity is the degree to which a substance can harm humans or animals. Section 11 lists the toxicity of the product and may go into detail about the chemical reactions the product has had on both humans and test animals. Be sure to read and understand the summary statements associated with the product.

2.1.12 Section 12—Ecological Information

Everyone must be careful not to harm the environment. Section 12 lists the environmentally damaging effects of the product. Diesel fuel, for example, can be toxic to the environment, animal habitats, and marine life.

2.1.13 Section 13—Disposal Considerations

Section 13 may be a little more important to heavy equipment operators than some of the other sections because it explains how to clean up and dispose of a given product. According to federal regulations, No. 2 diesel fuel, as produced, is not specifically listed as a hazardous waste. If discarded or disposed of and allowed to mix with water or some other substances, it may meet the criteria of hazardous waste. Check with federal, state, and local guidelines to determine the proper disposal methods.

2.1.14 Section 14—Transport Information

Different countries, and different states within the United States, have different guidelines on how to ship or transport given products. Section 14 of an SDS shows the product's proper shipping name, its hazard class, and how it should be packed for transport. When transporting products such as diesel fuel, the shipping vehicle must be labeled with the applicable UN/Identification number. That number allows anyone seeing the vehicle to know the hazard level of the product being transported.

Case History

At approximately midnight, the operator of a paving machine checked the level of fuel in his machine. He had difficulty seeing because it was dark, so he used his cigarette lighter to see the fuel level. The fuel vapors ignited, and the operator was burned badly enough to need hospitalization.

The Bottom Line: Use the information in the SDS. Stay alert. Think ahead.

Source: Occupational Safety and Health Administration (OSHA)

2.1.15 Section 15—Regulatory Information

Section 15 lists the various regulatory guidelines that may be associated with the product, as well as the different Right-to-Know regulations of different states and communities. If transporting a product such as No. 2 diesel across state lines, such information may be important to the shipper transporting the product.

2.1.16 Section 16—Other Information

Section 16 is a catchall section. It may contain additional information that the product supplier wants to add.

2.2.0 Material Labeling

Materials should be labeled according to any safety threat they pose. Two common systems are the National Fire Protection Association (NFPA) system (*Figure 26*) and the Hazard Material Identification System (*Figure 27*). Each system divides the potential safety hazard into categories (flammability, health, reactivity), while the NFPA system, sometimes referred to as the diamond, also adds a rating of the potential hazard from 0 (no danger) to 4.

Health	
4	Deadly
3	Extreme Danger
2	Dangerous
1	Slight Hazard
0	No Hazard
Reactivity	
4	May Detonate
3	Explosive
2	Unstable
1	Normally Stable
0	Stable
Specific Hazard	
OXY	May Detonate
ACID	Explosive
ALK	Unstable
COR	Normally Stable
₩	Stable
☢	Radiation

Figure 26 NFPA hazard labeling.

Name of Material	
3	Health
4	Flammatory
2	Physical Hazard
B	Protective Equipment

4	Severe Hazard
3	Serious Hazard
2	Moderate Hazard
1	Slight Hazard
0	Minimal Hazard
*	Chronic Hazard
1	Slight Hazard
0	Minimal Hazard

Figure 27 HMIS hazard labeling.

In 2016, OSHA's HazCom Standard required the completion of the phase-in of the Globally Harmonized System (GHS) of hazardous material labeling. While the purpose of the GHS labeling system is basically a combination of the NFPA and former HMIS systems, there are some key differences. For example, the GHS label has six information elements rather than four. As of this edition, employers may continue to use any of the systems as long as the resulting information and protections are not diminished.

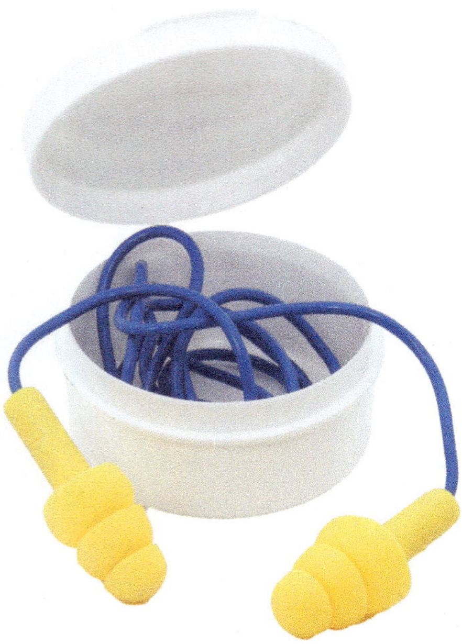

Figure 28 Hearing protection devices.
Source: Kedward/iStock/Getty Images

2.3.0 Noise

Noise on construction sites, and in closed spaces such as mines, makes it hard to hear voice commands and warning signals. When hearing protection devices (*Figure 28*) are used, care must be taken to ensure that the device does not mute sounds to the point that workers cannot hear safety signals, such as backup alarms and evacuation horns. In any area where hearing is difficult, workers must make eye contact with operators.

Long-term exposure to noise can cause noise-induced hearing loss or a permanent ringing in the ears, a condition called tinnitus. Employees are responsible for wearing hearing protection devices (HPD) and ensuring that they fit properly and offer adequate protection, but management must provide employees with the proper devices and train employees in their use and care. Additionally, it is management's responsibility to monitor noise levels and try to reduce excessive levels. Workers are required to wear hearing protection anytime there is a likelihood of exposure to loud noise. Even removing the device for a few minutes can seriously reduce its protective ability. If there are problems with hearing, or with the device, employees should seek help.

A personal hearing protection device (HPD) is anything that can be worn to reduce the level of sound entering the ear. Earmuffs, ear canal caps, and earplugs are the three principal types of devices. Each employee reacts individually to the use of these devices, and a successful hearing loss prevention program should respond to the needs of each employee. Making sure these devices protect hearing effectively requires the coordinated effort of management, the hearing loss prevention program operators, and the affected employees.

Industrial noise is measured using a certain method called A-weighting, which accounts for better human hearing in the middle and upper sound frequency ranges. Industrial noise level units are called A-weighted decibels, or dBA. Here are the dBA levels for some common everyday sounds:

- 0 dBA—Softest sound a person with normal hearing can hear
- 10 dBA—Normal breathing
- 30 dBA – Soft whisper
- 60 dBA—Normal conversation
- 110 dBA—Shouting in ear
- 120 dBA—Thunder
- 130 dBA—Stock car races
- 140 dBA—Jet plane (taking off)

The noise level on a typical construction site varies and is often irregular. Because most construction takes place outdoors, the loudness of the noise decreases rapidly as it travels away from the source. When heavy equipment is in use, the sound tends to be more constant. Heavy equipment operators have no choice but to stay close to the source of the noise, which can exceed 100 dBA.

Both OSHA and MSHA have hearing protection and noise level requirements. If the exposure equals or exceeds 105 dBA time-weighted average (TWA), or an 800-percent dosage, in addition to enrollment into the hearing conservation program, MSHA requires miners to use dual protection, including both earplugs and earmuffs.

To protect your hearing, avoid sources of loud noise as much as possible. Stay away from the area if not working there. This helps save your hearing. Limiting the number of workers around heavy equipment helps to protect hearing and also to prevent other accidents caused by contact with heavy equipment. Wear appropriate hearing protection devices but be sure to stay alert.

2.0.0 Section Review

1. The SDS for diesel fuel recommends the use of an air-purifying respirator with filter that is black or _____.
   a. green
   b. purple
   c. red
   d. yellow

2. A reactivity rating of 3 on a material label indicates that the substance _____.
   a. is stable
   b. is unstable
   c. is explosive
   d. may detonate

3. The noise threshold that establishes the need for double hearing protection is _____.
   a. 50 dBA
   b. 60 dBA
   c. 85 dBA
   d. 105 dBA

3.0.0 Heavy Equipment Safety

Objective

Identify and describe safety practices associated with heavy equipment operation.

a. List jobsite and equipment safety requirements.

b. Explain the safety hazards associated with weather.

Performance Task

3. Demonstrate how to use various types of personal protective equipment (PPE):

- Hard hat
- Safety glasses
- Ear protection
- Gloves
- Safety harness
- Reflective safety vest

Those who work around heavy equipment need a basic knowledge of all types of heavy equipment, as well as the hazards and safeguards of this equipment. The types of jobs performed by heavy equipment operators vary. The following are just a few of the jobs that use heavy equipment:

- Civil, residential, commercial, and industrial construction work
- Snow and ice control operations
- Maintaining road surfaces
- Loading, lashing, securing, and unloading equipment and materials
- Excavation and maintenance of public utilities

Construction workers may be expected to operate many types of heavy equipment (*Figure 29*). This equipment can vary greatly in size and weight. It is important to know how to use this equipment safely and to know the hazards and safeguards on the jobsite.

The first rule for heavy equipment operation is to never use a piece of equipment that you are not trained to use. Doing so not only puts you at risk of injury, but also puts your coworkers at risk. Current OSHA and MSHA regulations state that at the time of their initial assignment and then at least annually, employers must instruct every employee in the safe operation and servicing of all equipment that the employee will use.

(A) Bulldozer

(B) Scraper

(C) Excavator and Off-Road Truck

(D) Backhoe with Front Loader

(E) Motor Grader

(F) Soil Compactor

Figure 29 Examples of heavy equipment in use.

Sources: (A) Image courtesy of Komatsu America Corp.; (B) BanksPhotos/iStock/Getty Images; (C) Image courtesy of Komatsu America Corp.; (D, E) Courtesy of CASE Construction Equipment; (F) eni vidi...shoot/iStock/Getty Images

The following general points must be remembered when operating any heavy equipment:

- Before starting any equipment, study the manufacturer's operator manual.
- Learn the equipment's capacities and limitations.

- Learn the location and function of all controls, indicators, and warning devices.
- Be familiar with the safety devices on the equipment.
- Learn to recognize the equipment's warning and safety signals, such as high-temperature or low-pressure indicators.
- Securely fasten the seat belt.
- Do not permit others to ride in the equipment unless the equipment is specified for a rider.
- Operate the equipment smoothly—no jerky turns, starts, or stops.
- When towing something with the equipment, hitch only to points recommended by the manufacturer.
- When stopped, set brakes securely and use park lock if available.
- Maintain three-point contact when mounting or dismounting the equipment. Do not jump from the equipment.
- In case of contact with electrical lines, stay on the equipment unless there is a fire.

NOTE

OSHA and MSHA require that approved seat belts and rollover protection structures (ROPS) be installed on virtually all heavy equipment. (Old equipment must be retrofitted.) Do not use any heavy equipment that is not equipped with these safety devices.

3.1.0 Jobsite and Equipment Safety Requirements

Safety on the job is everyone's responsibility. The equipment operator has control over equipment operation, inspection, lubrication, and maintenance. Therefore, the operator must use good safety practices in these areas. Even though the operator has much of this responsibility, everyone on the site must work safely. They are also responsible for being aware of their coworkers and their actions.

Observe the following general job-safety rules to make sure everyone on the site is safe:

- Be alert. Watch out for moving equipment. If eye contact cannot be made with an equipment operator, assume the operator cannot see you.
- Usually, one person at a site is appointed signal person. Find out who this is on your jobsite and then watch for and obey their signals.
- Use retaining guards and safety devices on all equipment.
- Report all defective tools, machines, or other equipment, as well as all accidents and near-accidents to your supervisor.
- Make sure all others on the site are at a safe distance away from the equipment.
- Use traffic-control devices where required.
- Follow the manufacturer's safety rules and limitations for all equipment.

3.1.1 Power Lines

Equipment with tall booms, such as excavators and telescoping-boom forklifts, run the risk of coming into contact with live power lines. These lines carry high voltages that can kill anyone who comes into contact with them. For that reason, special care must be taken when it is necessary to work around power lines. A spotter should always be used in such cases. *Table 3* shows the minimum distance that must be maintained from power lines based on the voltage level.

WARNING!

Equipment does not have to come in direct contact with wires to act as a conductor. If operating near power lines, power to these lines should be shut off. Verify that the power is off with your supervisor.

TABLE 3 Power Line Clearances

Crane in Operation	
Power Line (kV)	**Boom or Mast Minimum Clearances (Feet)**
0 to 50	10
50 to 200	15
200 to 350	20
350 to 500	25
500 to 750	35
750 to 1,000	45

If the vehicle does come into contact with a power line, follow these guidelines:

- Stay inside the vehicle and do not touch any metal in the vehicle.
- If forced to leave the vehicle because of fire or to get help, keep your feet together and jump clear without touching any metal on the vehicle. To minimize the possibility of shock, hop to a safe area. Maintaining a single point of contact with the ground prevents current from finding a pathway through the body.

Downed power lines may still be live. When near downed power lines, follow these guidelines:

- Call emergency services for trained help.
- Do not allow other vehicles or people to enter the area.
- Never touch an overhead power line if it has been brought down by machinery or has fallen.
- Never assume power lines are dead; only the power company or its on-site representative can confirm that the system is de-energized.
- Treat anything in contact with a power line as energized. Anything the downed line touches and the ground surrounding it will also be energized.
- Never touch a person who is in contact with a live power line.
- If someone is trapped in a vehicle, stay well clear, at least 33 feet (more if high voltage), and try to keep the trapped person calm until help arrives.
- Do not attempt to help the victim; you are more likely to be electrocuted and may further endanger the victim.
- Wait for trained help to arrive.

Case History

Two employees were spreading concrete as it was being delivered by a concrete pumper truck boom. The truck was parked across the street from the worksite. Overhead power lines ran above the boom on the pumper truck.

One employee was moving the hose to pour the concrete when the boom of the pumper truck came in contact with the overhead power lines. These lines carried 7,200 volts. The employee was electrocuted and died immediately. He then fell on the employee who was assisting him. The second employee received a massive electrical shock and burns.

No one on the site had received the proper safety training. Otherwise, they would have known how dangerous it is to work under power lines.

The Bottom Line: Always check the jobsite for hazards before beginning work. Overhead and underground power lines can be deadly.

Source: The Occupational Safety and Health Administration (OSHA)

3.1.2 Equipment Safety

The identification of safety hazards begins with a daily machine-safety inspection. Before using any equipment, the operator should inspect around and under it for visible evidence of potential safety hazards or operational problems. These unsafe conditions must either be corrected immediately or reported. Heavy equipment operations should not proceed until the conditions are safe.

The following is a list of basic items that should be checked daily:

- Loose, worn, or leaky components
- Engine performance and gauges
- Oil pressure and levels
- Housekeeping tasks
- Fire extinguishers
- Audible warning devices
- Hydraulic lines and fluid
- Windshield, wipers, mirrors, and lights
- Batteries and charging system
- Visible hand signal and load charts
- Guards and clutches
- Outriggers, tracks, wheels, and tires
- Brakes, locks, and safety devices
- Drive chains, steering, and all rollers
- Sheaves, drums, and all cables
- Blocks and hooks
- Boom, gantry, jib, and extension
- Carrier assembly
- All controls, including hoisting, swing, travel, and boom

3.1.3 Personal Safety

Personal safety is as important as site safety. Your actions affect everyone on the site. Follow these guidelines to prevent accidents and injury:

- Always read any tool instruction manual before using the tool to understand safety information, required personal protective equipment, and proper operating techniques.
- Wear close-fitting clothing that is appropriate for the activity being performed.
- Tie back long hair before operating equipment.
- Wear safety glasses (preferably rated ANSI Z87.1), a suitable hard hat, goggles, hearing protection, gloves, and safety shoes where required (*Figure 30*).
- Wear an approved reflective vest.
- Fasten loose sleeves when working around machine tools or rotating equipment. Never wear gloves when working on these machines as gloves can get caught and pull hands into the machine.
- Remove rings and other jewelry when working.
- Be alert.
- Do not jump on or off moving equipment.

- Know the location of first-aid equipment, fire extinguishers, and emergency telephone numbers.
- Learn where to get assistance and how to use emergency communication equipment before you need it.

Figure 30 PPE.
Sources: (A, B) Courtesy of Stanley Black & Decker, Inc.; (C) Used with permission from MSA; (D) Courtesy of Stanley Black & Decker, Inc.

Never operate heavy equipment without a rollover protective structure (ROPS) or falling object protective structure (FOPS). The ROPS/FOPS should only be removed for maintenance and must be reinstalled before the equipment is operated again. Repairing or modifying ROPS/FOPS could weaken the structures and reduce your protection. See manufacturer's manuals for the equipment for complete instructions and inspection requirements.

3.1.4 Moving Equipment Safely

All heavy equipment is moved during the course of a job. It is moved to the site, around the site, and away from the site. Many injuries and deaths happen during the movement of heavy equipment. It is important to be especially safety conscious whenever equipment is moving. Be responsible when operating heavy equipment. A responsible operator must be a qualified and authorized operator. To be qualified, you must understand the instruction manual supplied by the manufacturer of the equipment, have training in the operation of the equipment, and know the safety rules and regulations for the worksite. Never operate any equipment under the influence of alcohol or drugs. If you are taking prescription or over-the-counter drugs, seek medical advice about whether you can safely operate machinery. Never knowingly allow a person who is impaired to operate heavy equipment.

Always observe the following guidelines when driving equipment on public roads:

- Drive slowly and never speed.
- Know the distance it takes to stop safely.
- Allow extra time to enter traffic.

- Know and follow the traffic control pattern.
- Travel with lights on.
- Use proper warning signs and flags.
- Secure all attachments and loose gear.
- Turn cautiously; allow for extensions or attachments and for structural clearances. Some equipment is top-heavy and will tip over if a turn is made too fast.
- Know the turning radius of the equipment.
- Know the swing radius of the equipment and make sure the swing path is clear of people and objects.
- Know and obey all state and local laws.

Always follow these guidelines for driving equipment on the jobsite:

- Never drive a machine in a congested area or around people without a spotter or flagger. The spotter or flagger is responsible for determining and controlling the driver's speed.
- Be sure everyone is in the clear while backing up, hooking up, or moving attachments.
- Never move buckets or shovels over the heads of other workers.
- If you cannot see the surrounding area clearly from the operator's seat, get a spotter or do not operate the equipment.
- Wait for an all-clear signal before moving.
- Signal a forward move with two blasts of the horn.
- Signal a reverse move with three blasts of the horn.
- Yield the right-of-way to loaded equipment on construction sites.
- Maintain a safe distance from all other vehicles.
- When moving, keep the equipment in gear at all times; never coast.
- Maintain a ground speed consistent with ground conditions and posted speed limits.
- Know where overhead electrical power lines are located.
- Ensure that buried pipes, powerlines, and gas lines have been located and marked.

Case History

A 49-year-old dozer operator was crushed to death while compacting earthen fill during the construction of an oil exploration island. The victim was operating the bulldozer on level ground, alternately in forward and reverse at half throttle. With the blade completely down, the victim shifted the transmission control lever toward neutral. The transmission was partially disengaged, but not fully in neutral.

He assumed, without checking, that the bulldozer transmission was in a stable neutral position. He exited the cab on the right side. When he did this, the transmission slipped back into the first gear of reverse, causing the bulldozer to suddenly move. As a result, the victim was pulled between the underside of the fender and the top of the track cleats. As the bulldozer continued in reverse, the victim was fatally crushed beneath the track cleats.

The Bottom Line: Always use parking brakes and double-check settings before exiting the equipment. Never leave equipment when it is still running.

Source: The National Institute for Occupational Safety and Health (NIOSH)

Pinch points: The area in which two moving equipment parts come together.

Each piece of heavy equipment has its own unique set of hazards. Be aware of its **pinch points** (*Figure 31*). Pinch points occur when there is motion between equipment parts. Serious injury or death can result from getting part of your body or your clothing caught in a pinch point.

(A) Loader Pinch Points

(B) Excavator Pinch Point

Figure 31 Pinch points.

Sources: (A) Vadym Terelyuk/iStock/Getty Images; (B) Svetlana Verbitskaya/iStock/Getty Images

Center of gravity: The point where all of an object's weight is evenly distributed.

Another danger that heavy equipment operators must face is tipping over. Tipping can be caused by improper loading, unsafe driving, or unstable terrain. These factors affect the **center of gravity**, which is the point at which the machine and load are in balance. To understand how to prevent tipping, you must first understand center of gravity. The center of gravity is the point where all of an object's weight is evenly distributed. The addition of a load causes a machine's center of gravity to move toward the load. This effect is very noticeable on a machine like a telescoping boom forklift. *Figure 32* is a load chart for such a forklift. Notice that moving the load up and down, forward, and back affects the load capacity of the forklift. These changes are, in effect, changing the machine's

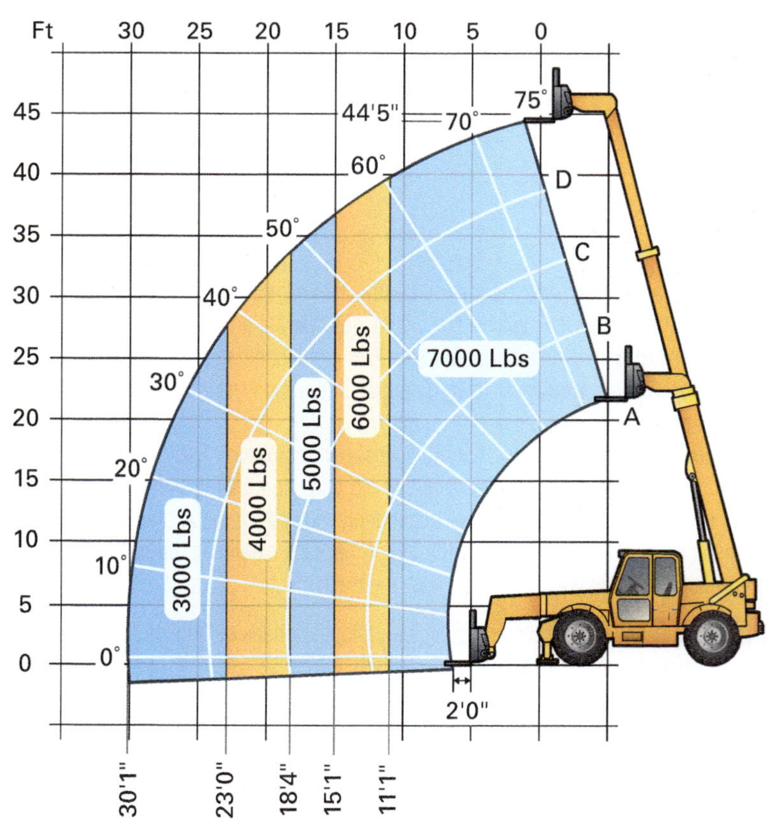

Figure 32 Effects of load position on load capacity.

center of gravity. If the forks are raised to their maximum height and extended the maximum distance forward, the load capacity is reduced by more than 50 percent (from 3,000 lb to 4,000 lb). If a 5,000-pound load were extended to the maximum height and distance, the forklift might tip over because its center of gravity shifted too far forward.

3.2.0 Weather-Related Safety Hazards

Under certain environmental conditions, such as extremely hot or cold weather, work can become uncomfortable and possibly dangerous. Be aware of the dangers when working under adverse conditions.

3.2.1 Cold Weather

The amount of injury caused by exposure to abnormally cold temperatures depends on wind speed, length of exposure, temperature, and humidity. Freezing is increased by wind and humidity or a combination of the two factors. Follow these guidelines to prevent injuries such as frostbite during extremely cold weather:

- Always wear the proper clothing.
- Limit exposure as much as possible.
- Take frequent, short rest periods.
- Keep moving to keep blood circulating throughout your body.
- Avoid icy surfaces.
- Do not expose yourself to extremely cold weather if any part of your clothing or body is wet.
- Do not smoke before exposure to cold. Breathing can be difficult in extremely cold air, and smoking can worsen the effect.
- Learn how to recognize the symptoms of overexposure and frostbite.
- Place cold hands under dry clothing against the body, such as in the armpits.
- Be careful of ice on ladders and walkway surfaces.
- Do not consume alcohol. Not only does this impair your function to operate heavy machinery, but it also makes you more susceptible to hypothermia and frostbite because your sensitivity to those things is dulled.

3.2.2 Cold Exposure Symptoms and Treatment

Spending long periods of time in the cold can be dangerous. It's important to know the symptoms of cold weather exposure and how to treat them. Symptoms of cold exposure include the following:

- Shivering
- Numbness
- Low body temperature
- Drowsiness
- Weak muscles

Follow these steps to treat cold exposure:

Step 1 Notify supervision.

Step 2 Get to a warm, inside area as quickly as possible.

Step 3 Remove wet or frozen clothing and anything that is binding, such as necklaces, watches, rings, and belts.

Step 4 Rewarm by adding dry clothing or wrapping in a blanket.

Step 5 Drink warm, not hot, liquids, and do not drink alcohol.

Step 6 Check for frostbite. If frostbite is found, seek medical help immediately.

Cold Weather Clothing Tips

Use the following tips to prevent injury due to cold weather:

- Dress in layers.
- Wear thermal-type woolen underwear.
- Wear outer clothing that will repel wind and moisture.
- Wear face protection and head and ear coverings.
- Carry an extra pair of dry socks when working in snowy or wet conditions.
- Wear warm boots and make sure they are not so tight that circulation becomes restricted.
- Wear wool-lined mittens or gloves covered with wind- and water-repellent material.

3.2.3 Symptoms and Treatment of Frostbite

NOTE

In advanced cases of cold weather exposure or hypothermia, mental confusion and poor judgment occur, the victim staggers, eyesight fails, and the victim falls and may pass out. Shock is evident, and breathing may cease. Death, if it occurs, is usually due to heart failure.

Frostbite is an injury resulting from exposure to cold elements. It happens when crystals form in the fluids and underlying soft tissues of the skin. The frozen area is generally small. The nose, cheeks, ears, fingers, and toes are usually affected. Affected skin may be slightly flushed just before frostbite sets in. Symptoms of frostbite include the following:

- Skin becomes white, gray, or waxy yellow. Color indicates deep tissue damage. Victims are often not aware of frostbite until someone else recognizes the pale, glossy skin.
- Skin tingles and then becomes numb.
- Pain in the affected area starts and stops.
- Blisters show up on the area.
- The area of frostbite swells and feels hard.

Use the following steps to treat frostbite:

Step 1 Protect the frozen area from refreezing. If there is a chance the affected area may get cold again, do not warm the skin. If the skin is warmed and then becomes cold again, it will cause severe tissue damage.

Step 2 Do not massage or apply heat to the frostbitten area.

Step 3 Get medical attention immediately. First aid for frostbite includes moving the victim to a warm area, removing wet clothing, and applying warm water of about 105°F (41°C) to the affected area. Do not pour the warm water directly onto the area, however.

3.2.4 Hot Weather

Hot weather can be as dangerous as cold weather. When someone is exposed to excessive heat, they run the risk of overheating. Some conditions associated with overheating include the following:

- Heat exhaustion
- Heat cramps
- Heat stroke

Heat exhaustion is characterized by pale, clammy skin; heavy sweating with nausea and possible vomiting; a fast, weak pulse; and possible fainting.

Heat cramps can occur after an attack of heat exhaustion. Cramps are characterized by abdominal pain, nausea, and dizziness. The skin becomes pale with heavy sweating, muscular twitching, and severe muscle cramps.

Heat stroke is an immediate, life-threatening emergency that requires urgent medical attention. Its symptoms are headache, nausea, and visual problems. Body temperature can reach as high as 106°F. This is accompanied by hot, flushed, dry skin; slow, deep breathing; possible convulsions; and loss of consciousness.

When working in hot weather, follow these guidelines to prevent heat exhaustion, cramps, or heat stroke:

- Drink plenty of water.
- Do not overexert yourself.
- Wear lightweight clothing.
- Keep your head covered and face shaded.
- Take frequent, short work breaks.
- Rest in the shade whenever possible.

3.0.0 Section Review

1. When working around a 200–350kV power line, how much clearance must the boom have between it and the line?
   a. 15 feet
   b. 20 feet
   c. 25 feet
   d. 50 feet

2. When operating any piece of equipment, what should the operator do before starting it up?
   a. Nothing, because the previous operator should have verified the piece of equipment was in operational condition before it was shut down.
   b. Walk around to see if there is any damage or anything that may be in the way of operation.
   c. The supervisor should walk around the piece of equipment before the operator comes to the jobsite.
   d. The piece of equipment will indicate to you if there is an issue that needs to be resolved before it will run.

3. When a person has frostbite, what is the first step you should take to treat it?
   a. Nothing, because it will fix itself over time.
   b. Massage the skin to bring back blood.
   c. Heat the area with hot water.
   d. Cover the affected area and protect it from refreezing.

4.0.0 Trenching and Excavation Safety

Objective

Identify safety practices associated with trenches and excavations.

a. Explain how soil composition relates to trench failures.
b. Describe the methods used to protect workers in trenches and excavations.
c. Identify personal safety hazards related to dust and soils.

Performance Tasks

There are no Performance Tasks in this section.

An excavation is any man-made cut, cavity, trench, or depression formed by removal of earth or soil. Sometimes the terms *excavation* and *trench* are used interchangeably, but there is a difference. A trench (*Figure 33*) is an excavation that is deeper than it is wide, and usually not wider than 15' (4.6 m). Nearly all trenches are dangerous if not protected. Because trenches are narrow, workers can easily become trapped. Trenches are typically dug to create footings, lay pipe, or place manholes. The soil that is removed from the ground is called spoil. When earth is removed from the ground, extreme pressures may be generated on the trench walls. If the walls are not properly secured by **shoring**, **sloping**, or **shielding**, they will collapse.

The collapse of unsupported trench walls can instantly crush and bury workers. This type of collapse happens because not enough material is available to support the walls of an excavation or too much stress is placed on the ground surrounding the trench, such as the weight of an excavator.

Any excavation project requires a great deal of planning by specially trained workers before it can safely begin. One of the tasks is to arrange for buried utilities in the excavation area to be located and marked. Each state has its own

Shoring: A support system designed to prevent the trench walls from collapsing.

Sloping: The angling of trench walls in order to reduce the risk of trench collapse.

Shielding: The use of structures that are inserted into a trench to protect workers.

Trenching

Figure 33 Trenching site.
Source: Svetlana Verbitskaya/iStock/Getty Images

agency to coordinate the location of buried utilities. They go by names such as One Call, Dig Line, and Dig Safely (*Figure 34*). There is now a nationwide 811 Dig Safely service that contractors and others can contact. The 811 service puts callers in touch with their state agency so they can arrange to have the jobsite surveyed and marked.

Figure 34 811 Dig Safely logo.
Source: Common Ground Alliance

Your company must individually contact any utility companies that do not subscribe to state agencies locating utilities. Utility workers will come to your worksite and mark their underground cables and pipes using the American Public Works Association (APWA) underground color codes shown in *Table 4*.

TABLE 4 APWA Underground Utility Color Codes

Color	Meaning
Red	Electric power lines, cables, conduit, and lighting cables
Yellow	Gas, oil, steam, petroleum, or gaseous material
Orange	Communication, alarm or signal lines, cables, or conduit
Blue	Potable water
Green	Sewers and drain lines
Purple	Reclaimed water, irrigation, and slurry lines
White	Proposed excavation
Pink	Temporary survey markings

The lead time for the One-Call agency to mark the jobsite varies by area and season. Do not start a digging job until the area has been cleared. Digging in uncleared areas can cause serious injuries, interruption of services, damage to the environment or equipment, or job delays.

Working around excavations and trenches is one of the most hazardous jobs. When heavy equipment is used around trenches, safety precautions must be exercised at all times to prevent injury to yourself and others.

Some of the hazards that may be encountered during an excavation include the following:

- Cave-ins due to trench failure
- Flooding from broken water or sewer mains
- Electrical shock from striking electrical cable in the trench or striking overhead lines
- Toxic liquid or gas leaks from nearby facilities, pipes, or idling vehicles
- Falling dirt or rocks from an excavator bucket
- Oxygen-depleted environment

Stresses and weaknesses can occur during excavations that affect the stability of trenches. The most common hazard during excavation is trench failure or cave-in. This happens when the walls of the trench are not strong enough to survive outside forces, causing the trench walls to collapse. Cave-ins can happen without warning and can bury workers in the trench.

4.1.0 Soil Composition and Trench Failures

Soil type is a major factor to consider in trenching operations. Only a company-assigned, qualified person has the experience, training, and education to determine if the soil in and around a trench is safe and stable. However, it is still your responsibility to know the basics about soil and its associated hazards.

Surface soil often contains some amount of organic material such as decaying plant matter and is a mixture of mineral grains coming from several kinds of rocks. Average soils are usually a mixture of two or three materials, such as sand and **silt** or silt and clay. The type of mixture determines the soil characteristics.

Silt: Soil deposited by running water.

Each of the various soil types behaves differently, depending on the condition of the soil at the time of the excavation. Sandy soil tends to collapse straight down. Wet clays and **loams** tend to slab off the side of the trench. Firm, dry clays and loams tend to crack. Wet sand and gravel tend to slide. Be aware of the type of soil you are working in and know how it behaves (*Figure 35*).

Loams: Soils containing clay, sand, and organic matter.

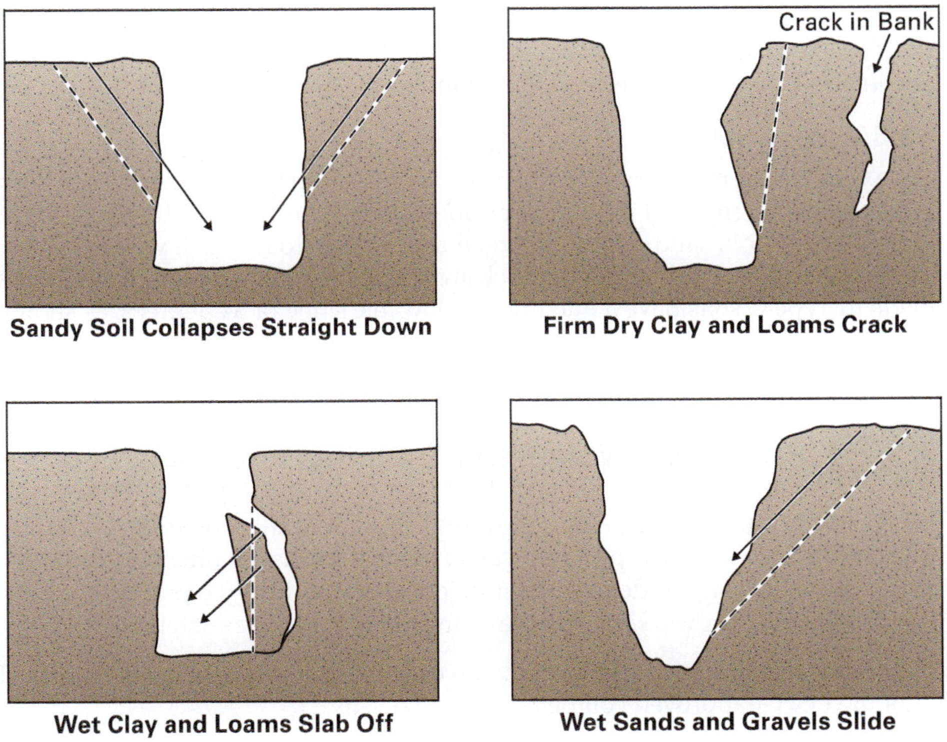

Figure 35 Soil behaviors.

When working in or around a trench, stay alert for changes in the trench. Watch for developing cracks, moisture, or small movements in the trench material. Alert the supervisor and coworkers to any changes. Small changes in trench walls may be an early sign of an impending cave-in.

Cave-ins are not the only cause of burial. Tons of dirt can be dumped onto workers when the excavated soil slides back into the trench. Such slides occur when the spoils pile is placed too close to the edge of the trench or when the ground beneath the pile gives way. OSHA requires a minimum of two feet between the trench wall and the spoils pile. This distance can vary depending on the type of soil and the height of the spoils pile. This area must also be kept free of heavy equipment, tools, and material.

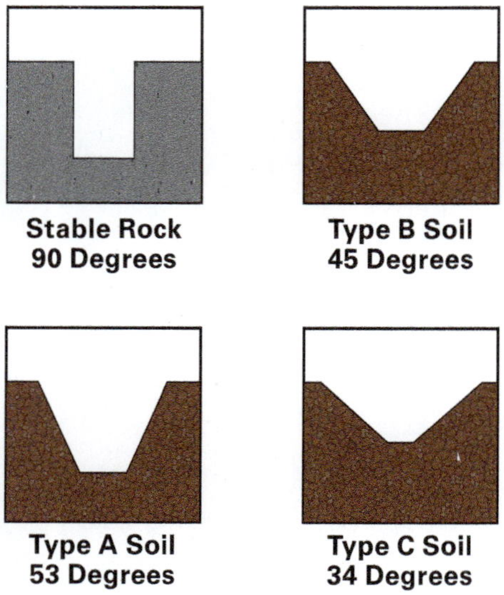

Figure 36 Maximum allowable slope for soil types.

4.1.1 Type A Soil

Type A soil refers to solid soil with good strength and cohesion. According to OSHA, examples of cohesive soils are clay, silty clay, sandy clay, and clay loam. Type A soil is more solid than any other type of soil with a compression strength that can withstand 1.5 tons or more per square foot. Excavations in Type A soil can have a maximum allowable slope of 53 degrees, as shown in *Figure 36*.

No soil can be considered Type A if any of the following OSHA outlined conditions exist:

- The soil is fissured. Fissured means a soil material has a tendency to break along definite planes of fracture with little resistance, or the material has cracks, such as tension cracks, in an exposed surface.
- The soil can be affected by vibration from heavy traffic, pile driving, or other similar activities.
- The soil has been previously disturbed. An exception to this rule occurs when the excavation is 12 feet deep or less and will remain open 24 hours or less.

4.1.2 Type B Soil

Type B soil refers to cohesive soils with compression strength of greater than 0.5 ton per square foot but less than 1.5 tons per square foot. According to OSHA, Type B soil is a granular soil, including angular gravel (similar to crushed rock), silt, sandy loam, unstable rock, and any unstable or fissured Type A soils. Type B soils also include previously disturbed soils, except those that would fall into the Type C classification. Excavations made in Type B soils have a maximum allowable slope of 45 degrees, as shown in *Figure 36*.

4.1.3 Type C Soil

Type C soil is the most unstable soil type. Type C soil refers to cohesive soil with compression strength of 0.5 ton per square foot or less. According to OSHA, gravel, loamy soil, sand, any submerged soil or soil from which water is freely seeping, and unstable submerged rock are considered Type C soils. Excavations made in Type C soils have a maximum allowable slope of 34 degrees, as shown in *Figure 36*.

4.2.0 Protecting Workers in Trenches and Excavations

When operating heavy equipment around any excavation or trench, you are responsible for your own personal safety as well as that of others in the work zone. The following guidelines must be observed by heavy equipment operators as well as other workers in the area to ensure everyone's safety:

- Never move heavy equipment into an excavation area without the approval of the OSHA-approved competent person on site.
- A competent person must inspect the excavation area daily, at the start of each shift, or when conditions change. Conditions that can affect an excavation include changes in the environment, such as rain, frost, severe vibration from heavy equipment, or change in soil conditions.
- Wear protective clothing and equipment, such as hard hats, safety glasses, work boots, and gloves. Use respirators and gas monitors if necessary.
- Do not walk under loads being handled by excavators, loaders, cranes, or hoists.
- Stay clear of any vehicle that is being loaded.
- Be alert. Watch and listen for possible dangers.
- Do not work above or below a coworker on a sloped or benched excavation wall.

- Barricade access to excavations to protect pedestrians and vehicles.
- Keep heavy equipment out of the area unless it is actively being used for work. Special attention should be given to cranes and crane matting near an excavation.
- Keep tools, equipment, and the excavated dirt (spoil) at least 2 feet from the edge of the excavation.
- Make sure sloping, shoring, trench boxes, or benching are used for excavations and trenches over 5 feet deep.
- Make sure means of exit, such as ladders or stairways, are placed within 25 feet of any employee in a trench with a depth of 4 feet or more.
- Stop work and immediately exit the trench if there is any potential for a cave-in. Make sure any problems are corrected before starting work again.

4.2.1 Making the Trench Safe

There are several ways to make a trenching site a safer place to work. Trenches are often reinforced with shoring or shielding systems. Other times, trench walls are cut at an angle away from the trench floor to relieve pressure and avoid cave-ins. This is called sloping. Shoring, shielding, benching, and sloping are different methods used to protect workers and equipment. It is important that you recognize the differences between them. *Figure 37* shows the difference between sloping and benching.

CAUTION

Protective systems are designed for even loads of earth. Heaving and squeezing can place uneven loads on the shielding system and may stress particular parts of the protective system.

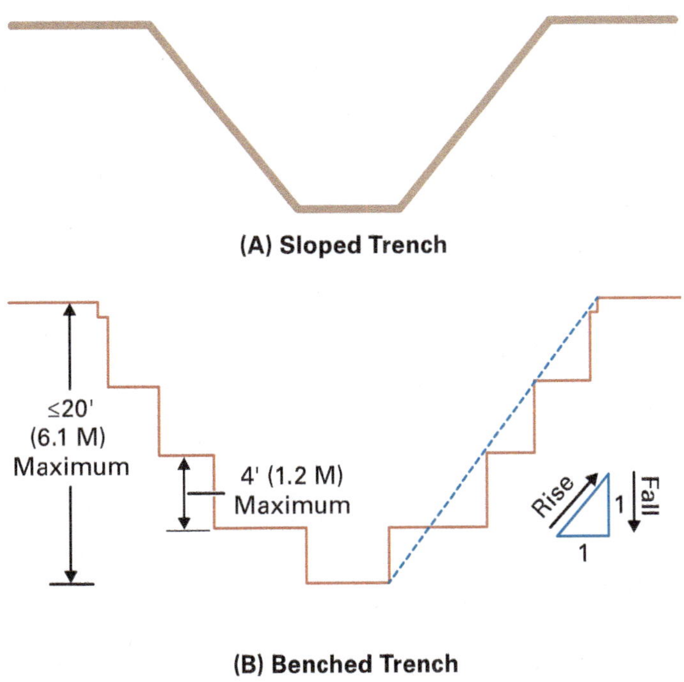

Figure 37 Sloping and benching Type B soil.

Shoring in a trench is placed against the trench walls to support them and prevent their movement and collapse (*Figure 38*). Shoring does more than provide a safe environment for workers in a trench. Because it restrains the movement of trench walls, shoring also stops the shifting of surrounding soil, which may contain buried utilities or support sidewalks, streets, building foundations, or other structures.

Trench shields, also called trench boxes (*Figure 39*), are placed in unshored trenches to protect personnel from trench wall collapse. They provide no support to trench walls or surrounding soil. For specific depths and soil conditions, trench shields withstand the side weight of a collapsing trench wall to protect workers in the event of a cave-in.

NOTE

Shielding systems are designed to protect workers in an excavation. They are not designed to prevent cave-ins.

Figure 38 Trench shoring.
Source: Kundel Industries

Figure 39 Trench box.
Source: Kundel Industries

When helping to place shielding systems, follow these safety guidelines:

- Be sure that the vertical walls of a trench box extend at least 18 inches above the lowest point where the excavation walls begin to slope. If the trench box is over a pipe, the box must extend 24 inches above the pipe.
- Make sure shielding systems are placed against the sides of the trench.
- Never permit workers to enter the trench box during installation or removal.
- Backfill the excavation as soon as the trench box has been removed.
- If a trench box is to be used in a pit or a hole, all four sides of the trench box must be protected.

WARNING!

Be aware of pinch points between trench boxes.

4.2.2 Ladders

There must be at least one method of entering and exiting all excavations over 4 feet deep. Ladders are generally used for this purpose. Ladders must be placed within 25 feet of any worker.

When ladders are used, the following requirements must be met:

- Ladder side rails must extend a minimum of 3 feet above the landing.
- Ladders must have nonconductive side rails if work will be performed near equipment or systems using electricity.
- Two or more ladders must be used where 25 or more workers are working in an excavation in which ladders serve as the primary means of entry and exit or where ladders are used for two-way traffic in and out of the trench.
- All ladders must be inspected before each use for signs of damage or defects.
- Damaged ladders should be labeled "Do Not Use" and removed from service until repaired.
- Use ladders only on stable or level surfaces.
- Secure ladders when they are used in any location where they can be displaced by excavation activities or traffic.
- While on a ladder, do not carry any object or load that could cause you to lose your balance.
- Exercise caution whenever using a trench ladder.

4.2.3 Sloping Systems

A sloping system is a method of protecting workers from cave-ins. Sloping is done by inclining the sides of an excavation. The angle of incline varies according to such factors as the soil type, environmental conditions of exposure, and nearby loads. There are three general classifications of soil types, and one type of rock. For each classification of soil type, OSHA defines maximum angles for the slope of the walls. MSHA uses the same guidelines for surface excavations but has special requirements for shafts and tunnels. The designation and selection of the proper sloping system is far more complex than described in this section. The depth of the trench, the amount of time the trench is to remain open, and other factors affect the maximum allowable slope. The important thing to understand is that a particular slope has been selected. When excavating slopes, follow the plans carefully.

4.3.0 Soil-Related Hazards

To stay safe, be aware of all hazards on the jobsite. When asked to list hazards of their profession, most construction workers would put heavy equipment, chemicals, and many others well ahead of hazards from soils, demolition dust, and sewage. While diseases from these sources are rare, they do occur. Most of these diseases are short-lived and mild, but they can sometimes be severe.

4.3.1 Soil and Demolition Dust

Excavation work generates a great deal of dust. The dust may contain animal droppings and other organisms that can cause diseases. Most of the illnesses that result from breathing this dust are mild, but a few can be serious. Fungi that occur naturally in the environment can cause serious illnesses when inhaled. Some diseases caused by bird droppings attack the lungs; these can be severe in people with pre-existing conditions such as asthma. Before demolition work is allowed to begin, a company safety inspector should have evaluated the site for materials that could be harmful. In some demolition situations, respirators and protective garments such as coveralls may be required.

WARNING!

Breathing asbestos fibers or getting them on your skin can cause serious illness. Proper PPE must be worn.

Asbestos was once commonly used in construction materials, such as insulation, ceiling tiles, siding, and flooring. Since that time, asbestos dust has been found to cause a variety of health problems ranging from simple skin irritation to cancer. Because of these hazards, asbestos use in this country is now extremely limited. Virtually all demolition work on asbestos material creates dust, thus exposing workers to this hazard. Asbestos removal may only be done by properly licensed companies. It requires specialized protective clothing, breathing apparatus, and containment equipment. Structures must be inspected by a qualified person for the presence of asbestos before demolition begins.

Another serious hazard is the silica encountered in concrete dust. Silica is a mineral found in concrete, masonry, and rock. During construction, silica may be found in a dust form. Silica dust can come from either the demolition of an existing structure or road (*Figure 40*), or from working in concrete-stabilized soil. The dust created when breaking up dry concrete has the same hazard as that of new cement.

Prolonged exposure to silica dust can cause silicosis, which is an incurable, and sometimes fatal, lung disease. The time it takes for silicosis to develop varies based on the length of time and the amount of silica someone is exposed to. In addition to silicosis, the inhalation of silica dust increases the risk of lung cancer and chronic obstructive pulmonary disease (COPD). Individuals working closely with materials containing silica, such as workers operating abrasive blasting (sandblasting) equipment, also risk the development of kidney disease.

WARNING!

When you are working in an area where silica dust is present, you must use the appropriate respiratory protection. Never work around silica dust without proper training, authorization, and PPE.

Figure 40 Concrete demolition.
Source: dalomo84/iStock/Getty Images

You can help to limit your exposure to soil contaminants and concrete dust by wearing a well-fitted respirator or proper PPE. Shower and change into clean clothes at your jobsite when possible. Avoid eating or drinking in the dusty area. Periodically spray the demolition area with water.

4.3.2 Sewage

Construction workers on an excavation site may uncover live sewers or septic tanks and should exercise caution when working with septic materials. Contact with raw sewage can place workers at risk for serious diseases such as *E. coli*, hepatitis, or pneumonia. Wear latex or similar gloves when working around sewage. Wash your hands frequently and avoid touching your face. Do not eat or drink in the area. If you do become ill after working on a job where you came in contact with sewage, tell your employer and doctor. Employers must offer the hepatitis vaccine to employees.

4.3.3 Soil Contamination

When working at a new construction site or performing demolition on a building, there is always the possibility of uncovering a contaminated site. Contaminants can be in the soil, buried underground, or stored in some containers such as a barrel. Identifying contaminants is a complicated and time-consuming procedure that requires advanced education and training. Heavy equipment operators are not expected to identify contaminants, but they are required to report any potential contaminations to a supervisor. Some warning signs of potential contamination are as follows:

- Puddles of fluid on the ground, especially fluid with an unusual color or odor
- Any fluid seeping out of the ground
- Unusual or foul odors
- Unmarked barrels or tanks—often buried or otherwise camouflaged

4.0.0 Section Review

1. The type of soil that is most likely to slab off is _____.
   a. sand
   b. dry clay
   c. wet clay
   d. gravel

2. Trenches must be inspected daily by _____.
   a. every worker
   b. the project manager
   c. a supervisor
   d. a competent person

3. The condition known as silicosis is caused by _____.
   a. coming into contact with sewage
   b. breathing concrete or cement dust
   c. handling bird droppings
   d. coming into contact with rotting vegetation

Module 22102 Review Questions

1. Caution signs with a yellow background and black letters are used to inform workers of _____.
   a. dangerous conditions
   b. potential hazards or unsafe conditions
   c. hazardous materials in the area
   d. weather conditions that could become hazardous

2. A sign identifying that an immediate hazard exists is a(n) _____.
   a. danger sign
   b. warning sign
   c. caution sign
   d. information sign

3. Barricades and barriers are used _____.
   a. only for traffic diversion on highway construction sites
   b. when there is not enough time to put up a sign
   c. to alert workers to potential hazards
   d. as a last resort to keep vehicles out of the work area

4. In areas where the ground is too soft to permit a machine to operate safely on a grade, it is a good idea to put up a _____.
   a. barrier
   b. flag
   c. caution sign
   d. danger sign

5. The spacing of temporary traffic control signs will depend on the road's _____.
   a. amount of traffic
   b. surface condition
   c. normal posted speed
   d. proximity to pedestrian traffic

6. Channeling devices are used in a temporary traffic control area _____.
   a. when flaggers are not available
   b. to guide traffic into a desired path
   c. to protect workers from speeding traffic
   d. to keep heavy equipment out of the road

7. The area next to the border that is designed to protect workers from closely passing vehicles is called the _____.
   a. buffer zone
   b. transition area
   c. pedestrian zone
   d. advance warning area

8. To use the Stop/Slow paddle, the flagger faces road users and _____.
   a. waves the flag rapidly across the lane of oncoming traffic
   b. steps into the lane of oncoming traffic and raises the flag
   c. repeatedly raises and lowers the flag to shoulder level
   d. extends the arm holding the paddle horizontally away from the body

9. For each hazardous substance in a shipment, OSHA and MSHA require that suppliers of hazardous chemicals provide _____.
   a. a respirator
   b. splash-proof goggles
   c. chemical-resistant gloves
   d. a safety data sheet (SDS)

10. The maximum amount of gasoline that OSHA allows to be stored outside an approved storage enclosure is _____.
    a. 5 gallons
    b. 10 gallons
    c. 25 gallons
    d. 50 gallons

11. What are the advanced symptoms of carbon monoxide poisoning?
    a. Dizziness
    b. Headaches
    c. Weakness and confusion
    d. Staggering and falling

12. Which of the following noise sources has the highest dBA level?
    a. Thunder
    b. Stock car races
    c. A jet plane taking off
    d. Shouting near the ear

13. Pale, clammy skin; heavy sweating with nausea and possible vomiting; a fast, weak pulse; and possible fainting are symptoms of _____.
    a. exposure to cold
    b. heat exhaustion
    c. hypothermia
    d. heat stroke

14. The only person who can decide if the soil in and around a trench is safe and stable is _____.
    a. a supervisor
    b. the engineer
    c. a company-assigned competent person
    d. the most experienced crew member

15. A device placed in an unshored trench to protect it from cave-in is called a _____.
    a. trench honeycomb
    b. trench shore
    c. trench box
    d. trench bench

16. Sloping is done to accomplish which of the following?
    a. To make the work area organized
    b. To protect the workers from cave-ins
    c. To provide jobsite control
    d. To allow the equipment to operate closer to the trench

Answers to odd-numbered review questions are found in the Module Review Answer Key at the back of this book.

Answers to Section Review Questions

Answer	Section	Objective
Section One		
1. c	1.1.1	1a
2. a	1.2.0	1b
Section Two		
1. d	2.1.8	2a
2. c	2.2.0	2b
3. d	2.3.0	2c
Section Three		
1. b	3.1.1	3a
2. b	3.1.2	3a
3. d	3.2.3	3b
Section Four		
1. c	4.1.0	4a
2. d	4.2.0	4b
3. b	4.3.1	4c

MODULE 22108

Equipment Operation, Inspection, and Readiness

Objectives

Successful completion of this module prepares you to do the following:

1. Describe the prestart procedures for heavy equipment.
   a. Describe how to safely mount a vehicle.
   b. Describe common operator prestart inspection procedures for equipment, attachments, and tires/tracks.
   c. Describe common operator prestart inspection procedures for vehicle drive systems.
   d. Describe common operator prestart inspection procedures for hydraulic systems.
   e. Describe common operator prestart inspection procedures for electrical systems.
2. Describe the procedures for starting, operating, and shutting down heavy equipment.
   a. Describe how to perform initial checks before starting the equipment.
   b. Identify basic maintenance procedures for the equipment.
   c. Explain how to properly shut down and secure equipment.
3. Describe the engine and auxiliary systems used in heavy equipment.
   a. Describe the primary engine types used in heavy equipment.
   b. Describe how to service auxiliary systems used with heavy equipment engines.
4. Describe hybrid and electric equipment preoperational checks.
   a. Describe what to check before operating hybrid-powered construction equipment.
   b. Describe what to check before operating electrically powered construction equipment.

Performance Task

Under supervision, you should be able to do the following:

1. Perform basic prestart inspection, startup, operational movement, and shutdown for three of the following types of equipment:
   - Utility tractors
   - Dozers
   - Loaders
   - Backhoes
   - Excavators
   - Compaction equipment
   - Motor graders
   - Scrapers
   - On-road dump trucks
   - Off-road dump trucks
   - Forklifts
   - Skid steers
   - Trenchers

Digital Resources for Heavy Equipment Operations

Scan this code using the camera on your phone or mobile device to view the digital resources related to this craft.

SCAN ME

Overview

Before operating any heavy equipment, it is important to conduct a set of preoperational checks to ensure the equipment is in operational condition. Maintaining equipment helps to keep it operational and ensures the operator is aware of potential problems and hazards. If a piece of equipment is not in serviceable condition, it must be put out of service until it can be repaired. Failure to conduct these types of inspections prior to operating a piece of equipment may result in injury or death.

This module covers some of the hazards and dangers that can occur when a piece of heavy equipment is not checked or inspected prior to operating it on a jobsite. It also provides information about preoperational checks and information related to evaluating the condition of heavy equipment. Ensuring a piece of equipment is in optimal condition decreases the possibility of an accident or having the piece of equipment fail on the jobsite.

NCCER **Industry-Recognized Credentials**

If you are training through an NCCER-accredited sponsor, you may be eligible for credentials from NCCER. The ID number for this module is 22108. Note that this module may have been used in other NCCER curricula and may apply to other level completions. Contact NCCER at 1.888.622.3720 or go to **www.nccer.org** for more information.

You can also show off your industry-recognized credentials online with NCCER's digital credentials. Transform your knowledge, skills, and achievements into credentials that you can share across social media platforms, send to your network, and add to your resume. For more information, visit **www.nccer.org**.

1.0.0 Prestart Procedures for Heavy Equipment

Performance Tasks

There are no Performance Tasks in this section.

Objective

Describe the prestart procedures for heavy equipment.

a. Describe how to safely mount a vehicle.
b. Describe common operator prestart inspection procedures for equipment, attachments, and tires/tracks.
c. Describe common operator prestart inspection procedures for vehicle drive systems.
d. Describe common operator prestart inspection procedures for hydraulic systems.
e. Describe common operator prestart inspection procedures for electrical systems.

Several of the prestart activities performed on heavy equipment machinery are the same on all machines. This section addresses the generic prestart activities that should be performed on any heavy equipment vehicle prior to starting it. Most companies have inspection checklists that specify which things need to be checked on a given machine. *Figure 1* shows an example of a shift inspection form used by construction companies. This type of inspection needs to be performed on all equipment.

The following are common areas that must be checked:

- Structural hardware (undercarriage, tires, tracks, protective structures, and body parts)
- Engine and power train
- Hydraulic system
- Electrical system for controls and instrumentation (gauges)
- Attachments

NOTE

For specific details on what needs to be checked on any given machine, refer to the vehicle-specific Operations and Maintenance (O&M) manual. The O&M manuals, also called the operator's manual, give specific procedures for any operator-performed maintenance activities. Most O&M manuals also give detailed pictures of all control panels, indicators, and control devices.

Shift Inspection List

Date______ Submitted by__________

Unit No./S.N. __________

Make/Model __________

Hour/Odometer Reading __________

Today's Use: Hours Machine...

- Operated __________
- Down (due to failure) __________
- Idle (machine not needed) __________

Use the daily inspection sheet as a guide to fill in the following information (if applicable). Explain any defects on the back side.

	OK F	OK R	Amount Added or *Needs Attention
Engine Coolant	___	___	______
Engine Oil	___	___	______
Transmission Oil	___	___	______
Pump Drive Oil	___	___	______
Pivot Shaft Oil	___	___	______
Hydraulic Oil	___	___	______
Fuel	___	___	______
Coolant Temp. Recorder Label	___	___	______
Engine Air Restriction Indicator	___	___	______
Ground Engaging Tools	___	___	______
Tires/Undercarriage	__/__	__/__	______
Fire Protection	___	___	______
Seat Belt	___	___	______
EMS/Gauges	___	___	______
Governor Control	___	___	______
Brakes	___	___	______
Transmission	___	___	______
Steering	___	___	______
Implements	___	___	______
Lights	___	___	______
Horn	___	___	______
Backup Alarm	___	___	______
Wipers/Washers	___	___	______

Repairs Needed__________

Additional Comments__________

Signature__________

Inform supervisor of any malfunction before operating machine

Figure 1 Example of a shift inspection form.

1.1.0 Inspecting the Jobsite and Mounting a Vehicle Safely

Most heavy equipment is large enough to kill a person who is struck by it. Before mounting a machine, and during operations, operators must be constantly aware of their surroundings. Anyone on the ground and near a heavy equipment machine must keep a safe distance, whether it is moving or not. Personnel near the machine should always assume the operator cannot see them. Some of the larger machines have a closed-circuit camera system that allows the operator to see blind spots that cannot normally be seen from the operator cab.

Operators must familiarize themselves with the day's operations and required tasks. They must review any job-related paperwork to make sure that they know what activities must be accomplished. They should also perform the following checks before mounting the equipment:

- Check for any warnings or notes in the job sheet.
- Perform a jobsite inspection, looking for any markers that show where to dig or fill.

- Walk around the area in which the work will be done, checking the ground for large holes, large rocks, soft or muddy areas, sudden drop-offs, and curbs. Driving over uneven terrain with a heavy load can cause the vehicle to tip.
- Observe the terrain, and plan how to move the equipment safely. Sharp objects can damage tires. If possible, move them or mark them.
- Look above the site for overhead power lines, trees, and other overhangs that could contact a raised **bucket** or boom. Be aware of any low clearances in the work area.

Bucket: A U-shaped, closed-end scoop attached to a front-end loader or backhoe.

WARNING!

Climbing into the vehicle can be dangerous. Slips and falls are common hazards, but they can be avoided. Clean the steps and handholds. Oil, grease, snow, ice, and mud can cause slips and falls. Make sure work shoes do not have any excessive amounts of mud or grease.

Mount the machine only at locations that are equipped with steps and/or handholds. Face the machine when mounting or dismounting. Always maintain three contact points with the machine. Observe the following safety rules when mounting heavy equipment:

- Do not use controls as handholds.
- Do not mount the machine while carrying tools or supplies.
- Do not mount a moving machine.
- Do not jump off a machine or allow others to do so.

Check that all safety equipment is securely in place and in operating condition. Clean and adjust the mirrors. Verify that the fire extinguisher is fully charged. Make sure that any signs, shields, or guards are in place. Check the service and parking brakes. Always fasten the seat belt before operating the machine. OSHA and MSHA require that any approved seatbelt and **rollover protective structure (ROPS)** be installed on virtually all heavy equipment. Check the ROPS bolts to make sure they are not loose or damaged. Old equipment must be retrofitted.

Rollover protective structure (ROPS): A metal structure attached to the equipment's cab designed to protect the operator should the equipment roll over.

Once in the operator cab, secure all personal items or other objects in the operator's area. Loose items can become flying objects, can block equipment controls, and can cause equipment damage or personal injury. Adjust the seat as needed to reach all the controls comfortably. Verify that the seat is locked in the correct position. Take time to become familiar with all the controls, warning devices, and gauges.

WARNING!

Do not use any heavy equipment that is not equipped with approved seat belts and a ROPS.

1.2.0 Inspecting Equipment, Attachments, and Tires

Heavy construction equipment is normally mounted on an undercarriage riding on either metal/rubber tracks or rubber tires. Structural hardware includes the undercarriage and all things mounted to it, as well as the machine's body. When inspecting these pieces of equipment, make sure to verify everything is tightened properly and any minor or major structure is in serviceable condition.

1.2.1 Tires

Some machines ride on inflatable tires while others ride on solid rubber tires. Check tires for bulges, cuts, punctures, and signs of excessive wear. Check the wheels to ensure that lug nuts are in place and properly torqued. Verify that the inflatable tires are properly inflated. Tire inflation pressures are often posted on the sidewall of each tire. The O&M manual is also a reliable source for proper tire inflation specifications.

Figure 2 shows examples of both inflatable and solid rubber tires found on construction equipment.

Inflatable Tire

Solid Rubber Tire

Figure 2 Construction equipment tires.

1.2.2 Tracks

Machines with metal tracks are a bit more complicated to inspect due to their construction. Metal tracks are made up of track shoes fastened to a track chain (*Figure 3*). The track chain is a drive chain that wraps around the drive sprocket (sometimes at the rear of the undercarriage) and the idler sprocket (closer to the front of the undercarriage). Between the sprockets, the chain is supported by rollers. The sprockets and rollers all ride on bearings that must be lubricated. Due to the environment in which most tracks operate, rocks, roots, and other debris may become trapped in between the track chain and the sprockets or rollers. Preoperational checks need to include making sure that all debris is removed from the tracks. Also, look for loose, damaged, or missing bolts or pins that hold the track shoes to the chain; check proper track tension per the manufacturer's instructions; and look for loose or damaged track shoes.

Some lighter-weight construction vehicles are equipped with rubber tracks. The rubber tracks are essentially one large rubber belt with molded cleats. Mechanically, the rubber tracks function the same as the metal tracks. A rubber track wraps around the drive and idler sprockets and is supported by rollers installed on the undercarriage frame. *Figure 4* shows an undercarriage with rubber tracks. Inspections of a rubber track should include making sure that the track appears to be properly installed and tensioned per the manufacturer's instructions. The track should not be excessively worn, ripped, or torn.

Figure 3 Metal tracks.

Figure 4 Rubber tracks.

1.2.3 Rotating Parts

It is important to keep rotating parts properly lubricated to get the most use out of them. Remember that rotating parts are riding on some form of bearing. Bearings must be either greased or bathed in a lubricant (oil). Refer to the O&M manual for lubrication instructions on any undercarriage components. If bearings need to be greased, make sure the grease fittings are cleaned before attempting to add grease to them. Make sure that the right kind of lubricant is being used. *Figure 5* shows a grease gun being used to grease a fitting on the bottom end of a hydraulic cylinder. All pivot points, such as the end of a hydraulic cylinder, have a grease fitting. If pivot points are not greased, the components will wear increasingly fast and will fail prematurely.

Figure 5 Example of a grease fitting.
Source: Welcomia/iStock/Getty Images

1.2.4 Frame and Body Panels

The frame of a heavy equipment machine is supported above the undercarriage of the vehicle. Keeping the main protective structure around and over the operator in good condition is critical to the safe operations of the machine. Anyone inspecting the machine's frame and body panels should be looking for cracks in the metal (especially the frames), loose connections (bolts or other types of fasteners), broken or cracked welds, and excessive corrosion. Dirt, mud, and moisture collect on the surfaces of the machine's body panels, ladders, and walking surfaces, and must be removed. Make sure ladders and walking surfaces are clean.

1.2.5 Safety Equipment

Check all installed or mounted safety equipment on the vehicle. This includes mirrors, seat belts, horn(s), and the fire extinguisher. Verify that lights are properly mounted and not damaged. If the machine is so equipped, ensure that all brush guards are properly mounted and undamaged. Make sure the windows are clean and undamaged.

1.3.0 Power Train Inspections and Checks

The power train of a vehicle starts with the engine. Heavy equipment machines are powered by either gasoline or diesel engines. The vehicle's engine is connected through a transmission or torque converter to the machine's drive axle. Regardless of engine type, the operator needs to visually check the machine's fuel supply system, all fluid levels, coolant system components, engine, transmission, drive axle(s), and hydraulic system for signs of leaks.

Going Green

Lubricants, Coolants, and Fuels

Lubricants, coolants, and fuels can harm the environment. Be extremely careful not to spill coolants and fuels when servicing a vehicle. Make sure that all lubricant leaks are contained, reported, and stopped. When oil must be changed in an engine or transmission, contain the used oil and recycle or dispose of it properly.

1.3.1 Fuel Supply Check

Engines must have the right type of fuel to operate. Gas engines must have only an approved grade of gasoline. Diesel engines must have only an approved grade of diesel fuel. Never use gas in a diesel engine, or diesel fuel in a gas engine. Most fuel tanks are clearly marked for gas or diesel fuel.

WARNING!

Never fill the fuel tank or work on any part of the fuel system with the engine running, while smoking, or near an open flame. Doing so may cause a fire or an explosion. Make sure a fire extinguisher is nearby.

Before the fuel reaches the engine from the fuel tank, it usually passes through a fuel filter. Those fuel filters must occasionally be changed. Fuel filters on heavy equipment machines are usually mounted near the engine's fuel injector (or carburetor), as shown in *Figure 6*. Check for visible water in the fuel-water separator (*Figure 7*) and drain it into an approved container.

Figure 6 Fuel filter.

Figure 7 Fuel-water separator.

1.3.2 Coolant System Check

Start the power train inspections at the engine's radiator. Make sure that the air paths into, through, and out of the radiator are free of debris. Some radiators have electric fans that increase the air flow over the fins of the radiator tubes. *Figure 8* shows a swing-out hydraulic fan on a large heavy equipment machine.

Figure 8 Radiator cooling fan.

CAUTION

Use the specified coolant. Do not mix coolant types (colors).

Figure 9 Radiator cap on a dump truck.

The hoses between the engine and the radiator must be securely fastened and in good condition. If the engine is cool, remove the radiator cap (*Figure 9*), and check the fluid level in the radiator. Some machines have coolant storage tanks for overflow coolant. If in doubt about any fluid level, review the O&M manual for the machine being inspected. Add coolant as needed and be careful not to overfill the cooling system or it will leak and boil over.

WARNING!

To prevent personal injury, never open the radiator cap on a hot engine. Always allow it to cool before relieving the pressure inside the radiator and opening the cap.

NOTE

Different machines have their radiators located in different places. The radiator cap and coolant storage tank are usually protected inside the body of the machine.

1.3.3 Engine Oil Check

Engines have dipsticks, which are used to measure the amount of oil in the engine (*Figure 10*). Dipsticks have markers on them that show where the oil level should be, and a level where oil needs to be added. Pull the dipstick out of its storage area and check the oil level of the engine. If oil must be added, make sure that only the oil called for in the machine's O&M manual (or company procedures) is used. Do not overfill the oil. Make sure that the dipstick is securely replaced.

Figure 10 Oil dipstick and engine oil fill area.
Source: Courtesy of JCB Inc.

Going Green

Oil Filter Disposal

To protect the environment, make sure that used and removed oil filters are disposed of in accordance with all environmental guidelines. Used oil filters may not be thrown into a landfill. Most communities have specific collection containers for used oil filters and empty oil containers.

While checking the oil, also locate and check the engine's oil filter(s). Some companies have their maintenance workers mark the oil filter(s) with a date when the filter was last changed. Maintenance records should also have that date. Try to determine if the machine's oil filter needs to be changed. Make sure the machine is serviced in accordance with the manufacturer's maintenance schedule.

1.3.4 Engine Air Filter Check

Due to the environments in which most heavy construction equipment is used, getting clean air to the engines is most critical. Most heavy equipment engines have large air filters housed inside metal canisters mounted somewhere near the engine. In most cases, a lid can be removed from the filter housing and the filter can be slipped out for visual inspection. In some cases, a filter can be cleaned if it is not too oily and dirty. If in doubt, replace any dirty filter. *Figure 11* shows an air filter being checked and replaced.

(A) Air Filter on a Compact Wheel Loader

(B) Servicing a Wheel Loader

Figure 11 Air filter replacement.
Source: Volvo Construction Equipment

1.3.5 Transmission and Drivetrain Checks

Transmissions and drive axles, regardless of type, need lubricants to protect the gears inside them. Any transmission, other than a standard shift transmission, also requires a specific transmission fluid to function as designed. Always check the O&M manual for the specific type of transmission fluid needed in the machine being inspected. Some transmissions have sight glasses that indicate the level of fluid in the transmission. Others may have a dip stick. Drive axles usually have inspection ports that can be opened to check the level of lubricant inside them.

WARNING!

Be careful when checking the fluid levels in transmissions and drive axles. If checking the fluid while it is hot, be careful to not get any of the hot fluid on exposed skin.

Some modern machines may have temperature and level sensors in or on both the transmission and drive axle(s). Monitoring a transmission's operating temperature is critical to the continued operation of the machine.

1.4.0 Hydraulic System Inspections and Checks

Almost all heavy equipment vehicles use some form of hydraulic system to control their attachments. Remember that a hydraulic system is a closed-loop system consisting of the reservoir, a pump, filters and strainers, control valves, hoses, and actuators (cylinders or motors). Heavy construction equipment may also use hydraulic power to drive cooling fans and lubrication pump motors. *Figure 12* shows a hydraulic power management system like the ones found on some heavy equipment.

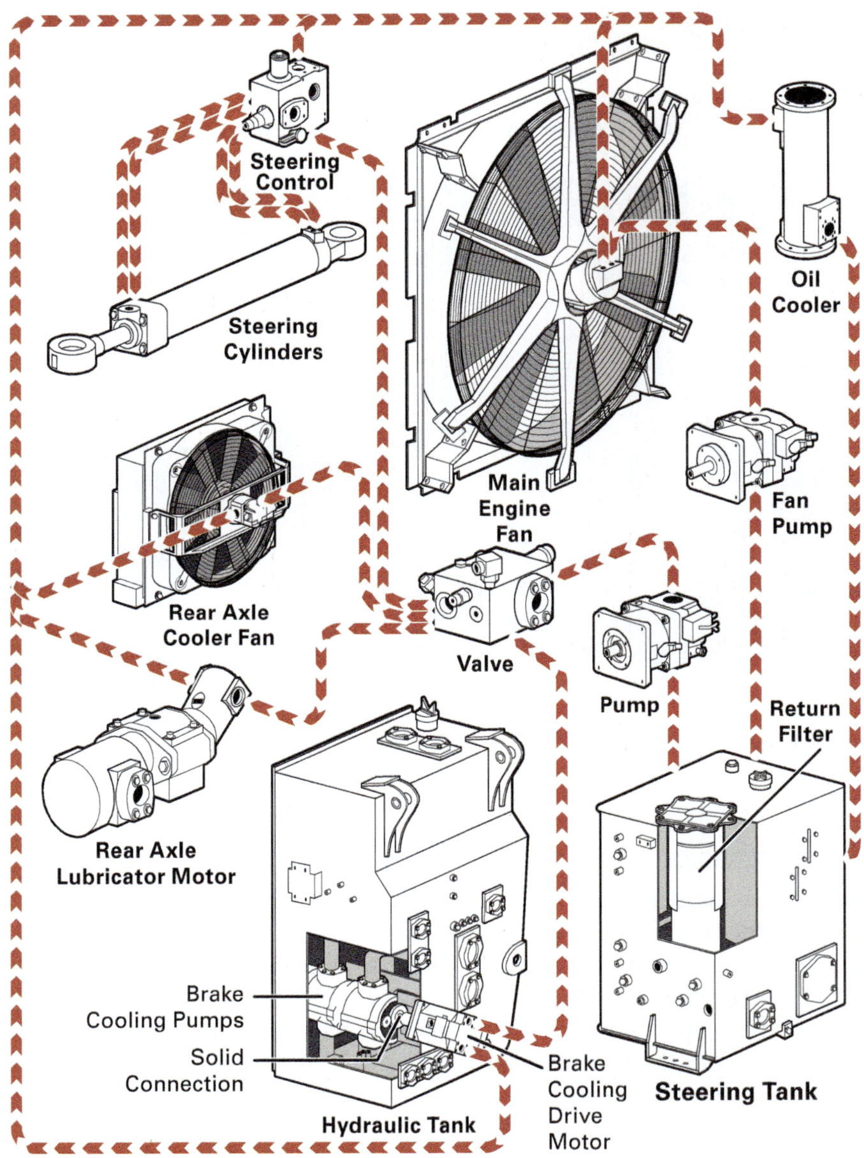

Figure 12 Hydraulic power management system components.

1.4.1 Hydraulic System Reservoir and Pump

The heart of a hydraulic system is its reservoir (storage tank) and the pump that moves the hydraulic fluid. Most hydraulic reservoirs have a sight glass that allows the level of fluid to be checked without opening the reservoir. If hydraulic fluid needs to be added, it can be poured in through the fill port located on the reservoir. Most fill ports have a strainer/filter inside them to help ensure that the fresh hydraulic fluid does not have trash in it. *Figure 13* shows the location of the reservoir's sight glass and fill port.

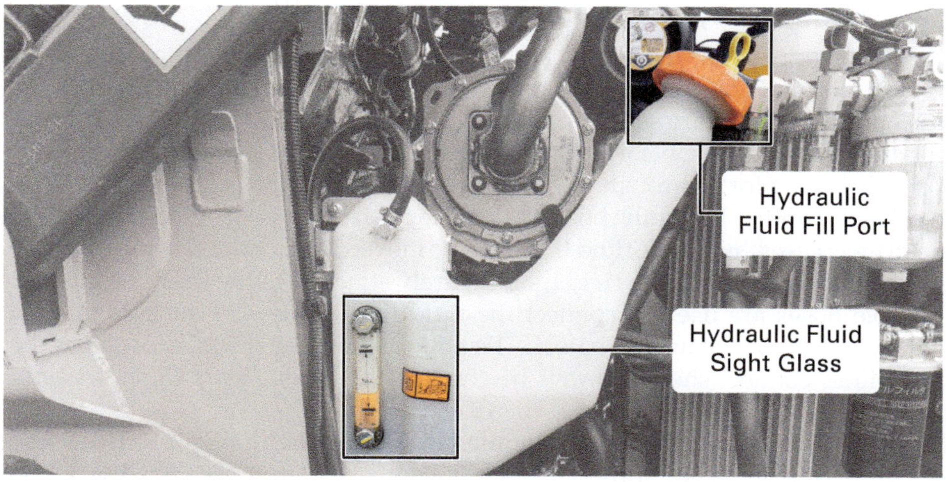

Figure 13 Typical hydraulic reservoir.

The temperature of hydraulic fluid rises as it is forced through the hoses and control valves to make the actuators perform work. After hydraulic fluid has done its job as an actuator, the fluid returns to the reservoir. It may or may not pass through a filter before returning to the reservoir. Internally, reservoirs are designed to cool the returning hydraulic fluid. In some cases, a heat exchanger may be used before the reservoir to provide additional cooling for the returning hydraulic fluid.

The size of the heavy equipment often determines the size and complexity of its hydraulic system. Hydraulic pumps may be driven by the same belt(s) used to run the engine's water pump and coolant fan, or they may be directly driven off a cam or shaft inside the machine's engine. *Figure 14* shows a hydraulic system pump mounted on the side of the machine's engine.

Figure 14 Hydraulic system pump.
Source: Jason Finn/Shutterstock

WARNING!

Do not use bare hands to check for fluid leaks on a vehicle. When checking for leaks, ensure that high-pressure oil isn't accidentally injected underneath your skin. Do not use your hand to "feel" for a leak. Use cardboard and deflect the oil away from your skin or eyes. Always use hand and eye protection to keep the fluid from contaminating your skin. Daily exposure to chemicals in this manner can cause health hazards.

Going Green

Hydraulic Leaks

Hydraulic systems tend to leak. To protect the environment, make sure any parked vehicles are not leaking onto the ground. Clean any leaking hydraulic fluid from the machine and dispose of the cleaning cloths properly. Report leaks so they can be repaired before the machine is used again.

1.4.2 Hydraulic System Hoses

When inspecting a hydraulic system pump, pay attention to the metal connectors that connect the hydraulic hoses to the pump's body. Ensure that the hose connectors are securely fastened at both ends. Always check the hoses for cracks and wear spots. Hydraulic hoses tend to move as the hydraulic fluid flows back and forth through them. If they happen to get too close to something else, even another hydraulic hose, the rubbing can wear a hole in the hose.

In most cases, hydraulic fluid that must be moved from one end of the machine to the other is run through rigid lines securely attached to the machine's frame. Flexible hoses are used to connect the rigid lines to devices such as pumps, control valves, pipes, and actuators. The points where the flexible hoses and the rigid lines connect should always be checked for leaks. *Figure 15* shows both flexible and rigid hydraulic lines.

The hydraulic lines and hoses mounted inside the machine's protective housing are less likely to be exposed to potential damage than those installed out on the boom and stick of a backhoe or excavator. When inspecting the hydraulic system hoses, check every hose or line that can be seen. The hoses and lines on the movable attachments are more likely to get damaged. *Figure 16* shows how the hydraulic hoses and lines on the boom and stick of an excavator are exposed. If you see any indication of leaking that shows on the ground, or dripping from the underside of the machine, you must follow the trail of the leakage to its source, which could even be at the top of the machine.

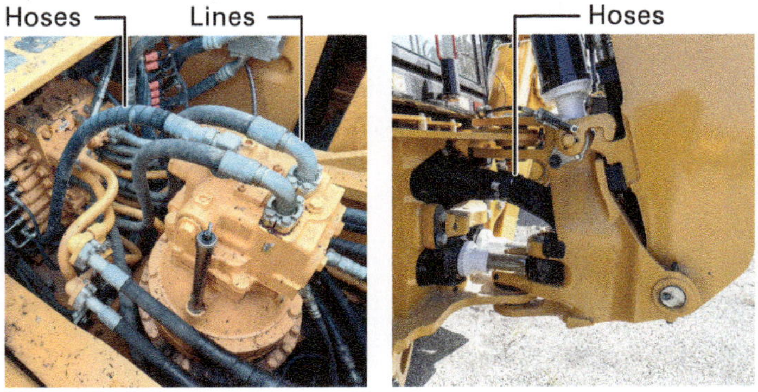

(A) Hydraulic Lines and Hoses in Engine Area

(B) Hydraulic Hoses at Base of Backhoe Boom

Figure 15 Hydraulic lines.
Source: (A) Zulfikar Ramadhan/iStock/Getty Images

Figure 16 Exposed hydraulic hoses and lines.
Source: Courtesy of CASE Construction Equipment

> **Going Green**
>
> **Hydraulic Lines**
>
> Hydraulic lines, especially those on construction equipment, are more likely to be damaged through daily use. Hydraulic lines, under intense pressure, occasionally rupture. If a line breaks, for whatever reason, the hydraulic pump continues to pump until shut off. Anytime there is a leak, the equipment should be shut down. Most companies have fluid containment kits located either on the machines or nearby. These containment kits include various devices and absorbent materials used to clean up oil spills. Always report any oil spill.

1.5.0 Electrical and Control Inspections and Checks

Modern heavy equipment vehicles have a variety of electrical sensors, controllers (computers), and indicators. The control panels in the operator's cab area are made up of switches, push buttons, analog gauges (gauges that have rotating pointers to show different values), digital readouts, and different colored indicator lights. Some control panels also have audio (sound) devices that give off different sounds for different situations.

Without turning the machine's On/Off switch On, it is difficult to tell if any of the control panel devices are operational or malfunctioning. Visually check the control panels for broken glass or gauges that appear to be inoperative. If allowed, turn the On/Off switch to On and recheck all the gauges and indicators on the control panel being inspected. *Figure 17* shows examples of operator control panels.

Figure 17 Operator control panels.
Source: Courtesy of CASE Construction Equipment

The typical charging and starting systems consist of a 12V or 24V starting system and an associated 12V **direct current (DC)** alternator, as shown in *Figure 18*. Direct current electrical systems are systems in which the electrical current flows in only one direction. DC voltage is the only type of voltage that can be stored

Direct current (DC): An electrical current that flows in only one direction.

within a battery. When looking at components on the piece of equipment make sure to verify if they are 12-volt or 24-volt type. Using the wrong voltage type of components will cause that component to fail when it is powered up.

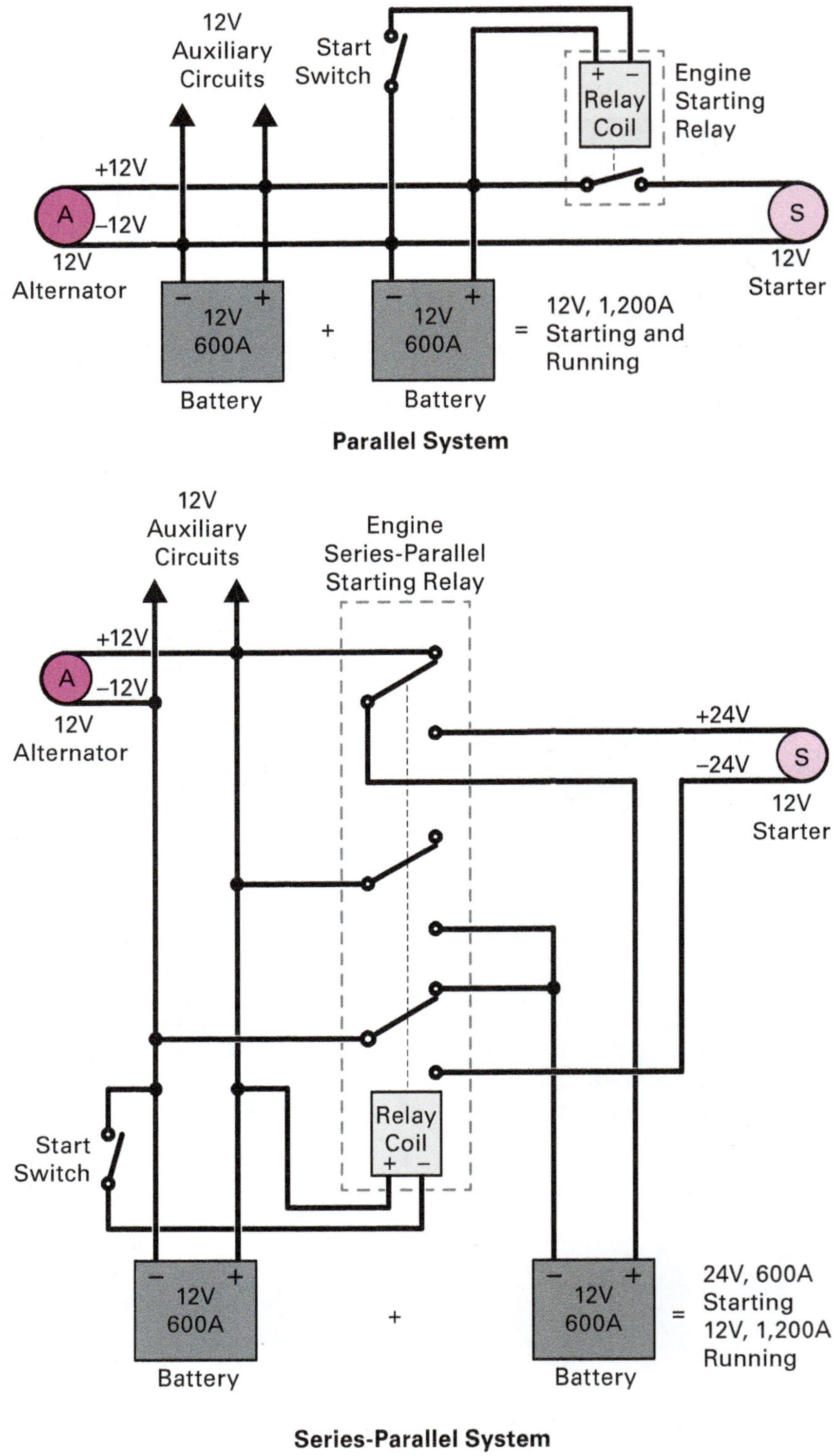

Figure 18 Typical charging and starting systems.

NOTE

Every machine has a slightly different control panel. Refer to the machine's O&M manual for details about the control panel for the specific machine.

WARNING!

When working on electrical systems and lead-acid batteries, wear proper eye and face protection. When charging, batteries release explosive hydrogen gas. A spark or flame can cause accumulated gas or batteries to explode. If battery acid gets on your skin, immediately flush the area with water and seek medical attention.

1.5.1 Batteries, Alternators, and Cables

The electrical system of a heavy equipment machine is not that different from the electrical system in a normal car or truck. The biggest difference is that the heavy equipment engines require more battery power to start, so the heavy equipment vehicles are equipped with multiple batteries. *Figure 19* shows a typical battery storage area in a heavy construction machine.

If a machine's batteries are dead (containing little or no electrical charge), do not attempt to charge or jump the batteries without reviewing the company's jump-starting procedures and the machine's O&M manual. Some companies do not allow the jumping of batteries.

WARNING!

Batteries contain acid. Most batteries are sealed but may still leak. Anyone working with or near any kind of vehicle battery needs to be careful. Keep battery acid away from bare skin. Do not allow any kind of metal object (such as a tool) to drop across the terminals of the batteries. If batteries must be changed out for new ones, review the machine's O&M manual for disconnecting and reconnecting batteries. If the electrolyte comes in contact with the skin or clothing, make sure to use water to flush the acid off the skin or material. Failure to do so could cause chemical abrasions on the skin and disintegrate any cloth material.

When inspecting a machine's electrical system, start at its batteries. Make sure that the ground cables are secured to the machine's chassis ground. Also be sure that the ground and positive cables are securely attached to the battery terminals. At the terminals, look for corrosion that could cause a bad connection between the cables and the battery terminals. Verify, as far as visually possible, that the battery cables appear to be undamaged.

Batteries must be recharged to continue working. An engine alternator (*Figure 20*), which is usually driven by a belt from the front of the engine, recharges the batteries. Verify that the alternator and all its terminals appear to be securely mounted and undamaged. When the engine is started, a current or amperage meter on the operator's console should indicate that the alternator is recharging the batteries. Until the engine is started, nothing more can be done to verify that the alternator is working. On later model applications, alternator control is commanded by the electronic control module (ECM) to coordinate it with the operation of the engine. This integration allows for the most efficient use of fuel and mechanical power.

Figure 19 Batteries on heavy construction equipment.

Figure 20 An engine alternator.
Source: Prasit Rodphan/iStock/Getty Images

Electrolyte: A solution of sulfuric acid and water, which is highly poisonous and can cause severe chemical burns.

Many batteries in heavy equipment machinery on a construction site are lead-acid batteries and contain an **electrolyte** solution of sulfuric acid and water. Often referred to as battery acid, this solution is highly poisonous and can cause severe chemical burns. To avoid injury, wear a full-face shield to prevent fluid contact with the face and eyes. Wear chemically resistant gloves and clothing to prevent this acid from damaging skin and regular clothing.

Explosive gases are always present when working around batteries. This is especially true when the battery is being charged. To avoid explosions, never smoke around batteries. Keep arcs, sparks, and open flames away from batteries, and always work on batteries in a well-ventilated area.

Follow the instructions in the manufacturer's manual for battery-boosted starting of engines (also known as jumping). Battery-boosted starting is a two-person job. One person needs to be in the driver's seat so that the equipment is under control when the engine starts. Be sure that the voltage of the replacement battery is compatible with the power source. If a battery needs to be replaced, the old battery must be disposed of in accordance with environmental regulations.

WARNING!

Always disconnect the battery before welding on any machine. If the battery is not disconnected, stray voltage could work its way through the electrical system and then cause any module or battery on the system to fail.

Jump-starting one street vehicle from another is done often but can be dangerous if improperly done. When dealing with the engines in larger heavy equipment machines, the danger increases because some of those machines require multiple batteries to start. Another danger is that different machine manufacturers wire their machines differently. Some use a positive ground for their electrical systems while others use a negative ground. Always refer to the manufacturer's operations and maintenance manual to see how the machine is wired.

The type of batteries used in these heavy machines also differs from year to year. Some older batteries give off hydrogen sulfide gas. That gas will sometimes accumulate inside battery compartments on hot days. A spark near that accumulated gas can set off an explosion, causing possible injury to anyone near the explosion. Sparks can be created by anything from a dropped tool to an improperly attached jumper cable.

Because most heavy equipment is located outside, the weather plays a major role in getting a large engine to start, especially in cold weather. Any engine is going to be hard to start when cold because the lubricants inside the engine are still stiff from the cold. Companies often install block heaters on engines operating in cold environments. The block heaters help keep the lubricants warm and thinned, which makes starting an engine easier. If the starting batteries have been frozen and need to be jumped, they may explode as a charge is introduced into the cells. Check with your supervisor on how to deal with frozen batteries.

For jump-starting procedures, refer to the manufacturer's operations and maintenance manual for each machine. Also, check company policies about jump-starting heavy equipment because jump-starting may not be allowed in some companies due to safety concerns.

WARNING!

Use extreme care if jump-starting is required. Improperly connected jumper cables can cause severe battery damage, starter damage, or an explosion. Make sure that the ground cable is connected to the chassis ground and not to the discharged battery ground.

1.5.2 Lights, Horn, and Alarms

Heavy equipment vehicles are often equipped with lights. Visually inspect any installed lights to make sure they are undamaged, clean, and securely fastened. Also, look at the physical positioning of the lights and make sure they appear to be pointing in the proper direction. Verify that the wiring into the lights is securely fastened and undamaged. Do the same thing for the machine's horn and any alarms that may be installed on the machine. These components allow for safe operation of the piece of equipment around the construction site and ensure those around the equipment will be aware of any changes of direction.

1.5.3 Sensors

Modern construction machinery relies heavily on computers and their related devices. A vehicle's computer relies on sensing devices to tell it things about the vehicle, such as water temperature and hydraulic oil temperature. On construction equipment, sensors also tell the computer how the devices attached to the machine are positioned (raised, lowered, or at some angle). *Figure 21* shows sensors used on modern graders.

(A) Grader Blade GPS Location **(B) Engine Temperature Sensor**

1 Housing for Rotational Sensor
2 Cable Connector to Sensor
3 Sensor Cabling
4 Housing for Tilt Sensor

Figure 21 Examples of sensors.
Source: (B) Aleksandr Kondratov/iStock/Getty Images

When inspecting sensors, always verify that the sensor itself is securely attached and properly positioned. Verify that the cabling connected to it is securely fastened, and that the cable connector and cable are undamaged. If a sensor or wiring is damaged, the ECM on the piece of equipment may put the equipment in limp mode, which will not allow it to operate as intended. Make sure to identify any areas of concern to your supervisor to minimize downtime on the site.

1.0.0 Section Review

1. Before mounting any machine to perform a job, what should the operator understand?
   a. Performance measurements
   b. Labor costs
   c. Daily schedule and break times
   d. The location of other nearby workers

2. The track chain wraps around which of the following?
   a. Tires
   b. Drive sprocket
   c. Drive gear
   d. Operational wheel

3. When looking at a fuel system on a piece of equipment, what should the operator know before operation?
   a. Average fuel mileage
   b. Fuel type
   c. Instrumentation maintenance schedule
   d. Fuel levels

4. Before operating equipment with a hydraulic system, what should you look for in your pre-operational check?
   a. The operational condition of the hydraulic pump
   b. The fluid level and if there are any visible signs of leakage
   c. The type of hydraulic fluid used with the system
   d. The size of the hydraulic pump

5. When checking whether the alternator is charging the batteries, what is the first step in evaluating its operation?
   a. The condition of the battery enclosure
   b. The condition of the wiring on the piece of equipment
   c. The condition of the dash indicator in the cab
   d. The condition of the drive belt and tensioner

2.0.0 Starting, Operating, and Shutting Down Heavy Equipment

Performance Task

1. Perform basic prestart inspection, startup, operational movement, and shutdown for three of the following types of equipment:
   - Utility tractors
   - Dozers
   - Loaders
   - Backhoes
   - Excavators
   - Compaction equipment
   - Motor graders
   - Scrapers
   - On-road dump trucks
   - Off-road dump trucks
   - Forklifts
   - Skid steers
   - Trenchers

Objective

Describe the procedures for starting, operating, and shutting down heavy equipment.

a. Describe how to perform initial checks before starting the equipment.
b. Identify basic maintenance procedures for the equipment.
c. Explain how to properly shut down and secure equipment.

Making sure equipment is safe to operate is important because unsafe equipment can cause downtime on the construction site and potentially injure the operator or other people on the site. When conducting prestart procedures, understand what you are trying to accomplish. Making sure the vehicle can operate as designed will keep you and the equipment working.

2.1.0 Initial Checks

Before starting the equipment, the operator must inspect it to verify that systems are working properly and that there is no physical damage to tires/tracks, wiring, hoses, and other equipment.

2.1.1 Indicator Checks

Before starting the engine, turn the key to the On position and check the instrument panel, but do not start the engine. Make sure that all gauges and indicators function in accordance with the operator's manual. The instrument panel should light for a few seconds. If an indicator or gauge does not light up, check the bulb. Repeat the test. If the indicator still fails to light, have the machine serviced.

2.1.2 Starting Aids

Gasoline and diesel engines are started differently. Most heavy equipment gasoline engines start the same way as gasoline engines in a street machine. These engines rarely need any kind of prestart aid. If the temperature is extremely cold, engine block heaters and other starting aids can be used. The startup and warmup of an engine must be done in accordance with the manufacturer's instructions.

After startup, the engine needs to warm up for a short period before beginning operations. Refer to the O&M manual for specific amount of warmup time.

WARNING!

Ether must never be used in combination with another starting aid. The manufacturer's instructions must be carefully followed. Ether or starting fluid can cause a backfire which can break engine components and cause catastrophic engine failure.

Glow Plugs

Prestart aids are often used with diesel engines. **Glow plugs** are frequently used to warm the engine prior to starting. Many diesel engines have glow plugs. These heating elements are used to heat up the combustion chamber and aid in igniting fuel in a cold engine. Current supplied to glow plugs causes them to produce heat. When the switch is activated, the plugs glow. When they stop glowing, the engine is warm enough to start.

Glow plugs: Heating elements used to preheat the combustion chamber and aid in igniting fuel in a cold engine.

Other models have a timing chart for how long the glow plugs should be activated. The activation time is based on the temperature surrounding the engine. An example of a timing chart is shown in *Table 1*. The chart for the machine is typically in the operator's manual or mounted on the dash.

TABLE 1 Timing Chart

Starting Aid Charge	
Starting Temperature	**Glow Plug Heat Time**
Above 60°F (16°C)	No
60°F (16°C) to 32°F (0°C)	1 Minute
32°F (0°C) to 0°F (–18°C)	2 Minutes
Below 0°F (–18°C)	3 Minutes

Block Heaters

Engine block heaters are used in colder climates. An engine block heater is an electrical heating element installed in the engine block water jacket. In cold weather, it keeps the engine water warm. This keeps the engine temperature warm enough to allow easy engine starts. Typically, the heater is plugged into a standard electrical outlet. Unplug the engine block heater before moving the machine. Properly secure the plug so it does not become fouled with ice and snow.

2.1.3 Starting

Before starting the machine, verify that all controls are properly set. Most vehicles must be started in neutral or park. This includes both the drive and the operational controls. Also check that the parking brakes, if provided, are properly set.

CAUTION

Do not push the starter button for more than 20 seconds. This can overheat and damage the starter. If the engine does not start after 20 seconds, wait two minutes before attempting to start again. Recheck all instrument settings and controls. Many controls automatically lock out the starter if not set properly. Some equipment ignitions require the key to be turned completely Off before attempting to turn them to Start a second time. You can break a key off in the switch if you're not careful.

Going Green

Oil and Grease

Lubrication oils and greases are either petroleum-based or synthetic. Heavy equipment is often serviced in the field. To protect the environment, make every effort to keep any oils or greases from getting onto the surrounding soils, and make sure to keep them away from any streams or bodies of water.

WARNING!

Before starting an engine or moving the machine, make sure that the area is cleared of personnel. Check that no one is underneath, around, or on the machine. Alert nearby personnel in accordance with site procedures.

Start the machine only after completing any prestart procedures and fastening the seat belt. Move the gearshift to neutral. Set the engine speed control as directed in the operator's manual. Sound the horn to alert people standing nearby that the machine is being started. Turn the key to the start position and/or depress the start button.

2.1.4 Warmup

It is important to let the machine warm up before operating. Operating the machine without warming it up can damage it. The warmup period is directly proportional to the temperature of the air surrounding the machine. The colder the temperature, the longer the warmup period is necessary. Refer to the O&M manual for any specific warm-up time before equipment operation.

Run the machine for the time listed in the O&M manual. When the engine is running, check that all gauges and instruments are working properly.

If the hydraulic oil is cold, the hydraulic system will be sluggish. Circulate warm hydraulic oil throughout the machine by cycling the hydraulic controls. With a backhoe, for example, lift the bucket to hood height and lower it. Extend and retract the boom. Continue to cycle the system until it moves freely. Once the machine is properly warmed up, operations can begin.

WARNING!

When checking the fluids in a hot engine, or when changing oil or filters, make sure the engine is stopped. Wear the proper eye, face, or hand protection as required because the oil, coolant, dipsticks, and filters will be very hot.

2.1.5 Air Filtration System

The air intake system to the engine must be monitored by the operator. This system allows the engine to draw in filtered air to mix with the fuel for combustion purposes. The air is filtered to prevent foreign material, which may cause damage to the engine, from entering the combustion chamber. *Figure 22* shows typical air filter systems. Some filters are equipped with dust cups or dust-unloading valves to rid the filter of some of the accumulated dust. Centrifugal action of the incoming air sweeps the dust into a dust cup or unloader valve.

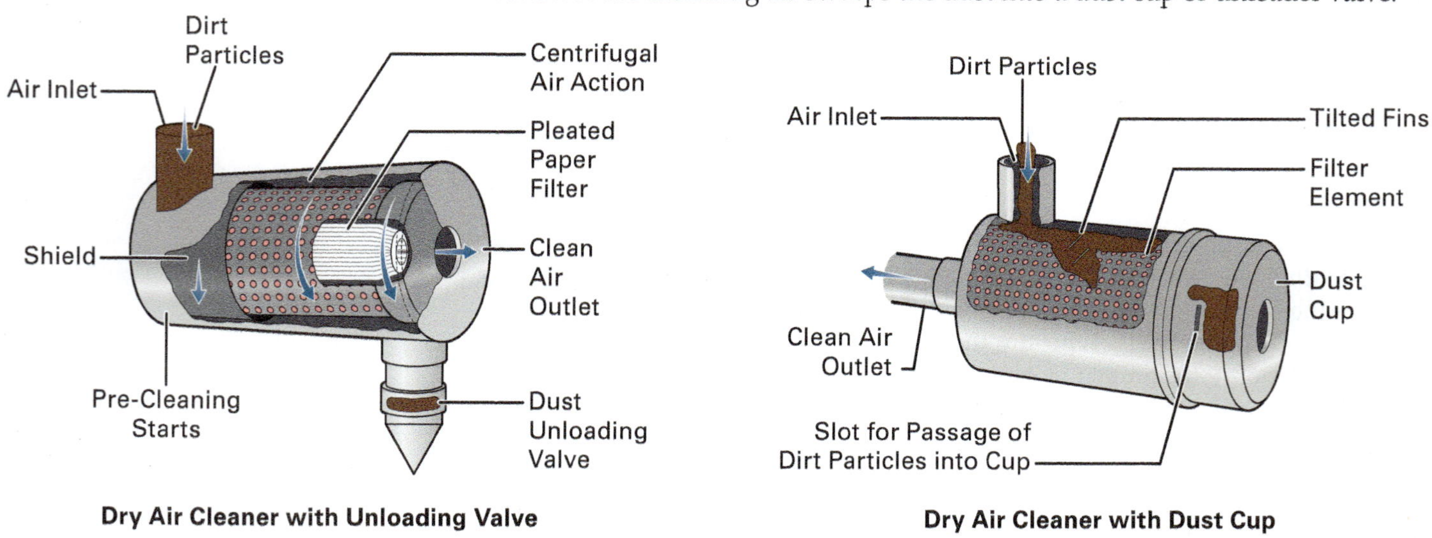

Figure 22 Typical air filters.

2.2.0 Equipment Maintenance Safety

Heavy equipment breakdowns can often be traced to improper maintenance or lack of regular maintenance. These malfunctions can cause unexpected failure of the equipment, resulting in scheduling delays and accidents. An operator's responsibility is to ensure that the equipment is in safe working order before using it. However, operators sometimes perform periodic maintenance.

Manufacturers supply detailed operating and maintenance manuals for equipment. Workers qualified and authorized to perform maintenance on the equipment should study the manufacturer's manuals, including maintenance schedules, lubrication charts, and parts lists, before starting the equipment. Maintenance can be dangerous unless performed properly. Workers should have the necessary skills, information, tools, and equipment to do the job correctly. Use parts, lubricants, and service techniques recommended by the manufacturer. If you do not know what you are doing—stop!

While performing maintenance on equipment, keep the following in mind:

- Disconnect power, then lock and tagout machines before performing maintenance.
- Use properly fitting wrenches on nuts and bolts.
- Keep the work area clear of scraps and litter.
- Dispose of combustible materials properly.
- Clean up any spilled liquids immediately.
- Store oily rags in self-closing metal containers.
- Never use compressed air to clean yourself or your clothing.

NOTE

Only people qualified to do so and authorized by the employer may perform maintenance on a vehicle or equipment.

2.2.1 Pre-Maintenance Safety

Before servicing equipment, prepare the service area. Begin by emptying your pockets of any items that could fall into machinery. Then collect any PPE that may be needed, such as aprons and gloves; safety glasses, goggles, or shields; safety shoes; welding protection equipment; and a filter mask or respirator. Wear all the protective clothing that the job requires.

If the worksite does not have a designated service area, select a clean, level area out of the flow of traffic. Make sure there is enough room to work, as well as adequate light and ventilation. Check overhead clearances if necessary. Before starting, clean the working surface, removing oil, grease, and water to eliminate slippery areas. Put sand or another absorbent material on these areas.

After moving the equipment to the service area, follow the necessary shutdown procedures. After the equipment is shut down, always attach a Do Not Operate tag or similar warning tag to the starter switch or steering lever before performing maintenance. Perform a lockout/tagout procedure to secure the piece of equipment. If the engine must not be started during maintenance, remove the ignition key and, if possible, use a lockout device. Many machines have ignition lockout switches or battery disconnects in or near the engine compartment for this purpose.

When performing maintenance, always use the correct tools. If guards or covers need to be removed during servicing, remove only those necessary to gain access and replace them as soon as possible. Never leave guards off while the equipment is unattended. Make sure any devices that are used during maintenance, like lifting and supporting devices, are strong enough to perform their jobs.

2.2.2 Engine Maintenance

Working on a running engine is dangerous. It must only be done by a qualified technician. If adjustments must be made with the engine running, always work as a two-person team with one person sitting in the operator's seat while the other works on the equipment. Never attempt to get around any safety devices or circuitry to start the equipment. Bypassing normal starting procedures may cause the machine to start and move suddenly, resulting in serious injury or death to anyone in its path.

If the engine needs to be running during maintenance, make sure there is proper ventilation in the service area because exhaust fumes can kill. Use an exhaust pipe extension if the engine is running in an enclosed area. If there is no exhaust pipe extension, make sure the doors are open to allow outside air into the area.

WARNING!

While performing any maintenance on any equipment, keep clear of all rotating components. Wrapping or entanglement may result in serious injury or death.

2.2.3 Hydraulic System Maintenance

The hydraulic system of the equipment is under pressure whenever the engine is running, and it can hold pressure even after being shut down. In some instances, cycle hydraulic controls such as steering after shutdown to relieve trapped pressure in the lines. This is not possible on every application. Whenever possible, remove tension from any attachments or devices. For example, lower the forks of a forklift to the ground. If this is not possible, install manufacturer-approved support devices or block the cylinders and equipment securely before working on the hydraulic system.

When venting or filling the hydraulic system, loosen the filler cap slowly and remove it gradually. If the system is equipped with an accumulator, see the manufacturer's manual for maintenance instructions. When filling the hydraulic system, using the proper type of hydraulic fluid is vital to pump, line, and actuator operation. Even after releasing the pressure on lines, remember that hot hydraulic fluid can cause severe burns. Wait for fluid to cool before disconnecting lines. Clean up spilled hydraulic fluid immediately.

Do not permit open flame around the hydraulic system. Only qualified technicians can recharge hydraulic accumulators with dry nitrogen gas only. Do not use oxygen or combustible gases. Use extreme caution when handling pressurized cylinders. Never reset any relief valve in a hydraulic system to a pressure higher than that recommended by the manufacturer. Do not bypass relief devices or close off overflow devices.

WARNING!

Hydraulic fluid under pressure can penetrate skin or eyes and can cause injury, blindness, or death. Fluid leaks under pressure may not be visible. Do not use your hand or any part of your body to check for a leak. Using cardboard or other flexible cellulose material to identify leaks will keep the operator's hands out of harm's way. Use proper PPE, including a face shield or safety goggles, for eye protection. If any fluid is injected under the skin, it must be removed within a few hours by a doctor familiar with this type of injury.

2.2.4 Cooling System Maintenance

Liquid cooling systems build pressure as the engine gets hot. Before removing the radiator cap, stop the engine and let the system cool. Wear protective clothing and safety glasses, goggles, or a shield. Remove the radiator cap only after the coolant is cold. Slowly turn the radiator cap to the first position to allow any residual pressure to escape before removing the cap entirely. For cooling systems with an overflow tank, the coolant can usually be checked at the tank without removing the radiator cap. See the manufacturer's instructions.

2.2.5 Fueling

Stop the engine and shut off electrical equipment while filling the fuel tank. Use extra caution when fueling a hot engine. Always ground the fuel nozzle or container spout against the filler neck to avoid sparks. Never overfill the fuel tank.

If any fuel is spilled, clean it up immediately. Keep sparks and open flames away from fuel. Never smoke while handling fuel or working on the fuel system. The fumes in an empty container are explosive. Never cut or weld on fuel lines, tanks, or containers. Do not walk away when fueling equipment. If the fill spout is elevated, climb up and then have someone hand up the fuel hose. When fueling the piece of equipment, make sure to use the proper cetane level of fuel that matches the operational requirements of the piece of equipment. Some applications can operate on up to B20 low sulfur diesel fuel. Others, including most older equipment, cannot operate on this mixture of biodiesel and regular petroleum-based diesel fuel.

WARNING!

Do not grind, flame-cut, braze, or weld any fuel tanks or lines. Doing so can cause residual vapors to explode, causing serious injuries or even death.

Don't Get Burned

The following are some facts about gasoline:

- One gallon of gasoline contains the same explosive force as 14 sticks of dynamite.
- Gasoline vapors are heavier than air, can travel several feet to an ignition source, and can ignite at a temperature of –45°F. To be safe, keep open gasoline containers well removed from all potential ignition sources.
- Gasoline has a low electrical conductivity. As a result, a charge of static electricity builds up on gasoline as it flows through a pipe or hose. Getting into and out of a vehicle during refueling can build up a static charge in your clothing, especially during dry weather. That charge can cause a spark that can ignite gasoline vapors near the fuel nozzle.

Source: US Department of Energy

2.2.6 Tire Maintenance

The stability and handling of the equipment can be dramatically affected by underinflated tires or those with inadequate ballast fill. Examine the tires and wheels on the equipment before each use and ensure they are in accordance with manufacturer's specifications.

Tires must be serviced by a qualified tire service center or by an authorized person who has been properly trained in the procedures and equipment used during servicing. The types of rims and tires used on heavy equipment usually require that special care be taken during maintenance to prevent serious injury or death. The following are important points to remember:

- Inflate tires from a distance using a self-attaching chuck. Never stand to the side of a tire when adjusting tire pressure, because the rim ring or bead may fail as the tire is inflated.
- Do not inflate a tire with flammable gases or from a system using an alcohol injector.
- Never exceed the maximum pressure shown on the sidewall. The tire can explode.
- Punctures that have allowed the ballast to leak out must be repaired and the ballast refilled before the equipment may be placed back in operation.
- Never reinflate a tire that has been run flat or seriously underinflated without removing the tire from the wheel. Closely inspect the tire and rim for damage before remounting.
- Clean the area around all wheel lug nuts or bolts, and periodically check the torque per the manufacturer's instructions until the torque value stabilizes. Then check at regularly scheduled intervals.
- Never weld on a wheel rim with the tire in place. The tire could explode.

- Check for and remove any debris and rocks between dual tires.
- Check tire sidewalls and tread for cuts or exposed metal belts. If found, remove the machine from service until the tire is replaced.

2.2.7 Air Conditioning System Maintenance

Only qualified technicians may work on air conditioning equipment. When servicing air conditioning systems, refer to the refrigerant gas container for proper use. Keep in mind that the use of air conditioning refrigerant in the United States is governed by the Environmental Protection Agency (EPA), which strictly prohibits venting any type of refrigerant into the atmosphere. Therefore, EPA-approved recovery and recycling systems must be used to avoid accidental venting of refrigerant. To help maintain the air conditioning system, keep the windows closed when it is in operation and check filters daily.

WARNING!

Do not smoke or have open flames in the area when servicing the air conditioning system—refrigerant gases may be present. Inhaling refrigerant gas through a cigarette or other smoking material or inhaling fumes released when refrigerant gas is burned can cause bodily harm or death. Also note that contact with the refrigerant stream can cause severe frost burns.

2.2.8 Post-Repair Maintenance

Complete all servicing and repairs before releasing any equipment. Tighten all bolts, fittings, and connections to torques specified by the manufacturer. Reinstall all guards, covers, and shields after servicing. Replace or repair any damaged ones. Refill and recharge pressure systems only with manufacturer-approved or recommended fluids. Start the engine and check for leaks. Operate all controls to make sure that they are working properly. Road-test the equipment if necessary. After testing, shut down the engine and check the work you performed. Look for any missing hardware like cotter pins, washers, and locknuts. Recheck all fluid levels before releasing the equipment for operation.

2.3.0 Equipment Shutdown Procedures

Before turning off a piece of heavy equipment, allow it to slowly cool down and idle for a few minutes. Listen for any abnormal noise and look at the gauges to make sure everything is within specification. If you notice any questionable sounds, lights, or gauge readings, note them and report them to your supervisor, who can put in a trouble ticket with the equipment repair department. After you have successfully powered down the piece of equipment, make sure to conduct a post-operation inspection. This includes checking all fluids; inspecting all safety devices, tires, and tracks; and identifying any areas that need to be looked at before the next day. Make sure the equipment is fueled up and ready for the next day. Clean the piece of equipment of any debris or mud that has built up on it. It is much easier to remove fresh buildup than it is to remove caked-on, hardened dirt. Keeping the machine ready to go for the next day will increase productivity and help to keep the equipment operating smoothly.

2.0.0 Section Review

1. The operator should fasten the seat belt in the cab and then do which of the following before equipment startup?
   a. Turn the ignition switch to Start and start working.
   b. Check the oil pressure.
   c. Turn the ignition on and inspect all instrument panel gauges.
   d. Turn the ignition on and leave the cab to verify engine fluid levels.

2. A diesel engine requires which of the following components to warm the cylinder and help start a cold engine?
   a. Spark plugs
   b. Glow plugs
   c. Ignition plugs
   d. Starting aids

Figure SR01
Source: Brunohitam/Shutterstock

3. The arrows in *Figure SR01* are pointing at the _____.
   a. engine oil fill port
   b. fuel fill port
   c. coolant fill port
   d. hydraulic fluid fill port

3.0.0 Vehicle Engine Systems in Heavy Equipment

Objective

Describe the engine and auxiliary systems used in heavy equipment.

a. Describe the primary engine types used in heavy equipment.
b. Describe how to service auxiliary systems used with heavy equipment engines.

Performance Tasks

There are no Performance Tasks in this section.

The mechanical operation of heavy equipment relies on two separate systems—the vehicle drive system and the hydraulic system. This section focuses on the vehicle drive systems and emissions-related systems on equipment. Understanding how these systems operate, along with the engine support systems, will allow the operator to determine if there is an issue when operating a piece of equipment.

3.1.0 Engine Types

The two primary types of engines used in construction machines are gasoline and diesel. However, more companies are building hybrid-powered (electric and fuel) machines that are strong enough to be used in heavy construction. Although such hybrid-powered machines may be on the market, this module focuses on the conventional, internal-combustion engines still used in most construction machines.

Gasoline engines are occasionally used in some smaller machines, but most machines operated on construction sites use diesel engines. Diesel engines (*Figure 23*) operate at twice the compression ratio of gasoline engines, and therefore deliver much more power to the drive mechanism. This increased compression ratio also increases the engine's efficiency, increasing the amount of work it can accomplish.

Figure 23 Diesel engine.
Source: 3D Horse/iStock/Getty Images

CAUTION

Do not put gasoline in a diesel engine. It will damage the engine.

Both gasoline and diesel engines are internal combustion engines. That is, their drive comes from the explosion of a fuel-air mixture in a combustion chamber. The main difference between diesel and gasoline engines is in the way they are fired. In the gasoline engine, air and gasoline are mixed, and then introduced into the combustion chamber where the mixture is compressed. A spark plug then emits a spark that causes the compressed air-fuel mixture to explode. A diesel engine differs in that only air is drawn into the cylinder through the intake valve. As the piston compresses air to top dead center, a small amount of fuel is injected directly into the combustion chamber. The high temperatures and pressures at this point of injection cause the highly atomized fuel to spontaneously combust, thus starting the combustion event. After the combustion event in a gas or diesel engine, the pistons transmit the force to the crankshaft and through the drivetrain to propel the piece of equipment.

All new diesel engines must be built to comply with the clean engine standards established by environmental agencies all over the world. Those standards especially apply to all new diesel- and gas-powered vehicles built for both on-road and off-road uses.

Environmental agencies in different parts of the world have their own guidelines for all the different types of vehicles. Manufacturers of such vehicles must

Sustainability

Komatsu manufactures a hybrid excavator powered primarily by electric drive motors. The excavator still uses a combustion engine, but the engine now drives a generator that produces electricity, and that electricity powers the electric drive motors. These hybrid-powered machines emit less exhaust into the atmosphere, and thus are more environmentally friendly. Other companies are building similar systems for other types of construction equipment.

The Diesel Engine

The diesel engine was patented by Rudolph Diesel in 1898. This engine revolutionized industry, which until that time had used less efficient steam engines. Today, locomotives, heavy construction equipment, trucks, and many other applications depend on the diesel engine. In a diesel engine, air and diesel fuel are not mixed before being injected, and there is no spark plug. Air and fuel are injected directly into the piston chamber and compressed. The extreme compression creates enough heat to cause the mixture to ignite.

build machines that comply with the standards of each country. In the US and Canada, the environmental standards that apply to new diesel engines used in off-road equipment are known as Tier 4 standards. While the Tier 4 standards have been in existence for several years, they were more strictly applied to all new off-road (construction) equipment being built during and after 2011. The purpose of the Tier 4 standards is to lower the amount of nitrogen oxides (NO_x) and particulate matter (PM) being emitted from engines. Particulate matter is the small solid or liquid matter that remains in gas or liquid emissions being released into the atmosphere. Nitrogen oxides and particulate matter are pollutants that must be controlled.

To meet the Tier 4 standards, new diesel engines must be built to burn cleaner ultra-low sulfur diesel fuel that contains no more than 15 parts per million (ppm) of sulfur. To burn the cleaner fuel more efficiently, the new engines are equipped with computer-controlled components that better manage the fuel and air mixture coming into the engine, and the combustion processes taking place inside the engine. All those efforts are to optimize engine performance while minimizing the harmful emissions. On the exhaust side of the engines, changes are being made to the exhaust systems themselves. Even the design and location of the exhaust pipes are being changed to better control the exhausts being released into the atmosphere. Inside the exhaust systems, improved technologies are being used to better clean the exhaust gases passing through the exhausts.

3.2.0 Auxiliary Systems

The engine requires a number of supporting systems, including the fuel, exhaust, cooling, electrical, lubricating, and air filtration systems. Without these auxiliary systems, the engine could not function.

3.2.1 Fuel System

The fuel for a diesel engine is injected under high pressure into the combustion cylinder (*Figure 24*). Some engines use high-pressure fuel injection pumps

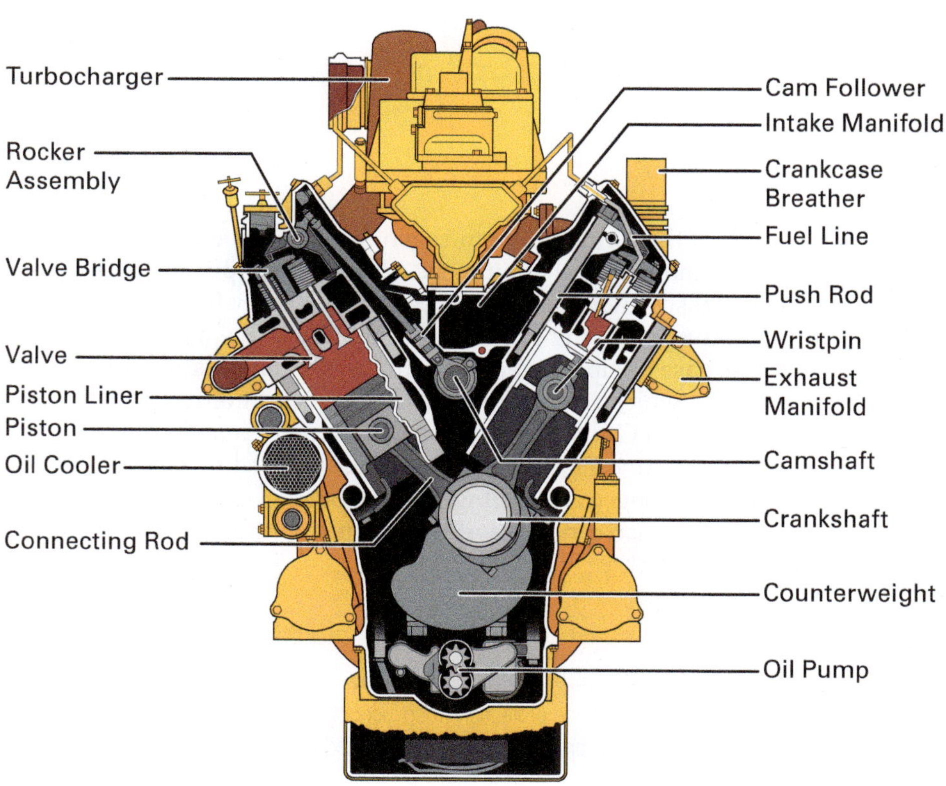

Figure 24 Diesel engine cutaway.

(20,000 to 28,000 psi). Other engines use lower-pressure pumps and cam-operated injectors to pressurize the fuel to the pressure required during injection. To reduce pollution and increase efficiency, many newer engines control the injection time using electronic controls that monitor a variety of engine functions and conditions during operation.

Some engines are also equipped with turbochargers. Turbochargers are designed to increase the amount of air delivered to each of the engine's cylinders. This increases the compression ratio and produces more power. When the compressed air is admitted into the cylinder, the fuel is injected as the air is being compressed by the piston, and compression continues until a detonation takes place when the piston is near top-dead center of its compression stroke. The explosion creates the piston's power stroke, which in turn rotates a crankshaft (*Figure 25*), producing crankshaft torque to the engine's load.

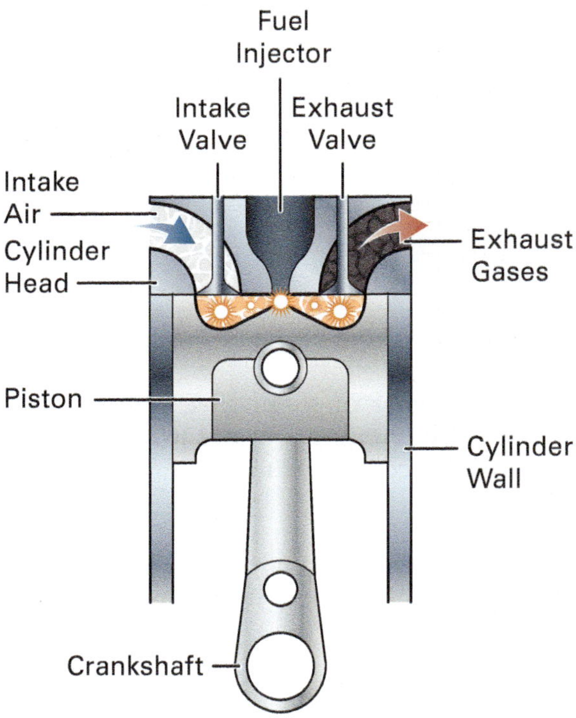

Figure 25 Cylinder and crankshaft.
Source: US Department of Energy

One way to reduce the formation of nitrogen oxides in an engine is through Exhaust Gas Recirculation (EGR) equipment. The EGR equipment takes a portion of the exhaust gases and runs them through a cooler before routing them to the air supply returning to the combustion chamber of the engine. Inducing the cooler exhaust gases helps lower the combustion temperature inside the combustion chamber, which in turn reduces the formation of nitrogen oxides. Engines equipped with EGR equipment have additional manifolds and extra plumbing around the engine. *Figure 26* shows a side view of an engine with EGR equipment installed. Along with cooling the combustion charge entering the engine, the EGR system also can help with burning any unburnt hydrocarbons from the engine's cylinders.

WARNING!

Always wear the proper eye, face, and hand protection when working on diesel engines. High-pressure fuel leaks can cause severe injury. Only properly trained and qualified technicians are permitted to work on diesel engines.

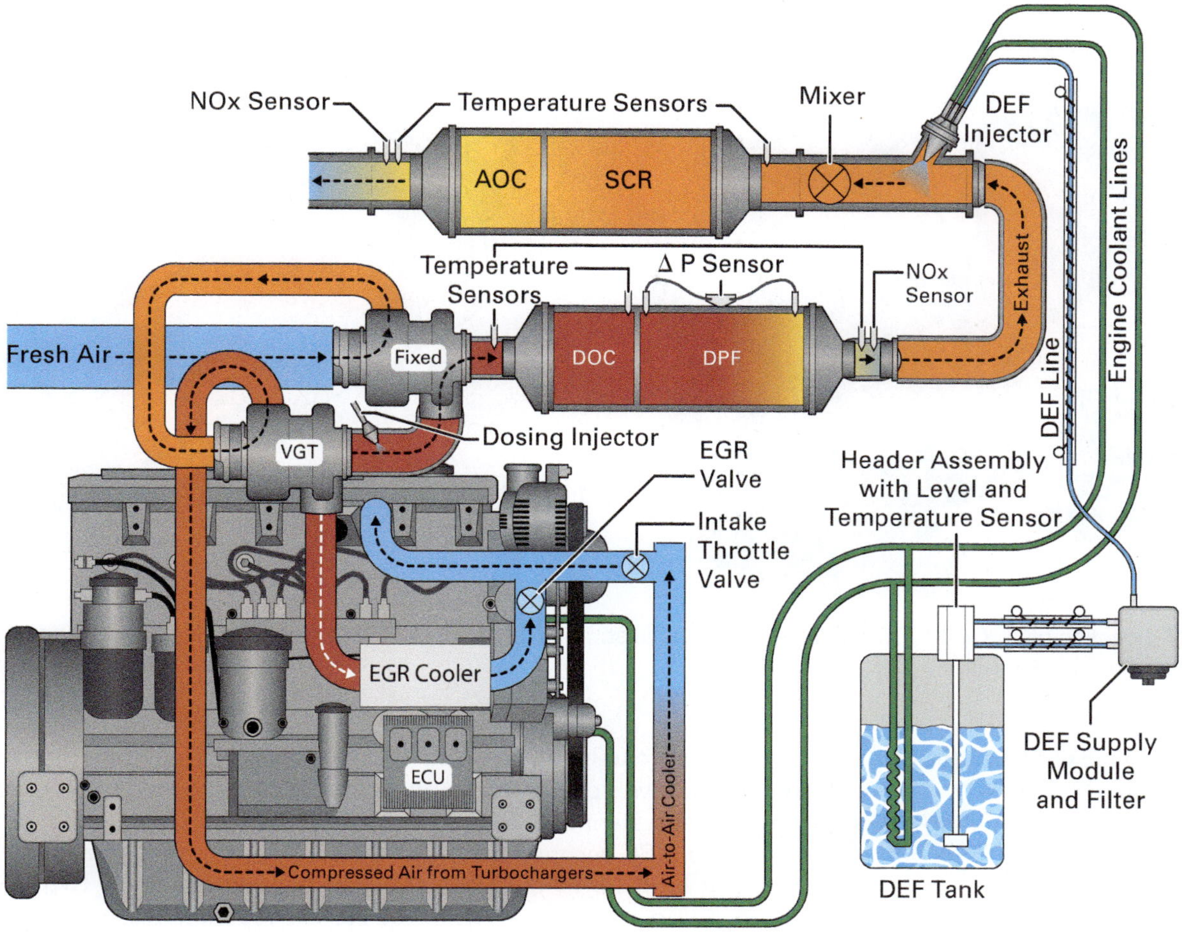

Figure 26 Location of EGR on a diesel engine.
Source: OEM Off-Highway

3.2.2 Exhaust System

When combustion engines operate, they emit gases that exit the vehicle through its exhaust system. All modern vehicles have fuel systems that are directly related to the vehicle's exhaust system. Sensors in the exhaust system monitor the exhaust gases and tell the fuel system controller if the fuel is burning correctly or not. If not, the fuel system controller tries to adjust the combustion process to clean up the exhaust fumes as much as it can.

The major pollutants coming out of a combustion engine are hydrocarbons (HC), carbon monoxide (CO), and nitrogen oxides (NO_x). Modern exhaust systems are designed to change the composition and volume of these pollutants as they pass through catalytic converters installed in the exhaust system.

The three types of catalytic converters used in today's exhaust systems include:

Diesel oxidating catalyst (DOC) - This is typically the first converter in the exhaust system. The DOC starts breaking down harmful emissions chemically by oxidizing CO and HC.

Diesel particulate filter (DPF) - This filter is typically the second stage of the exhaust system and is designed to trap remaining exhaust particles, known as soot, that the DOC was unable to oxidize. These particles are a result of an incomplete combustion process. Prior to 2007, these particles were released into the atmosphere. That same year, the US government put restrictions on the release of emissions out of internal combustion engines into the atmosphere. As the engine is operated, the particulate matter is combusted in the cylinder and then sent through the exhaust manifold through the DPF (*Figure 27*). As the particles travel through the DPF, the matter sticks to the honeycomb present in the DPF (*Figure 28*) and accumulates until the temperature gets hot enough to fully combust the buildup.

Diesel particulate filter (DPF): A filter that catches particulate matter and burns it up before it exits the engine into the atmosphere.

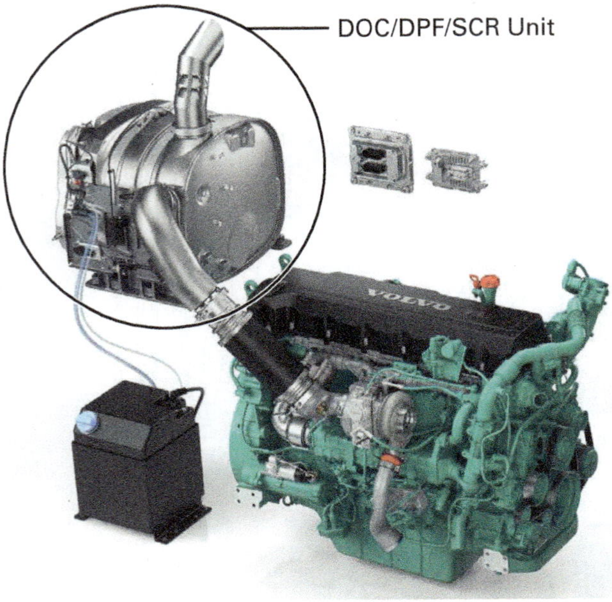

Figure 27 Location of DOC/DPF/SCR on an engine's exhaust.
Source: Volvo Construction Equipment

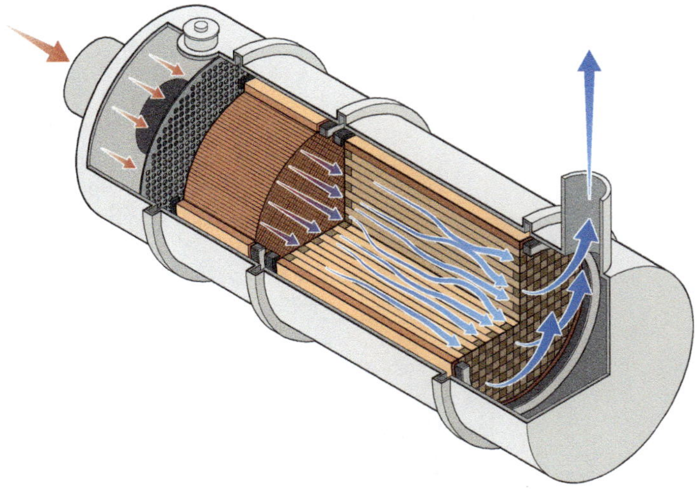

Figure 28 A DPF unit catches and burns all particulate matter.

Selective catalytic reduction (SCR) system: A system that allows the introduction of urea fluid into the exhaust stream to react with the NO_x present in the exhaust emissions.

Selective catalytic reduction (SCR) system – This converter uses DEF to reduce the levels of NO_x emissions in the exhaust. *Figure 27* shows a side view of an exhaust system with SCR equipment installed.

As the DPF becomes full, the system must be regenerated to remove all material built up in the DPF. This can happen in one of two ways: passive regeneration or active regeneration. Passive regeneration occurs when the diesel engine is operated at a high enough revolutions per minute (rpm) to create the conditions necessary in the DPF to ignite the matter to fully combust the material. This is assisted, in some applications, with a slightly rich fuel mixture or dedicated injector, which provides a little bit of fuel along with the high temperatures to start the DPF on fire. Once the temperature within the DPF meets the required criterion, all the material is burnt off over time and the DPF is ready to collect particulate matter again. If the engine is not operated at a high enough rpm for the proper amount of time, passive regeneration cannot occur because the exhaust system will not get hot enough to support the cleaning process. In this situation, active regeneration must be conducted.

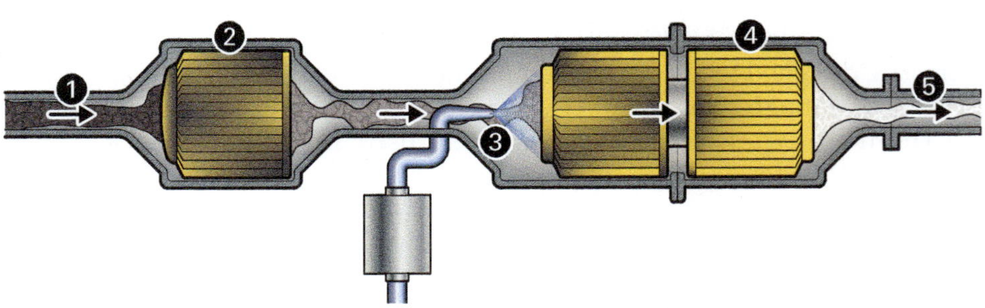

1. Exhaust gases coming from diesel oxidating catalyst (DOC).
2. Particulate matter (PM) is trapped in the diesel particulate filter (DPF).
3. Diesel exhaust fluid (DEF) is injected into the exhaust.
4. NO_x reduction occurs in the SCR catalyst.
5. Reduced volume of emissions leaves tailpipe. PM and NO_x near zero levels.

Figure 29 Location of SCR on an engine's exhaust.
Source: Diesel Technology Forum *www.dieselforum.org*

Active regeneration is a forced regeneration based on the DPF's soot load. If the engine's electronic control module (ECM) determines that the DPF is full and needs to be regenerated, it will indicate to the operator that an active regeneration must occur. Depending on the application, the operator must put the piece of equipment away from any flammable materials, isolate the piece of equipment from other equipment and then conduct the active regeneration process. When the ECM is commanded to regenerate the DPF, the rpm is raised to get the exhaust temperatures up. At the proper time, the ECM will inject fuel into the DPF to start the cleaning process. When this happens, the exhaust system will become very hot and increased smoke output will occur. This process can take anywhere from 20 minutes to over an hour depending on the soot load within the DPF. The ECM will utilize pyrometers to determine if the DPF is getting to the proper temperature and it will determine when the DPF is clean. When operating a piece of equipment on a jobsite, this type of regeneration must be planned because the piece of equipment will be out of service until the forced regeneration process is completed. If the regeneration process is ignored and the piece of equipment continues to be operated, the ECM will disable the engine until it can be forced into regeneration. This will cause downtime and can delay the project until a trained equipment technician can get to the jobsite to repair the issue.

Before the exhaust of the engine reaches the SCR device, a dosing system injects periodic sprays of a mist of a chemical reagent called a **diesel exhaust fluid (DEF)**. The DEF solution is a nontoxic solution made up of approximately 68 percent purified water and 32 percent urea. Urea is an organic product created from animal waste. When injected into the exhaust system and heated, the DEF creates ammonia that flows through the SCR catalyst and reacts with the engine gases to form harmless nitrogen and water vapor that make up the engine's exhaust. The DEF fluid reacts with the NO_x present in the exhaust emissions. When the conversion process occurs, the nitrogen oxides are broken down into harmless H_20, CO_2, and N_2 that are sent out of the exhaust pipe. Because DEF is mostly purified water, it will freeze in weather below 32°F. To prevent this from occurring, the DEF tank has heaters that help keep the fluid from freezing when the equipment is operating.

Diesel exhaust fluid (DEF): Urea fluid that promotes the conversion of oxides of nitrogen into harmless gases.

SCR-equipped vehicles are equipped with a DEF storage tank and DEF indicator light (*Figure 30*). The vehicle operator refills the DEF tank as needed. Because of the ongoing efforts to reduce pollutants from all combustion engines, heavy equipment used in construction will eventually be equipped with similar SCR systems using DEF. Operators are also responsible for visually inspecting the EGR and SCR systems installed on the engines of their machines. If the engine is operated on an empty DEF system, the engine's ECM will provide a countdown until the engine will no longer be able to be run. Manufacturers have put safeguards in place to make sure the piece of equipment will not operate unless the operator continues to use the DEF system.

Figure 30 The DEF indicator light shows if the system fluid is low.

WARNING!

Modern engines produce extreme temperatures with their exhaust systems. Be careful working around them and be aware that they can start fires if they are parked on grassy surfaces.

3.2.3 Cooling System

Depending on the system's design, a diesel engine is either air-cooled or water-cooled. If the engine is water-cooled, it will have a system similar to the one shown in *Figure 31*, which removes the excess heat generated during combustion.

When working on liquid coolant systems, make sure that the engine is stopped and wear proper eye, face, and hand protection. Hot coolant systems operate under pressure. To prevent an eruption of steam and coolant from the radiator, slowly depressurize the radiator before removing the radiator cap.

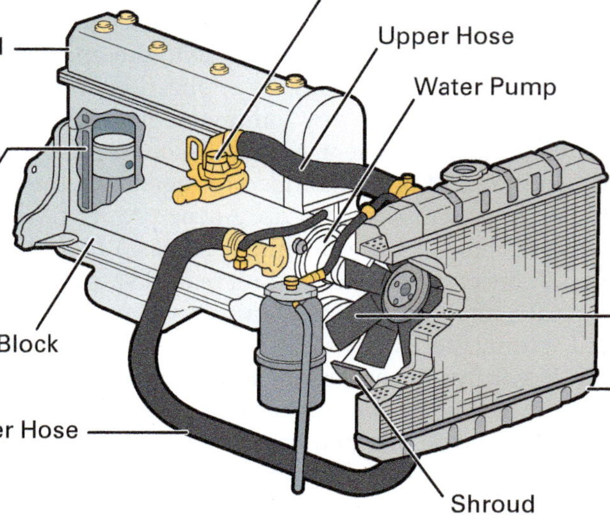

Figure 31 Typical cooling system.
Source: United States Navy

WARNING!

Do not attempt to add cold water to an overheated liquid coolant system. Allow the system to cool before depressurizing the system and adding coolant and/or water. Besides the danger of opening an overheated system, damage to the engine can be caused by adding cold coolant and/or water.

Figure 32 Radiator surge tank cap.
Source: Thissatan/iStock/Getty Images

Components of this system are typical of most water-cooled engines. The coolant, radiator, radiator surge tank cap (*Figure 32*), fan, pressure cap, engine-cooling circuit, connecting hoses, and thermostat work together to keep the engine at the proper operating temperature. This is accomplished by the circulation of pressurized coolant (antifreeze) using a water pump.

Pressurizing the cooling system increases operating range of the cooling system without boiling. Water boils at 212°F at sea level with atmospheric pressure. For every 1 psi increase of pressure on the closed system, the boiling point of the water increases by 3°F. That means a 15-psi radiator cap will make the cooling system capable of reaching 257°F before the water starts to boil. The thermostat should have opened long before that temperature is reached. Cooling fans should have been pulling air through the radiator before then. These multiple fail-safe systems allow operators to know they can shut down the engine before anything catastrophic can occur.

Going Green

Antifreeze

When antifreeze is used as a coolant, identify the source of any leaks and repair the leaks. Also, try to control any spills. Antifreeze can harm animals and may pollute any soil it contacts. It must be kept away from any streams or water sources. Check local regulatory guidelines for instructions on how to handle coolant spills. Remove and properly dispose of any contaminated soil.

Water Pumps and Coolants

The water pump is normally shaft-driven or turned by a belt connected to the crankshaft of the engine. As the coolant circulates, it is drawn from the bottom of the radiator by the pump. The pump forces the coolant into and through the water jacket of the engine. Coolant flowing through the water jacket controls the temperature of the engine. A thermostat regulates the flow of the coolant through the water jacket. After the coolant exits the water jacket and thermostat, it flows to the top of the radiator. The coolant flows down through thin tubes inside the radiator. The tubes are made with thin cooling fins mounted along their outer surfaces. Air passing over the tubes and fins extracts heat from the coolant. A cooling fan is used to increase the air flowing through the radiator. The more air moving past the radiator tubes, the better cooling for the engine.

Each day, before starting operations, the machine's engine belts, hoses, and coolant level should be checked. A problem in this area could result in expensive repairs if the engine overheats. The coolant level should only be checked when the system is cool. Opening a closed cooling system when the coolant is hot can cause personal injuries. Always allow the system to cool before adding coolant. Verify that the coolant is at the level recommended by the manufacturer.

Figure 33 Typical heavy equipment coolant.
Source: Dr. Nick Goodnight — Ivy Tech Community College

Checking Coolants

After an engine has been operating for 30 minutes or so, always check the temperature of the coolant. Check the temperature often when the engine is running. If an engine is showing signs of overheating, verify that the cooling fan is operating and the air paths into and out of the radiator are clear. If the fan is running and the air paths are unblocked, the next thing to do is to verify that the radiator has the proper level of coolant. Verify that the radiator and cooling system are not leaking coolant before attempting to check the level of coolant inside the radiator. If coolant does need to be added to the cooling system, make sure you are adding the proper type of coolant (*Figure 33*). The correct coolant is vital to ensuring the system will operate properly and the components will not prematurely wear out. Mixing coolants or using nondistilled water in the cooling system will cause the coolant in the system to degrade and fail to protect the system as designed.

WARNING!

Always allow an overheating engine to cool before checking or adding coolant.

3.2.4 Lubrication System

Another important system associated with the gasoline and diesel engine is the lubrication system. This system provides lubrication and cooling to the engine bearings and other friction surfaces inside the engine.

This system can be one of two types: full-flow and bypass (*Figure 34*). In some cases, both systems may be used on an engine. These systems consist of a pump, an oil sump, a pressure-regulating valve, and filter(s). In the bypass system, the pump pulls oil from the oil sump and sends it to the engine bearings under pressure. Some of the oil is also sent continuously through a filter to remove contaminants. The pressure-regulating valve bypasses a small amount of pressurized oil from the system back to the sump if the system pressure gets too high. In the full-flow system, all the oil is sent through a filter to the engine bearings. The filter is equipped with a bypass-relief valve in case the filter clogs. Clogging of the oil filter can be attributed to poor engine maintenance or operating in a highly dusty environment. Dirt entering the lubrication system will minimize the life of the internal engine components and cause the engine to fail.

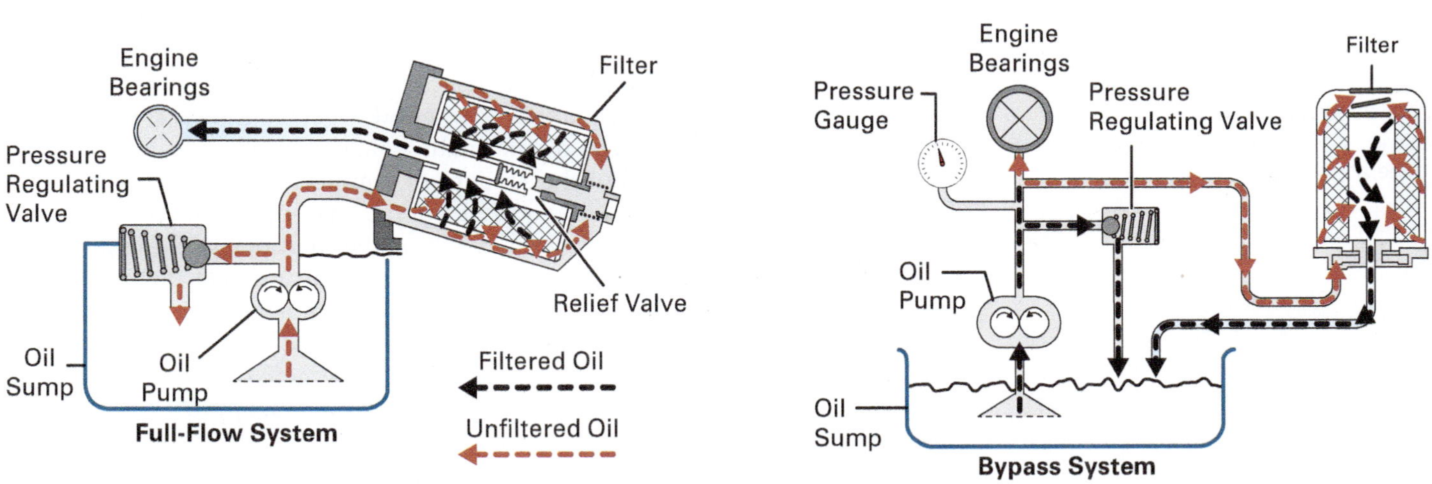

Figure 34 Lubrication systems.
Source: United States Navy

3.0.0 Section Review

1. What is the purpose of Tier 4 diesel engine standards?
   a. Increase engine horsepower.
   b. Reduce carbon dioxide emissions.
   c. Allow engines to burn kerosene.
   d. Lower nitrogen oxide emissions.

2. What is the primary function of the DPF?
   a. Increasing engine power
   b. Catching any particulate matter in the exhaust
   c. Storing exhaust fumes
   d. Increasing turbocharger speed

4.0.0 Hybrid and Electric-Powered Equipment Maintenance

Performance Tasks

There are no Performance Tasks in this section.

Objective

Describe hybrid and electric equipment preoperational checks.

a. Describe what to check before operating hybrid-powered construction equipment.
b. Describe what to check before operating electrically powered construction equipment.

As the world transitions to heavy equipment that is more climate friendly, hybrid and electric-driven construction equipment will become more commonplace. Hybrid and electrically powered equipment require different types of preventive maintenance. The use of these types of equipment will continue to increase on the jobsite as owners realize the increased efficiency and lower environmental impact that these applications provide.

When working around these types of equipment, understand that there are many differences between a conventionally powered and a hybrid or electric-driven piece of equipment. Some of these differences include making sure that certain pieces of equipment are plugged in and charging before leaving the site. In some cases, the operator may even need to make sure there are no electrical faults in the system. Knowing the differences will help the operator and those on the site to maintain safety at all times.

4.1.0 Basic Hybrid Equipment Operational Checks

Hybrid electric equipment (HEE): A type of application that has an internal combustion engine coupled with a power generation unit.

Hybrid electric equipment (HEE), like the excavator shown in *Figure 35*, has an internal combustion engine coupled with a power generation unit. The internal combustion engine still needs regular maintenance like a conventionally powered piece of equipment, but the amount of maintenance is less. Oil, coolant, fuel, and filter changes are still required, but the amount is reduced. When performing a preoperational inspection, checking the condition of the engine's fluids and related equipment is vital to minimizing equipment downtime. The hybrid drive feature has an additional maintenance schedule that should be evaluated before operation. Depending on the piece of equipment, the type of hybrid drive system could be as simple as a power generation unit that helps electrify some part of the drive system to a more integral approach. It is recommended that you become familiar with the make and model of the piece of equipment so you understand what you should be looking for when running through your pre- and postoperational equipment checks.

Figure 35 Hybrid electric excavator.
Source: Image courtesy of Komatsu America Corp.

4.2.0 Electrically Powered Equipment Operational Checks

When inspecting **electrically powered equipment (EPE)**, such as the one shown in *Figure 36*, the technician should look for any wiring hanging free from the piece of equipment. An EPE does not have an internal combustion engine. Instead, these types of equipment are powered by batteries. Any damaged high voltage wiring that appears to be compromised could potentially be deadly.

Electrically powered equipment (EPE): Types of equipment that are solely battery powered.

Figure 36 A typical EPE.
Source: Courtesy of Doosan Bobcat North America, Inc. All Rights Reserved.

CAUTION

Fluorescent orange wiring indicates high voltage is present and should not be touched without the proper high voltage PPE. Serious injury or death may result if the proper PPE is not used.

Along with the electrical wiring, cooling system loops should be inspected for any coolant leaks that could cause an overheating condition within the power electronics. When the system detects a machine fault it can disable the drive system and make the piece of equipment not operate until the issue is resolved. When power cycling the machine, pay attention to any indicator lights that come on. These lights keep you aware of any potential issues with the machine before it is required on the jobsite. If a failure indicator is illuminated, notify the supervisor and lockout the machine so no one else will attempt to operate it.

4.0.0 Section Review

1. What is the purpose of employing a hybrid or electrically powered piece of equipment?
   a. Increased operator requirements
   b. Reduced environmental impact
   c. Increased fuel usage
   d. Decreased operational efficiencies

2. Which of the following needs to be checked before operating hybrid electric equipment?
   a. Nothing, because it is a totally sealed system and there is nothing that is serviceable.
   b. Engine fluids, because they still wear like a conventional piece of equipment.
   c. The battery electrolyte in the battery pack, because it can deplete during operation.
   d. Ensure that the battery has been changed every day prior to operation.

Module 22108 Review Questions

1. What drives the pump in heavy equipment's hydraulic system?
   a. Battery
   b. Transmission
   c. Engine
   d. Alternator

2. Which direction does electrical current flow in a direct current system?
   a. Current flows in both directions in the circuit.
   b. Current flows in only one direction in the circuit.
   c. Current alternates in two directions within the circuit.
   d. Current alternates in one direction within the circuit.

3. How much time is needed to heat a glow plug in temperatures above 60°F (16°C)?
   a. Heating time is not required.
   b. 1 minutes
   c. 2 minutes
   d. 3 minutes

4. With regard to equipment maintenance, what is the operator's miminum responsibility?
   a. To ensure it is in safe working order before using it.
   b. To have it inspected by a supervisor before using it.
   c. To ensure the equipment is completely refueled before using it.
   d. To wash and clean it before it is used.

5. What is the procedure called when you take equipment out of service due to failure?
   a. Field servicing
   b. Lockout/tagout
   c. Swap out
   d. Tagout failure

6. What causes the explosive reaction between fuel and air in a diesel engine?
   a. The fuel
   b. An agitator
   c. Extreme compression
   d. An electrical spark

7. Tier 4 diesel engine standards apply to which of the following vehicle groups?
   a. Street-legal vehicles only
   b. Off-road vehicles only
   c. Company vehicles
   d. All new diesel engines

8. Which of the following actions is correct when working on a cooling system?
   a. Add coolant before depressurizing the system.
   b. Allow the system to cool before adding coolant.
   c. Add cold water before depressurizing the system.
   d. Remove the radiator cap to allow the system to depressurize.

9. True or False: Maintenance on hybrid electric equipment includes checking oil, coolant, fuel, and filters.
   a. True
   b. False

10. The high voltage circuits on an electrically powered piece of equipment are which of the following colors?
    a. Black
    b. Orange
    c. Green
    d. Red

Answers to odd-numbered review questions are found in the Module Review Answer Key at the back of this book.

Answers to Section Review Questions

Answer	Section	Objective
Section One		
1. d	1.1.0	1a
2. b	1.2.2	1b
3. d	1.3.0	1c
4. b	1.4.1	1d
5. d	1.5.0	1e
Section Two		
1. c	2.1.1	2a
2. b	2.1.2	2a
3. d	2.2.3	2b
Section Three		
1. d	3.1.0	3a
2. b	3.2.2	3b
Section Four		
1. b	4.0.0	4a
2. b	4.1.0	4a

MODULE 22201

Introduction to Earthmoving

Source: Image courtesy of Komatsu America Corp.

Objectives

Successful completion of this module prepares you to do the following:

1. Describe earthmoving operations.
   a. Describe the general approach to excavations and related planning.
   b. Describe the characteristics of different types of soils.
   c. Explain how plans and specifications are used in earthmoving.
   d. Describe methods used to lay out slopes and grades.
   e. Explain how to set up and coordinate operations, including setting up staging areas, determining cycle times, and site preparation.
   f. Describe site drainage requirements.
   g. Explain the safety practices related to earthmoving operations.
2. Describe the equipment and methods used in excavating.
   a. Describe how to measure production for various earthmoving activities.
   b. Identify the equipment and methods used in loading excavated soil.
   c. Identify the equipment and methods used in hauling excavated soil.
   d. Identify the equipment and methods used in dumping excavated soil.
   e. Describe fill, backfill, and embankment construction.
3. Describe the different methods used in stabilizing soils.
   a. Identify the different types of binders.
   b. Explain how to apply binders to soil.
   c. Explain soil compacting.

Performance Tasks

Under supervision, you should be able to do the following:

1. Lay out a basic earthmoving operation.
2. Identify and select the proper equipment for a given earthmoving operation.
3. Draw a plan for basic earthmoving operations:
   - Clearing and grubbing
   - Excavating the foundation
   - Constructing embankments
   - Backfilling
   - Compacting

Digital Resources for Heavy Equipment Operations

SCAN ME

Scan this code using the camera on your phone or mobile device to view the digital resources related to this craft.

Overview

One of the main purposes of heavy equipment is to move earth to excavate, fill, and shape the land's surface. This module provides an overview of the various aspects of earthmoving operations. Some of these topics include planning and safety measures, typical equipment used in excavation, grading methods, worksite practices, soil properties, and stabilization.

NCCER **Industry-Recognized Credentials**

If you are training through an NCCER-accredited sponsor, you may be eligible for credentials from NCCER. The ID number for this module is 22201. Note that this module may have been used in other NCCER curricula and may apply to other level completions. Contact NCCER at 1.888.622.3720 or go to **www.nccer.org** for more information.

You can also show off your industry-recognized credentials online with NCCER's digital credentials. Transform your knowledge, skills, and achievements into credentials that you can share across social media platforms, send to your network, and add to your resume. For more information, visit **www.nccer.org**.

1.0.0 Earthmoving Operations

Performance Task

1. Lay out a basic earthmoving operation.

Objective

Describe earthmoving operations.

a. Describe the general approach to excavations and related planning.
b. Describe the characteristics of different types of soils.
c. Explain how plans and specifications are used in earthmoving.
d. Describe methods used to lay out slopes and grades.
e. Explain how to set up and coordinate operations, including setting up staging areas, determining cycle times, and site preparation.
f. Describe site drainage requirements.
g. Explain the safety practices related to earthmoving operations.

Pit: An open excavation of some depth that usually does not require vertical shoring or bracing.

Embankment: Material piled in a uniform manner so as to build up the elevation of an area. Usually, the material is placed in long narrow strips.

Groundwater: Water that resides and moves beneath the surface of the ground.

Trench: A long, narrow, temporary excavation that workers fill in when work is completed.

Earthmoving is any activity involving digging, loading, hauling, and dumping any earth material that is needed for construction or is in the way of construction. The work may also include digging and removing material from a **pit** or mine and processing the material. The primary earthmoving operations are soil or rock excavation, backfill, and **embankment** construction. Earthmoving may also include the transportation of materials, stabilization of embankments, and control of **groundwater**.

Earthmoving consists of moving material, usually soil or rock, from one place to another. This can be a short distance, as in the case of excavating a **trench**, or a longer distance, such as a cut and fill operation for construction of a highway or building site.

1.1.0 Excavation Planning

Stakes: Markers that are used to mark cut, fill, or grade information; underground utilities; or other survey points.

Grade: The surface of a road, channel, or natural ground area. In civil engineering, the term usually refers to the surface level required by the plans or specifications.

Several activities are required before starting any excavation and embankment work. The type, quantity, and location of material on the site determine the kinds and numbers of pieces of equipment needed to do the job. Each project has a set of grading plans that shows how the site should look after the earthmoving is complete. Survey crews transfer the information on the plans to the actual building site using **stakes** specific to each **grade** so that workers know what areas require excavation and fill.

Projects next to public highways or other areas where people pass by require traffic and pedestrian control. The controls must be in place before construction begins. The project drawings and specifications usually show the required signs and layout.

The term *earth* as used in earthmoving refers to any soil or **aggregate** material in the earth's crust that is above **bedrock**. Bedrock is the crust's foundation and consists of generally hard, dense rock materials. Rock itself may be of **igneous** or **sedimentary** origin and can be **crystalline** or noncrystalline. Soil is a mixture of **organic** and **inorganic** particles, as well as large and small pieces of rock. Weathered bedrock will produce soil, large boulders, and aggregate materials, which often lie on top of the parent bedrock in distinct layers.

The methods of excavating and removing rock differ from those required for removing soil. They are also more expensive. Bedrock removal usually involves drilling and blasting, and/or heavy equipment especially configured for excavating solid rock. Because of this cost, contractors must first define what makes up rock excavation, and then consider everything else to be soil excavation. The common practice is to call all excavation unclassified until the earthmoving contractor defines how much of the excavation is rock and how much is soil. This allows the contractor to determine what equipment is needed and thus the cost of the excavation.

Aggregate: Manufactured or natural rock or sand that has a specific size characteristic. Used in foundations and subgrade work.

Bedrock: The solid layer of rock under Earth's surface. Solid rock, as distinguished from boulders.

Igneous: Identifies earth materials that were formed from formerly molten (melted) rocks.

Sedimentary: Identifies rock materials that were originally deposited by wind or water.

Crystalline: Containing a mixture of closely adjoining mineral crystals. If the rock does not contain visible crystals, then it is noncrystalline.

Organic: When considering earth materials, any substance found in living organisms (plants, animals, or microorganisms) or produced by them.

Inorganic: When considering earth materials, any substance that does not come from living organisms.

1.1.1 Mining Operations

Mining is a type of earthmoving work and is a form of excavation. Mining requires special equipment and methods, so it is briefly described here separately. The Mine Safety and Health Administration (MSHA) governs mining safety and operations. It establishes regulations to prevent death, disease, and injury from mining.

In mining, the term *pit* describes any open excavation that is made to obtain materials of value, such as coal, mineral ore, select fill material, gravel, or quarry rock. Pit operation usually involves not only the use of heavy equipment, but also blasting, processing, and drainage. The size of pits varies from very small project pits to quarry **select material** for fill, to large open-pit mining operations such as those used for extracting iron, copper, or other ores (*Figure 1*). Quarries that yield construction materials, such as sand, limestone, and granite, are located all over the United States. Some are set up temporarily for the purpose of supplying material to a specific project; others are large permanent business ventures.

Select material: Earth or manufactured material that meets a predetermined specification as to some physical property such as size, shape, or hardness.

The type of heavy equipment used for a pit operation depends on the size of the pit and the material being excavated. For small pits, dozers, front-end loaders, and dump trucks are enough. For large or more permanent operations,

Figure 1 The Kennecott copper mine in Utah.
Source: Dale Chadwick

Figure 2 Dragline.
Source: Liebherr USA, Co.

Draglines: Crane-like excavating machines configured to drop a bucket from a boom and drag it toward the machine by a cable.

draglines like the one shown in *Figure 2* may be more economical. In larger pit operations, there may still be a need for larger specialized equipment to perform jobs such as heavy duty hauling (*Figure 3*), the construction of haul roads, drainage work, and stockpiling.

Figure 3 Rigid-frame truck.
Source: Image courtesy of Komatsu America Corp.

1.1.2 Emergency Services

If the job is expected to last for any length of time, or is separate from an existing facility or site, certain preparations are required for emergency services. Most jurisdictions have strict requirements for emergency vehicle access to construction sites. Usually, the first step is to set up a 911 (or local emergency number) address. This is a unique address given to a site that allows police, fire, and other emergency vehicles to find the location easily. Many areas require that the number portion of the 911 address be clearly visible from the road. The numbers are usually a specified height and are made of reflective material.

In some cases, local authorities may call for construction of special roads that allow emergency vehicles easy access to the site. For example, California, which is subject to wildfires, requires an access road at least 20' wide with a minimum $13\frac{1}{2}$' vertical clearance and a suitable load-bearing capability to ensure easy access for fire department equipment. Some jurisdictions may require standpipe or fire hydrant installations to supply water for firefighting equipment.

1.2.0 Soil Characteristics

The success and stability of any construction project depend upon the ground on which it is built. Most people give little thought to soil, but there is nothing more important in determining the suitability of a site for building. Site selection involves a series of complicated and rigorous tests performed by specially trained workers to determine the site's soil composition and properties.

There are many types of soils; each has unique characteristics. Soil composition varies by region, with most areas having a combination of two or more types. It is very rare to find a site that is composed of a single soil type. There are several classification systems for soil. A general description of the different types of soils follows:

- *Gravel* — Rock pieces greater than 0.08" and less than 10" in diameter. Larger particles are called cobbles or stones, and those larger than 10" are called boulders. Gravel occurs naturally, or it can be made by crushing rock. Natural gravel is usually rounded from the effects of being transported by flowing water, while crushed rock is usually angular.
- *Sand (coarse and fine)* — Mineral grains measuring 0.005" to 0.08". Sand is made from grinding or weathering rock. It usually contains a high amount of quartz. It is called granular material because it separates easily, giving it almost no **cohesive** strength. Coarse sand commonly consists of rounded grains like gravel and is often found mixed with gravel. Fine sand is usually more angular.
- *Inorganic silt* — Very fine sand with particles that are 0.005" or less. Silt consists of sand that has been ground very fine. It has a dusty appearance and powdery texture when dry but sticks together when wet. Silt has almost no cohesive strength, so dried lumps are easily crushed. It tends to absorb moisture by capillary action, which means that moisture wicks up through the soil. This makes it unstable in areas where the **water table** depth is shallow.
- *Clay* — Clay results from the weathering of certain types of minerals. Clay particles are tiny and plate-like. The particles have a strong attraction to each other, so clay is very cohesive. When wet, clay soils feel like putty and can be easily molded and rolled into ribbons. This property is called plasticity. When dry, clay is strong, and clumps are difficult to crush. A small amount of clay in the soil mixture at a building site makes it ideal for building, but clay is an **expansive soil**—it swells and shrinks with moisture changes. For that reason, pure clay is not suitable for building.
- *Organic matter* — Organic soils contain partly decomposed vegetable and animal material. Decayed organic earths, such as peat, are usually soft and fibrous and odorous when warm. This material is not suitable for building or as fill because it loses volume as it decays. This causes air pockets and makes the ground unstable. Very fine organic material is humus, a desirable component of topsoil.
- *Colloids* — Colloids are very fine particles less than 40-millionths of an inch in diameter. Colloidal organic and clay materials in most soils have a significant effect on soil properties and can enhance soil stability in small percentages. They promote cohesiveness and act as binders to other inorganic soil components. However, soils containing a large percentage of colloidal clays are susceptible to swelling and shrinking, so they are not suitable for building.

Soil testing is required at many jobsites. Exceptions are some small foundation excavation, small trenching, utility work, and landscaping. For large projects, specialists perform testing before any earthmoving work begins. This information gives the earthmoving contractor a good idea of the kinds of equipment required and how to plan the work.

On some jobs, it may be necessary to examine the soils or determine rock surfaces by digging a **test pit**. Although the pit furnishes the best information on the different layers of soil and rock, digging pits is slow and costly work. Pits are usually dug only where special foundation conditions need investigation. **Test boring** by drilling is the preferred method where any depth is required. **Core samples** taken from the borings provide the needed information about the various soil types.

Site plans are developed based in part on the information gathered during soil testing, so it is important for equipment operators to understand and accurately follow the project plans and specifications. If in doubt, consult the supervisor for guidance.

Cohesive: The property of a material that sticks or clumps to itself.

Water table: The depth below the ground's surface at which the soil is saturated with groundwater.

Expansive soil: A clay-like soil that swells with an increase in moisture content and shrinks as it dries out.

Soil testing: Any mechanical or electronic test used to determine the density, moisture content, and other properties of the soil that determines the amount of compaction required.

Test pit: Similar in purpose to a test boring but is a larger opening in the earth that provides a better understanding of the underlying ground structure and soil characteristics.

Test boring: A drilled or excavated hole created to take samples of the earth materials from different layers beneath the surface.

Core samples: Samples of earth taken from a cylindrical test boring.

1.3.0 Plans and Specifications

Dewatering: Removing standing water from a low area using a drain or pump.

Construction personnel need to understand what is included in project plans and specifications and how that information is transformed into directions for doing the work.

Most new construction work requires a single set of plans. Other work, such as major reconstruction or rehabilitation of roads and vertical structures, has two sets of plans: the as-built plans documenting the existing roads or structures, and a set of plans for the new work. As-built plans can be very helpful in determining what kind of problems the operator might have in excavating, **dewatering**, or demolishing the existing site.

Project specifications are important documents that provide directions, provisions, and requirements for performing the work illustrated and described in the plans. The items in the specifications cover the methods of performing the work or describe the quality and quantity of materials and labor to be furnished under the contract. Addenda are issued as approved additions and revisions to the original specifications.

The specifications normally incorporate other documents. These documents include standards for material specifications, test procedures, safety regulations, and traffic control devices.

Plans and specifications should be accurate, complete, and leave little room for assumptions or interpretation. They should also define specific responsibilities under the contract. If there is a discrepancy, the contract documents usually establish the authority hierarchy. When in doubt, consult the supervisor.

A typical set of plans consists of pages of drawings and notes that provide information about the dimensions, materials, and construction sequence of the project. Information in plans varies, depending on the type of project. Plans for horizontal construction projects, such as roads, canals, and airport runways, normally contain many drawings showing cross-section details, and plan and profile sheets. For vertical construction, such as warehouses, office buildings, and houses, the plan set has foundation drawings and any special shoring requirements for excavating the foundation, as well as the normal building design drawings.

Heavy equipment operators are often involved in the cut and fill work that needs to be done in preparing a jobsite for construction. The plans that guide cut and fill operations are part of the drawing set that defines a project. *Figure 4* shows a three-dimensional view of a jobsite. The coloring on the plan has significance. Green represents the site as it exists, blue indicates fill required, and red/yellow shows where cuts are needed.

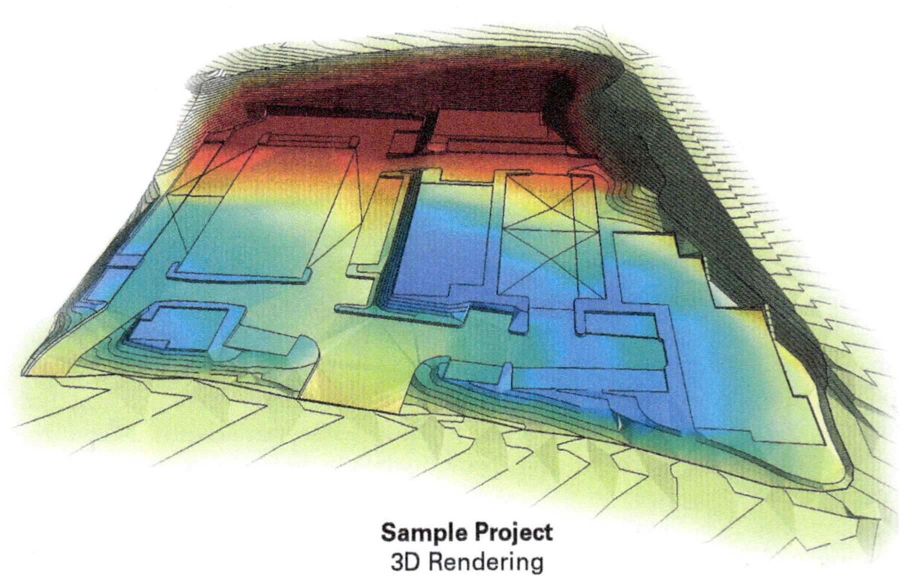

Figure 4 Three-dimensional view of a jobsite.
Source: Mark Jones

Figure 5 shows cross sections of the same site. The red lines on the charts represent locations of rock, the blue lines are finish grade, and the green lines

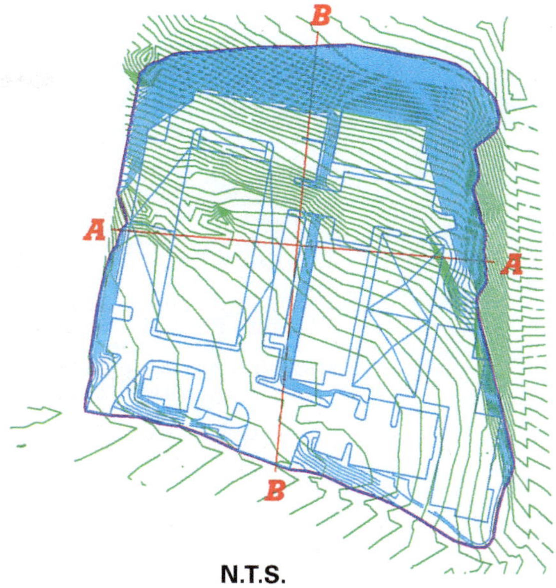

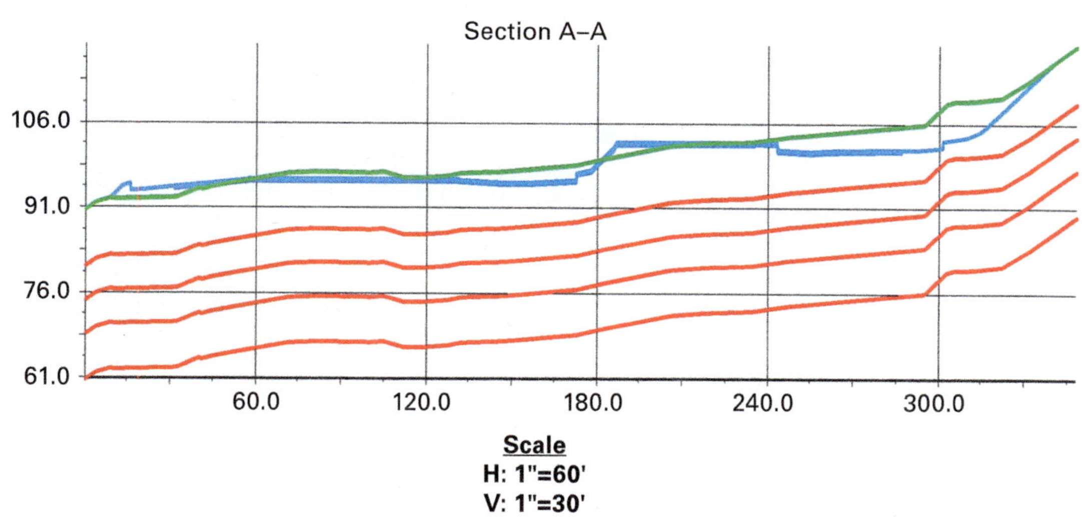

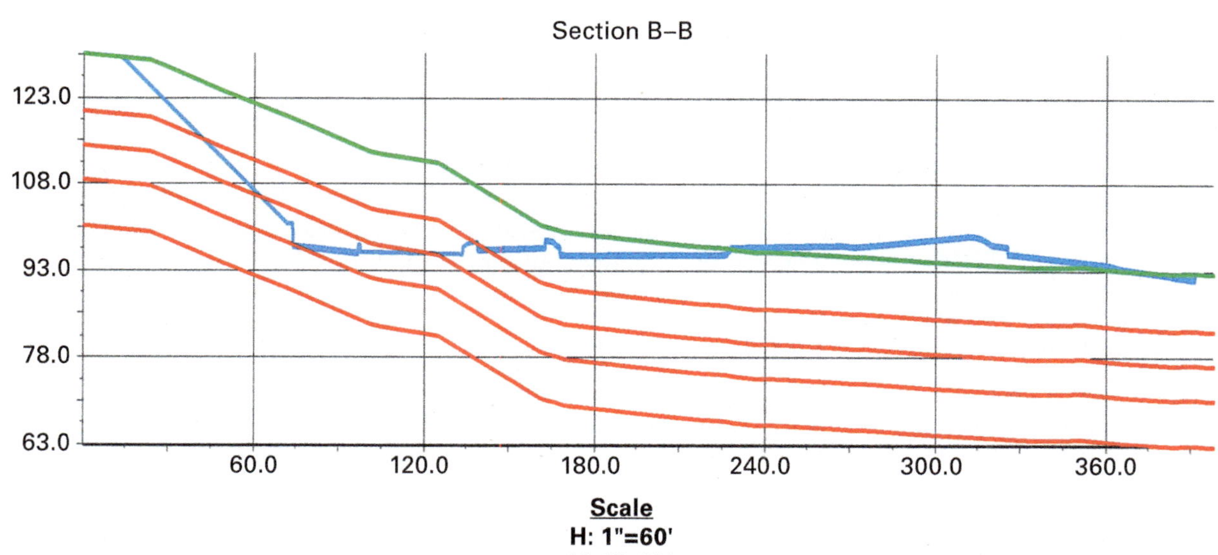

Figure 5 Jobsite cross section.

Source: Mark Jones

are existing grade. *Figure 6* shows how much cut and fill work is needed to complete the project. F represents fill, and each F bubble indicates how much fill is needed. For example, F1+72 means fill 1' plus 0.72'. The C (cut) bubbles are interpreted the same way.

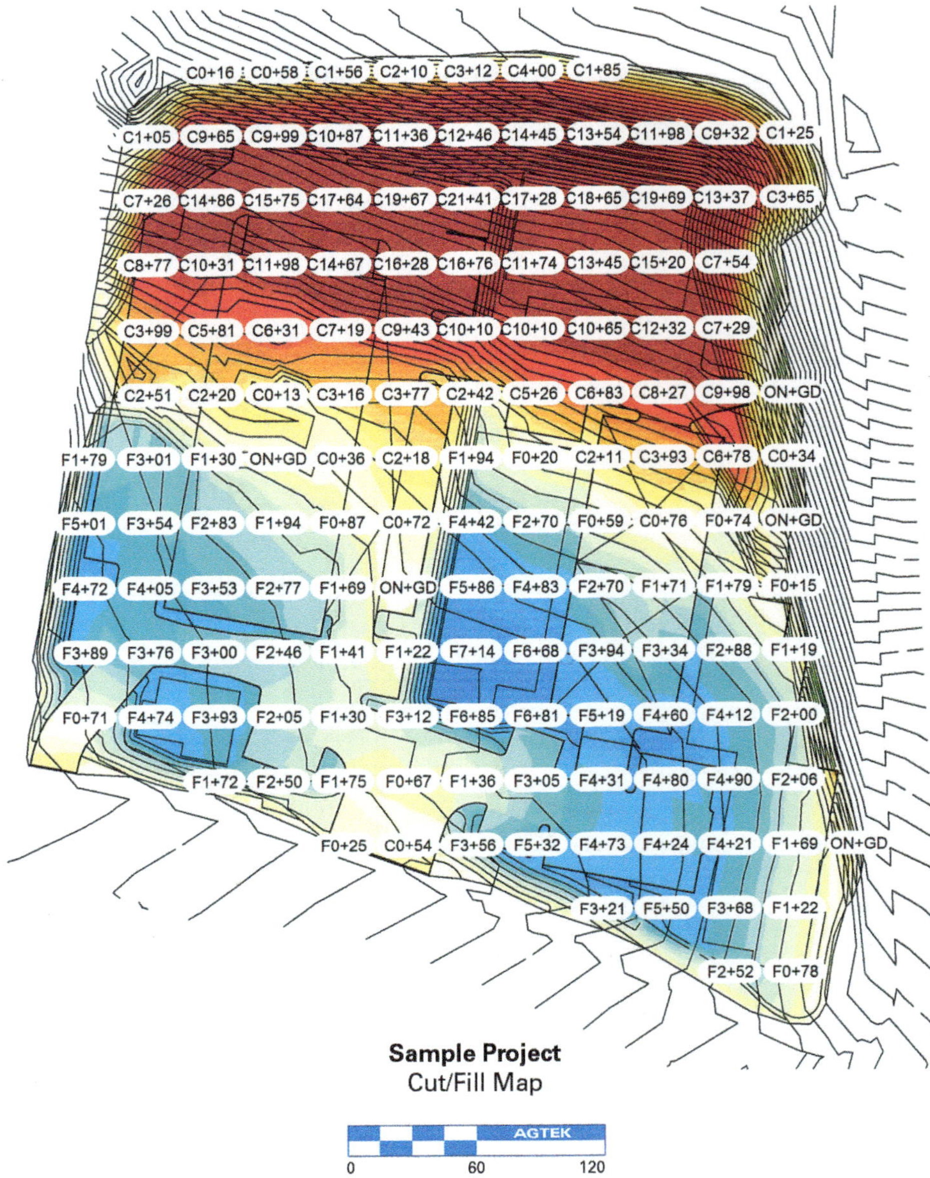

Figure 6 Cut and fill map.
Source: Mark Jones

Figure 7 is an example of a site plan for part of a subdivision. It shows property line boundaries, existing roadways, and building foundations. In addition, it shows the existing contours of the ground in dashed lines, and the planned grades in solid lines.

Benchmarks: Permanent points of known or assumed elevation from which surveyors can establish grades.

Slope: Ground that forms a natural or artificial incline upward or downward.

Slope stakes: Surveyor's stakes placed at the intersection of a cut or fill slope with the natural ground. Instructions include the slope ratio to meet the hinge point of a roadway or other reference point. Also given the terms *cut* or *fill stakes*.

1.3.1 Layout Control

Layout controls are established by accurately setting specific stakes to use as permanent reference points. These are called **benchmarks**. In turn, the benchmarks are used for placing other stakes, such as **slope** and grade control stakes. All of these points are shown on the site construction plans.

1.3.2 Slope Control

During excavation and fill activities, offset and **slope stakes** control grade slope. The survey crew places them at the edge of the construction limits. They are the

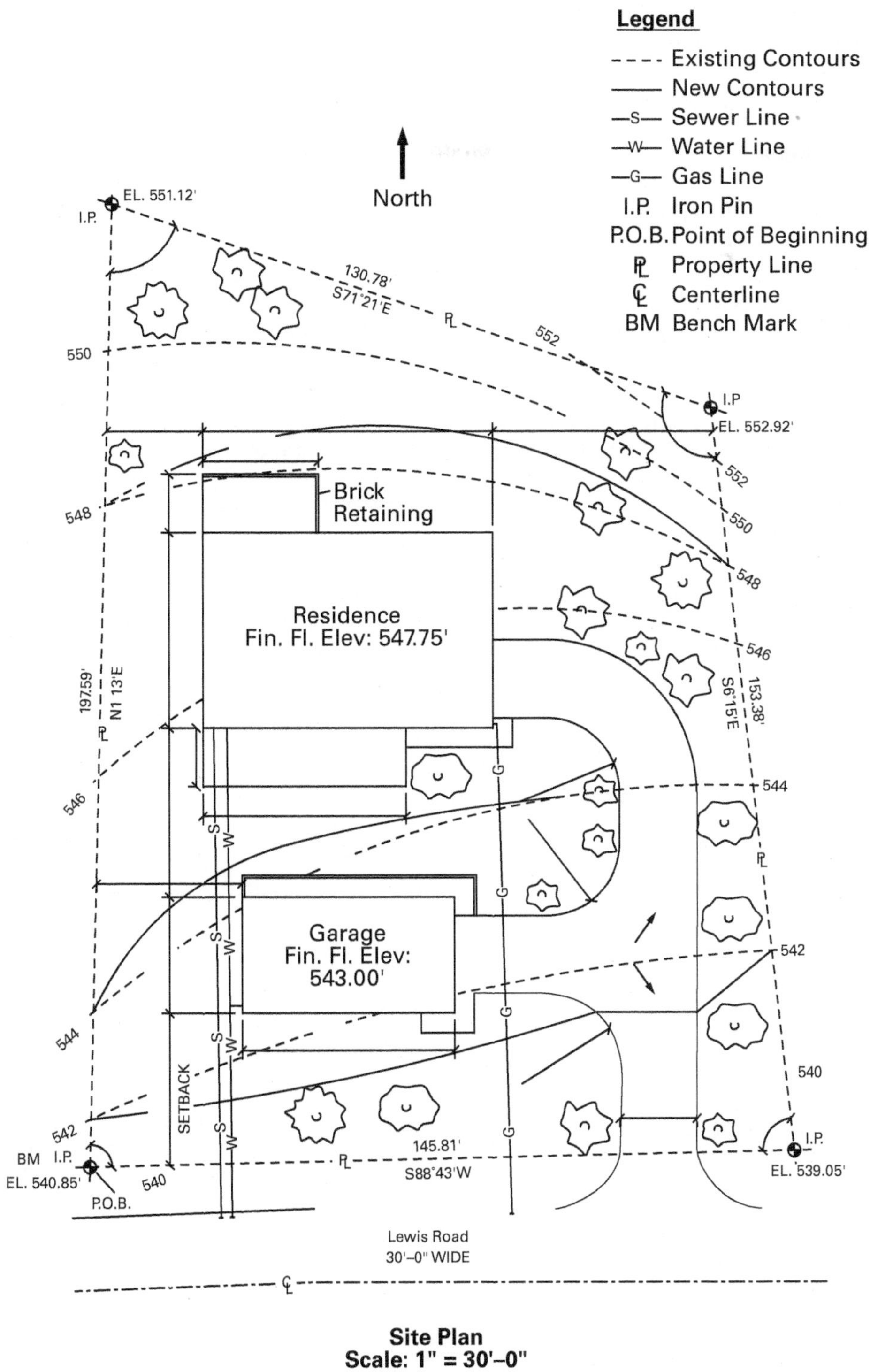

Figure 7 Site layout for part of a subdivision.
Source: John Hoerlein

principal guides for grading operations. It is important that construction equipment not disturb them.

Although this is usually the responsibility of the contractor, the owner's representative also checks them to be sure the slope is correct. Slopes should be checked in the following situations:

- At the completion of equipment operations but before removal from the area.
- When there is an indication that the slope is flatter or steeper than specified on the plans, even though the slope agrees with the stakes. Potential human error should be considered when checking the slopes.

Backslope: The area between the bottom of a ditch adjacent to a roadbed and the top of the cut into the existing natural grade.

Foreslope: The area of the roadway between the shoulder/hinge point and the bottom of the adjacent ditch or natural grade.

Cross slope: The pitch or grade of a road's surface perpendicular to its centerline that helps drain water from the road as well as improve vehicle stability on curves.

Centerline: The line that marks the center of a roadway or other construction project. This is marked on the plans by a line and on the ground by stakes.

- While finishing a **backslope**, **foreslope**, or **cross slope** of highway projects. Checking these during the process prevents extra work later.

1.3.3 Grade Control

The grade is the rise and fall of the constructed surface relative to the horizontal plane. The profile sheets in the project plans show the basic grade requirements. Slope stakes provide the primary information needed to control longitudinal grades for roadway projects. These stakes normally show the vertical distance from the stakes to the subgrade shoulder points, along with cut and fill information. This information maintains grade control during rough grading.

The survey crew places finish grade stakes, called blue tops, to guide finishing work and trimming of the subgrade. Surveyors drive these blue tops so that their tops lie at the proposed grade lines along each edge of pavement and the **centerline**. Heavy equipment operators are responsible for cutting or filling areas to the prescribed grade as stated on the grade stake.

1.4.0 Laying Out Slopes and Grades

NOTE

You should not enter the area beyond the limit of construction unless noted on the plans. It may be private property.

The layout of construction sites varies according to the kind of construction required, the terrain, and the location. Surveyors establish the layout using wooden and metal stakes placed in the ground. Spray paint and flagging attached to some solid object in the construction area may also be used. Right-of-way stakes determine the limits of highway and road projects.

The stakes set by the survey crew (*Figure 8*) contain written grade requirements for the site. These stakes tell workers about various cut and fill requirements. All workers must understand the importance of stakes and avoid damaging or removing stakes as work progresses. Resetting stakes is a costly effort that can delay the completion of a job and cause cost overruns.

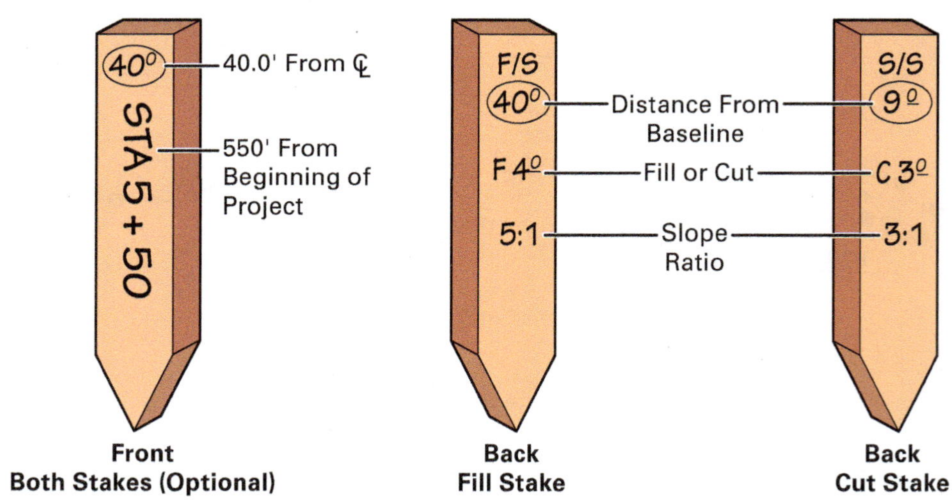

Figure 8 Examples of cut and fill stakes.

1.5.0 Setting Up and Coordinating Operations

Before beginning any earthwork, the contractor must set up the job. The setup depends on the type of project. A highway project is typically a long, narrow area where access may only be available from one or both ends. To get to intermediate points along the route, the project may require access roads so that equipment can be carried to the jobsite.

Setting up for a commercial or industrial site is different from setting up for highway work. Because existing roads provide access to these sites, equipment may be easily transported to the site and unloaded. These areas are often located in populated or high-traffic areas, so security and site access may be a more significant issue than for highway work. In addition, these sites may have added complications of excavating for parking areas, landscape areas, and drainage. Different requirements for excavating and hauling may affect the type of equipment to use. For example, the following questions may need to be considered:

- Are there requirements to fill planters and other landscape areas with topsoil?
- How much contour grading is required?
- Is there any topsoil that must be stripped and saved, or does it have to be hauled away?
- If there is any vertical excavation, how much shoring is required?
- Has project planning accounted for all underground utilities?

Regardless of the type of project, preparation for construction usually includes a preconstruction conference held after the completion of contract negotiations and before any work begins. At the preconstruction conference, the owner's representative, the designer, the construction manager, and the contractor discuss work tasks. The main topics may include the following:

- Foreseeable problems and unusual conditions
- Structures designated for removal
- Erosion, sediment, and **stormwater** runoff control
- Traffic control
- Utility locations
- Emergency plan setup
- Debris disposal
- Signing requirements
- The contractor's work plan and schedule of operation
- Project requirements and specifications
- Equal Employment Opportunity Commission (EEOC) regulations
- OSHA regulations, and MSHA regulations, if applicable
- Any applicable local, state, and federal rules or regulations
- Permits required before work can begin

Stormwater: Water directly resulting from rain or snow flowing over the ground's surface, or in natural or man-made water courses.

The purpose of the meeting is to bring the contractor and the owner to the same understanding of the project requirements.

1.5.1 Staging Areas, Equipment Parking Areas, and Site Access

During the preconstruction conference, participants may review requirements for equipment staging and parking areas. On smaller jobs, parking may not be a problem because there may be some space that is not being excavated or filled. On highway projects where larger equipment such as scrapers and dozers are used, keeping idle equipment out of the way may be a problem. Most contractors are very concerned about the condition and security of their equipment. They prefer to move it to an area around the worksite or to a fenced location when it is not in use. This may not be practical for highway jobs because of the distance a machine must be transported to get it to a secured area. For highway work, operators often shut down the equipment and park it on some level spot close to where the machine is working.

Site access is a concern for many contractors because of the liability that can result from accidents. Security of the equipment must be considered because of vandalism and theft. For highway projects where the work is strung out over several miles, it is difficult to secure the area with any kind of fencing or enclosures. Depending on local conditions, the contractor may decide to park equipment in only one or two locations that can be guarded. For commercial and industrial construction sites, the area is usually smaller, and the boundaries are well defined.

Equipment operators should know the policy and procedures for securing their equipment during nonworking periods. This includes what can be left in and on the equipment after shutdown, as well as where the equipment is to be parked.

Access to the site should be limited to authorized persons at all times, so operators should be alert to any unauthorized persons on the jobsite. This includes people who wander onto the site by accident, people who are trying to take shortcuts, and children who are curious about the equipment. Normally, requirements for being on the jobsite include the use of hard hats and safety vests, so it is relatively easy to spot someone who does not belong there. If anyone looks as if they do not belong on the site, notify the supervisor.

1.5.2 Construction Signs

In some cases, it may be necessary to post construction signs to warn the public of the construction activity. Flaggers or other methods of traffic control may be required, especially where construction equipment must cross the road or when the work is very close to a public area, such as a commercial building site in an urban area. Work may not begin until all signs and warning devices are installed according to the project plans and applicable government regulations.

1.5.3 Clearing and Grubbing

The worksite must first be stripped of all vegetation, obstructions, and unwanted structures before excavation or other construction activity. The removal of this material is referred to as clearing and grubbing, or right-of-way preparation. It also includes the preservation of trees and other objects that are to be saved. Generally, the distinction between the two operations is as follows:

- Clearing is the removal of objects above the original ground.
- Grubbing is the removal of objects below ground.

The contractor is generally required to clear and grub all areas of the site. In the cross section shown in *Figure 9*, the area within the property lines is to be cleared and grubbed of all trees, roots, stumps, and other protruding objects not designated to remain. The depth of cover for objects usually determines whether they need to be removed.

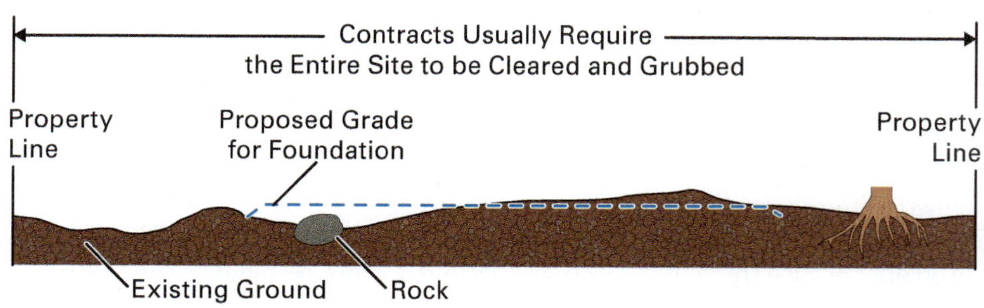

Figure 9 Clearing and grubbing.

Clearing and grubbing may include the removal of structures and obstructions, including buildings, fences, old pavements, and abandoned pipe or culverts. After the area has been cleared and grubbed, it is scalped of brush, roots, and grass. The topsoil is generally removed and stockpiled during this phase.

If the area is undeveloped, some vegetation may need to be cleared. Vegetation is usually removed by scraping it off with earthmoving equipment and placing it in a designated area. When clearing areas, remember the following points:

- When using a scraper to remove vegetation, remove shanks and teeth from the scraper's cutting edge to reduce the amount of soil taken with the vegetation.

- Do not leave any vegetation in the construction area. Anything that grows is unsuitable for compaction because it loses volume as it decays and causes a depression to form on the excavation surface.
- Use the proper equipment for grubbing. Large trees or boulders need to be removed before a scraper or grader can start pushing the soil around. Depending on the size, an excavator, dozer, or track loader can be used to remove such obstructions.

There are several ways to dispose of debris. It may be hauled away from the construction area, burned (if allowed), buried, or mulched. High-quality topsoil may be stockpiled to be used during landscaping at the end of the project. Heavy equipment operators should be briefed on the desired disposal method to avoid costly errors.

1.5.4 Maintaining Haul Roads

Well-maintained haul roads are important for efficient operation. On long-haul roads, a grader and water truck should make a pass over the road periodically to keep it in good condition. This reduces the wear and tear on equipment as well as **cycle time**.

On small jobs, a grader may not be available for use on the haul road. In this case, the scraper operators can maintain the road. If the road is being watered to keep the dust down, the scraper operators must be careful. Wet soil can cause a scraper to skid out of control.

Cycle time: The time it takes for a piece of equipment to complete a repetitive operation. This normally would include loading, hauling, dumping, and then returning to the starting point.

1.6.0 Drainage Requirements

Excavation descriptions in this module are based on the movement of dry material. However, most materials are not completely dry in their natural environment. Even in a desert, the first couple of feet of soil are dry, but as the digging goes deeper, the earth feels cool and moist to the touch. Millions of yards of material are moved every year without major problems with water. Rain may slow or temporarily halt operations because of its effect on the surface, but it usually does not make excavation impossible. Many construction sites have plans to control the flow of stormwater over the site to prevent erosion and water pollution. The federal government requires these plans, so it is important to understand and follow them. Other drainage problems on construction sites that are related to earthmoving are the control of groundwater and the allowance for drainage from embankment materials.

1.6.1 Control of Stormwater

The Environmental Protection Agency (EPA) requires that contractors performing any activity that disturbs one or more acres of land obtain a National Pollutant Discharge Elimination System (NPDES) stormwater permit before work is started at the site. The purpose of the permit is to ensure the contractor has a workable plan to prevent pollution and control stormwater runoff from the site, as well as to prevent erosion.

NOTE

Construction sites that are smaller than one acre, but are part of a larger development, also need to have an NPDES stormwater permit.

Contractors need to use adequate measures to prevent erosion and sedimentation and to control stormwater runoff. Erosion is the removal of soil by water or wind. Sediment refers to soil that has been moved from its original place by wind, water, or other means. Stormwater runoff is water from rain and snow that flows over the ground surface rather than being absorbed. Heavy equipment operators need to know and understand the site's erosion, sediment, and runoff prevention plans.

Every site is different, so each has its own erosion and sediment control plan. The best way to prevent erosion is to avoid disturbing the surface of the soil, so it is important to follow plans closely and disturb only the earth that needs to be graded. Other measures include reducing the steepness of slopes, covering disturbed soil with mulch, erosion cloth, or **riprap**, and contouring the grade so

Riprap: Loose pieces of rock or manufactured materials placed on the slope of an embankment in order to stabilize the soil from the effects of erosional processes.

that water is absorbed rather than shed. A slope graded with a rough surface slows the flow of water and promotes absorption of water rather than runoff, thus helping to prevent erosion.

Because sediment is earth that has been moved from its original place, the best way to avoid sediment is to prevent erosion. In earthmoving, some erosion is unavoidable, so most sites use devices such as silt fences to trap displaced soil. Other ways to help prevent sedimentation include the following: avoid walking or driving across disturbed soil, wash soil off truck tires before leaving the site, and confine vehicle operation to approved haul roads.

As rainwater flows over a construction site, it picks up loose soil and debris, as well as gasoline, oil, and other chemicals from soil or paved areas. These chemicals can pollute water and degrade ecosystems. The first step in preventing such pollution is to prevent the release of pollutants into the environment. One way to do this is by being very careful not to spill solvents, chemicals, oil, grease, paints, and gasoline onto the ground. Employers usually have designated areas for equipment maintenance and wash-downs and a protected storage facility for chemicals, paints, solvents, and other potentially toxic materials. Workers can also help prevent pollution by placing trash in the proper receptacle and using the sanitary facilities provided on the site.

1.6.2 Control of Groundwater

Water existing below the ground surface is typically divided into two zones. The zone of aeration contains varying amounts of moisture and air in the spaces between soil particles. Beneath it, water completely fills the spaces, forming the zone of saturation. The water in this lower area is groundwater. The top surface of groundwater is the water table. *Figure 10* shows a cross section of the zones of aeration and saturation.

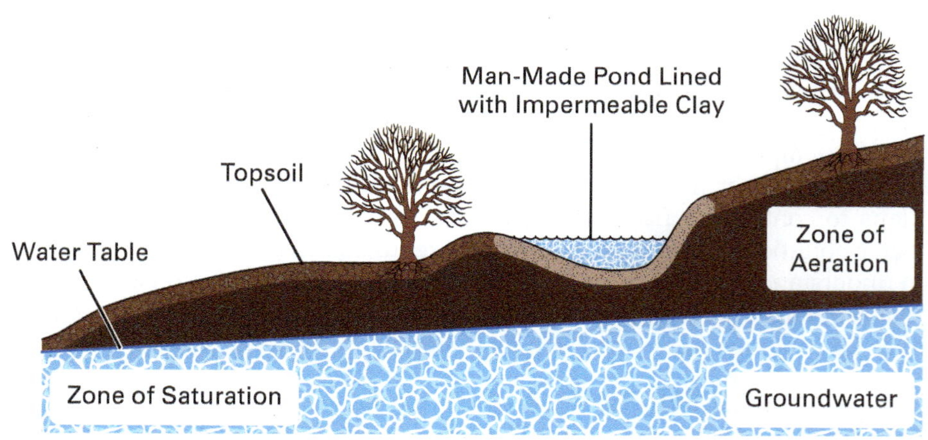

Figure 10 Cross section of a water table.

The depth of the water table in an area can vary significantly by location and time of year. It depends on the amount of annual precipitation, the terrain, the underlying geology, and the amount of extraction due to irrigation and other uses of the water.

The depth of the water table and the amount of groundwater may be determined by soil borings. Borings are made at specific locations over the jobsite to plot the profile of the water table. From this profile, the need for any type of special dewatering activity, such as sumps or various types of drains, can be established.

Sumps are open pits or holes built to collect groundwater. Sumps are then pumped out mechanically. The requirements for sumps vary with the depth and flow rate of the groundwater.

Drains are various types and sizes of ditches that collect and channel the flow of water away from the excavation area. They should be limited in width and depth to the minimum requirements and should intercept the groundwater as close to the bottom of the excavation as possible. The layout of these temporary drains is determined by soil conditions and the location, direction, and quantity of flow.

1.6.3 Drainage of Stockpiles

A typical problem with stockpiles of topsoil or other products is the storage of wet earth materials. Wet material does not stack well and tends to spread out or flow. The ease of water drainage from the stockpile depends on the type of material.

Water that drains off the stockpile must be controlled. When building a large stockpile of select material, a shallow channel should be cut around the stockpile to collect the runoff and drain it away from the storage area to a holding area. Silt fences must be used to slow water flow and the movement of sediment.

1.7.0 Safety Guidelines

Once workers are comfortable reading the construction plans for the project, it is important that they relate them to the physical building site. Safety is everyone's first priority on the job, and workers must take responsibility for their own safety. Once workers are familiar with the site plans, they should walk the construction area to get a firm idea of the work to be performed and be able to identify potential safety hazards. Report any potential problems to the supervisor. Never assume that the project engineer or supervisor is aware of a potentially unsafe condition.

1.7.1 Personal Safety

Before starting any earthwork, walk the site if possible. Be alert for large depressions or other obstacles that are not shown on the plans. Large animal burrows, boulders, and decayed stumps may be present, especially in undeveloped areas. Driving over unexpectedly rough terrain can throw the equipment off balance and put it in danger of tipping over. These areas may need to be corrected during clearing and grubbing.

In the warm weather, look for swarming insects that could mean a nest or hive in the area. Many people are highly allergic to insect stings and need to carry special medication to counteract the sting effects. Notify the supervisor of such findings. Removal of the nest may not be possible, but other workers should be notified of its existence.

When it is not possible to walk the site before starting work, be especially careful during the first drive over a section of land. Always scan the terrain looking for potential hazards. Drive slowly and be ready to stop.

1.7.2 Site Contamination

Working at a new construction site or an industrial site, or performing demolition on a building, always presents the possibility of uncovering a contaminated site. While walking around the site, look for signs of illegal dumping and other contamination. Contaminants can be in the soil, buried underground, or stored in a container such as a barrel. Warning signs of potential contamination include the following:

- Puddles of liquid on the ground, especially liquids with an unusual color or odor
- Rainbowlike reflections on surface puddles
- Any non-water liquid seeping out of the ground
- Unusual or foul odors
- Unmarked barrels or tanks—often buried or otherwise obscured
- An unexplained and remarkably bare patch of land in an area that is otherwise heavy with vegetation

Identifying contaminants is a complicated and time-consuming procedure that requires advanced education and training. Heavy equipment operators are not expected to identify contaminants but are required to report any potential contamination to the supervisor.

Workers can limit exposure to soil contaminants by wearing proper PPE, showering and changing into clean clothes at the jobsite when possible, avoiding eating or drinking in the dusty area, and periodically spraying the area with water.

1.7.3 Sewage

Construction workers at an excavation site may accidentally uncover live sewers or septic tanks. Contact with raw sewage is not only unpleasant but also can place workers at risk for disease. One of the most common diseases is hepatitis A. Hepatitis B is less common. Hepatitis is a viral disease that attacks the liver. It is contracted when infected human waste or blood is swallowed or passed on by person-to-person contact. To prevent infection, wear latex or similar gloves when contacting sewage, wash hands frequently, avoid touching the face, and do not eat or drink in the area.

WARNING!

Some people are very allergic to latex. Some symptoms can be mild, such as an itchy red rash where the latex contacts the skin. Other symptoms are life threatening, such as anaphylactic shock. This involves swelling of the airway, causing shortness of breath. If any of these symptoms occur while or after wearing latex gloves, remove them immediately, wash the affected area, and seek medical assistance if necessary.

If illness occurs after working on a job that included exposure to sewage, tell the doctor. Hepatitis is not often fatal, but there is no cure for hepatitis A once contracted. Most otherwise healthy individuals can recover from hepatitis B. There have been cases in which otherwise healthy individuals have died from hepatitis. Long-term infection can require liver transplants. If a job requires frequent contact with sewage, talk to your supervisor about being vaccinated against this disease before exposure or symptoms appear.

1.7.4 Safeguarding Property

As employees, workers are responsible to their company for their actions on the job. When workers think about safeguarding property, operating equipment usually comes to mind. The operator's duties include operating the equipment responsibly, reporting defects, and securing the equipment when not in use. In addition, workers must treat the building site responsibly because the site owners have contracted with employers to construct a project to certain specifications described in the plans.

Workers should treat the jobsite responsibly by following the plans and being careful not to unduly damage the site. This is especially important for small jobs where a designated equipment staging and maintenance site has not been selected. It may be up to the worker to select a site; it should be done with the construction plans in mind. For example, it would be hard to maintain a flowerbed after construction if oil and hydraulic fluid soaked into the soil during construction. If the jobsite has no designated areas for loading, unloading, parking, and maintenance, the worker should select a site well away from the building.

Know the general plans well enough to avoid damage that heavy equipment can cause. Locate any existing structures, roads, or walkways to ensure that the operator can avoid driving over them. Know the location of underground structures, such as culverts, pipes, and septic systems at the site. Driving over these structures with heavy equipment can crush them and cause the equipment to be thrown off balance, causing an accident.

Never cross property lines without permission. Not only is this trespassing, but heavy equipment can also damage the terrain and cause the property owner to consider taking legal action against the employer. Further, there may be underground structures on the adjacent property that the equipment could damage.

When the plans call for preserving certain trees, workers must know their locations and avoid guesswork. The supervisor is available to help when needed. Trees and bushes are good places to look for survey markers or benchmarks (*Figure 11*), because they indicate the ground hasn't been disturbed for some reason. Never destroy or move these markers. Notify the project engineer or supervisor of the locations of any markers not shown on the site plan.

Figure 11 Tree-posted notice locating a benchmark.

1.0.0 Section Review

1. The primary earthmoving operations are excavation, backfill, and _____.
   a. trenching
   b. unneeded material disposal
   c. embankment construction
   d. grubbing

2. A very cohesive soil includes a large percentage of _____.
   a. silt
   b. clay
   c. gravel
   d. organic matter

3. Labels that include the letter F and some digits distributed over a construction site plan indicate locations for _____.
   a. fill application
   b. cutting excavation
   c. spoils areas
   d. freight hauling lane width

4. Stakes set by the survey crews are important because they _____.
   a. identify the property owners
   b. specify the kind of fill to use
   c. show compass directions
   d. contain written grade requirements

5. An important concern to most contractors at worksites is _____.
   a. water content of the soil
   b. security of their equipment
   c. the route they take to access the site
   d. whether towns are close by

6. Which of the following methods would *not* be very effective for controlling stormwater runoff?
   a. Reducing the steepness of a slope
   b. Applying riprap to a slope
   c. Cutting gullies down the face of an embankment
   d. Contouring the grade to slow the flow of water

7. The greatest hazard of dealing with raw sewage in an excavation is _____.
   a. the stench of human waste materials
   b. contracting hepatitis
   c. decreased ground stability
   d. having to compensate the owners of the damaged septic system

2.0.0 Site Excavation Practices

Performance Task

2. Identify and select the proper equipment for a given earthmoving operation.

Objective

Describe the equipment and methods used in excavating.

a. Describe how to measure production for various earthmoving activities.
b. Identify the equipment and methods used in loading excavated soil.
c. Identify the equipment and methods used in hauling excavated soil.
d. Identify the equipment and methods used in dumping excavated soil.
e. Describe fill, backfill, and embankment construction.

NOTE

Before you dig on a jobsite, you should ensure a planner, engineer, or supervisor has identified underground utilities. The *Call Before You Dig* program can help workers contact the right agencies in their state who can assist in locating underground utilities and mark them before digging on a jobsite. A worker simply dials 811 and is connected to a local call center where they can explain the location and situation and the operator can assist them.

Potholing

A method known as potholing is used when it is necessary to locate the exact position of a buried utility by exposing it. Potholes can be dug with mechanical equipment, such as air blasters or excavation vacuums, or they can be dug by hand.

Planning is important in excavation work. Plan work so that the equipment never stands idle while grades are being checked or while a decision is being made on where to start the next cut or fill. The operator should study the plans and stakes carefully before starting work. Contractors need to take time to analyze the type of equipment needed. For example, is a scraper sufficient to cut the soil, or is the soil hard enough that some type of ripping is needed to loosen it before using the scraper?

Before breaking ground, it is important to determine if any utilities or other structures are buried on the site. In most areas a special utility information phone number or website has been established so that contractors can arrange to have a survey conducted to locate and mark buried utilities. This service requires advance notice. Flags are commonly used to mark the locations and types of utilities. These markers are color-coded according to the standard American Public Works Association (APWA) underground utility color codes found in *Table 1*. When workers see these markers, they should act accordingly. Check the plans to make sure no utilities were missed. If in doubt about the proper actions to take, workers should ask the supervisor.

Several types of equipment are used for excavating. The machine chosen for a particular job must fit the requirements of the work and have a cycle time that allows the work to be done cost effectively. Typically, the equipment is matched to the type of material to be excavated. The equipment must be

TABLE 1 APWA Underground Utility Color Codes

Color	Meaning
Red	Electric power lines, cables, conduit, and lighting cables
Yellow	Gas, oil, steam, petroleum, or gaseous material
Orange	Communication, alarm, or signal lines, cables, or conduit
Blue	Potable water
Green	Sewers and drain lines
Purple	Reclaimed water, irrigation, and slurry lines
White	Proposed excavation
Pink	Temporary survey markings

capable of operating in the environment without being unstable or inefficient. In general, the following types of heavy equipment are used for excavation:

- Scrapers (*Figure 12*)
- Excavators (*Figure 13*)
- Motor graders (*Figure 14*)
- Dozers (*Figure 15*)
- Wheel and track loaders (*Figure 16*)
- Backhoe loaders (*Figure 17*)

Each type of excavation presents its own challenges. Some projects require work at several different excavation sites, including the following:

- Highway/roadway excavation
- Several types of bulk excavations
- Limited-area vertical excavation
- Trenching

Figure 12 Scraper.
Source: agnormark/iStock/Getty Images

Figure 13 Excavator loading a rigid-frame, off-road dump truck.
Source: Mark Jones

Figure 14 Motor grader.
Source: Image courtesy of Komatsu America Corp.

Figure 15 Dozer.
Source: Image courtesy of Komatsu America Corp.

Figure 16 Wheel loader.
Source: Image courtesy of Komatsu America Corp.

Figure 17 Backhoe loader.

2.1.0 Production Measurement

Although equipment operators usually do not get involved in the details of planning the overall project work, they need to understand how it is done. At some point, they may also provide suggestions or input based on their experience with different types of soils and the handling of different machines under various conditions.

2.1.1 Mass Haul Diagram

The mass haul diagram is a graphic method of analyzing the movement of soil from one loading area to the dumping site. It is mainly used in highway, large commercial, or industrial projects where the cut and fill work is spread out in a long, narrow band. The mass haul diagram relates the proposed grade to the existing grade and is useful in determining the location and volume of excavations and the haul distances to points in the fill.

This analysis identifies any areas where it would be better to obtain material from a place outside the project limits rather than hauling material a very long

Spoils: Excavated materials reserved for future use or unsuitable for construction that have been stockpiled.

distance. Also, it shows where to dump surplus excavated material in a **spoils** area rather than haul it to a fill area that is further away. These decisions depend on the cost of the equipment and the materials being used.

2.1.2 Cycle Times

Cycle time is the measurement of how long it takes to perform a repetitive operation, such as the time for a piece of equipment to load, haul, dump, and return to the loading point to start the next cycle. This allows the contractor to estimate how long it will take to complete a specific activity, like rough grading or excavating a ditch.

Cycle times vary for different types of equipment. For scrapers that haul and dump as well as load, the cycle time includes picking up the material from the ground, hauling a full bowl to the dump site, spreading the material, and then driving back to the excavation area to scrape another load. In the case of a front-end loader, the cycle time is measured from the time the loader begins digging with the bucket, through maneuvering to load the truck or stockpile, loading (or dumping the bucket), and then returning to the point of digging.

Most equipment manufacturers provide information about average cycle times, which can be used to estimate the time and cost of doing the work. Contractors also use cycle time measurements to determine the operator's efficiency.

2.1.3 Roadway Excavation

Roadway excavation is a leveling process used in highway construction and airfield grading. It often involves the movement of material from high ground to low areas. This type of job is called balanced excavation (*Figure 18*) because the soil needed for fill areas is obtained from cut areas and there is no need to haul material in from outside the project. In other cases, cut material may be wasted because it is not needed or is unsuitable for fill. Material moved from a borrow area to fill low spots for embankment construction may also be included.

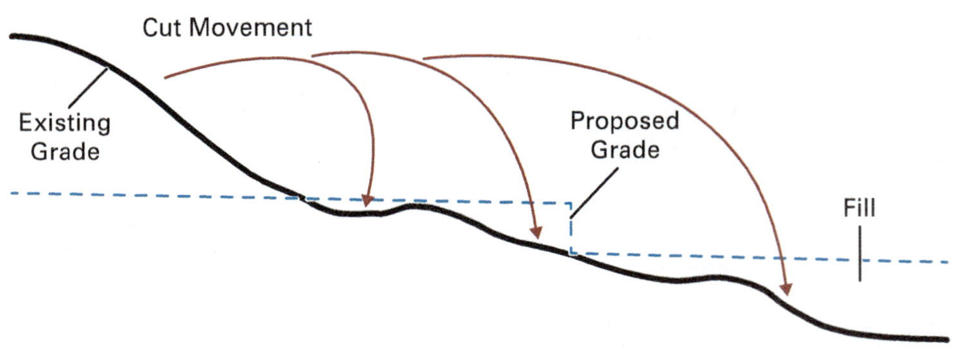

Figure 18 Balanced excavation.

2.1.4 Bulk Pit Excavation

Bulk pit excavation involves digging and loading material of great depth and volume and then hauling the excavated material to another site. The pit may be at the project site, such as the construction of a building with a basement, or it may be outside the worksite, such as at a borrow pit.

As discussed, the cycle time of any equipment is the measurement of how long it takes to perform a specific function. When excavated material is loaded onto a truck for removal from the site, it is important to minimize the movement of the excavator or loader for the shortest cycle time. *Figure 19* shows methods of loading trucks. In Method 1, the truck is positioned so that the excavator boom swings between the excavation face and the truck bed without having to move the excavator. In Method 2, the loader must move to the excavation face for loading, then back up and move into position to dump its load into the truck bed.

To keep the cycle time at a minimum, the loader picks up a load, backs up a short distance, turns 45 degrees, and raises its bucket. The truck then backs under the raised bucket.

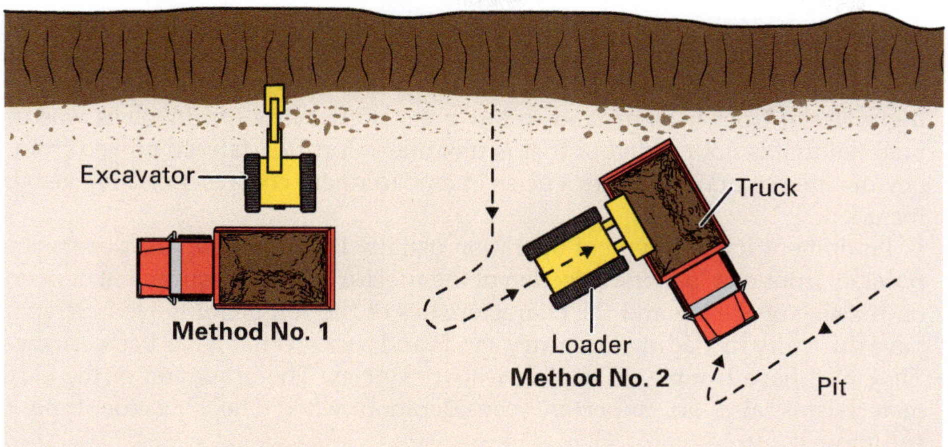

Figure 19 Loading a truck at the digging face.

A third method of moving excavations away from the excavation site is used in special conditions where there is no need to do more than push the excavated material aside. In this method, a crawler-mounted loader or dozer uses its bucket or blade to dig and collect material on the front, and then pushes beyond the access ramp to the fill site.

In the scenario in *Figure 20*, both the entering and exiting traffic are using the paved road below the site. A more efficient traffic flow would use the paved road above the site for entry and the one below the site for exiting vehicles. Even when workers are fortunate enough to experience this scenario on large jobs, local permitting may limit the construction traffic flow to a single point. In addition, the construction of two access roads would increase the cost of the job.

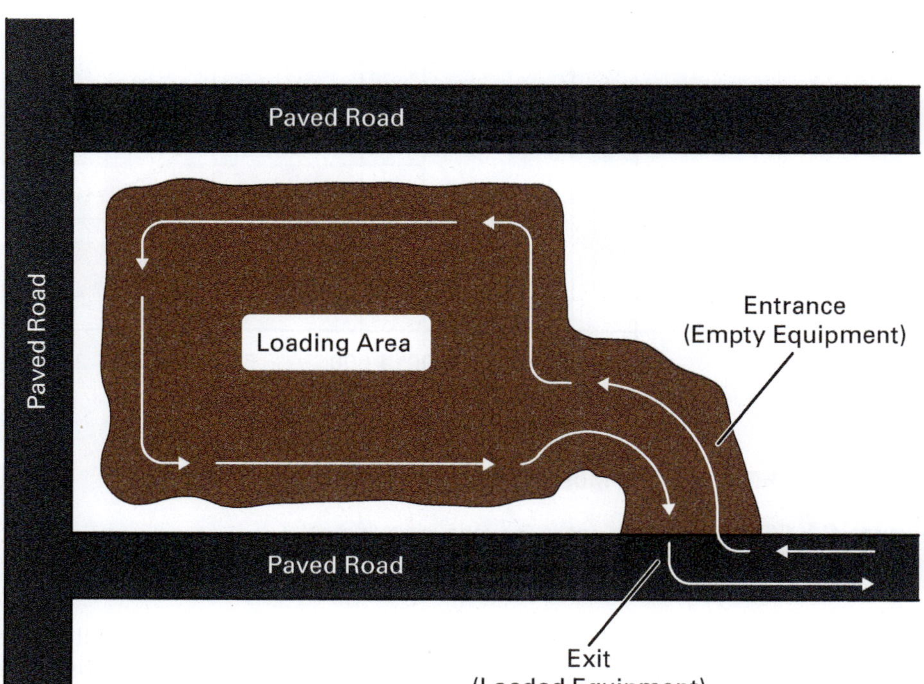

Figure 20 Bulk pit excavation.

2.1.5 Bulk Wide-Area Excavation

Bulk wide-area excavation is similar to bulk pit excavation, but this type of excavation permits access to the area from many directions, making it easier to access and leave the excavation site. Bulk wide-area excavation is usually shallower in depth but larger in area than bulk pit excavation.

Bulk wide-area excavation is used in highway construction, airfield runway grading, the removal of top layers of soil (as in quarrying or strip mining), and the building of earthen dams and basins. In some cases, it only involves moving material from high ground to low areas. Other times, it involves wasting material that is not needed or that is unsuitable for fill. Material moved from a borrow area to make up deficiencies in embankment construction may also be included.

Equipment for this type of operation usually involves scrapers, dozers, and possibly front-end loaders. Selection of the specific pieces of equipment depends on the size of the job and the characteristics of the soil being moved. Scrapers have difficulty in loading loose, dry sands and rock, even if it has been crushed. They also have trouble unloading wet sticky clay. The **gradation** of the earth materials is also an important consideration when choosing the type of equipment.

Gradation: The classification of soils into different particle sizes.

In this type of excavation, the ideal location ensures short haul distances. For short distances, dozers and graders can be used to scrape the material, push it the required distance, and then level it as required. If haul distances are longer, scrapers should be used when the material can be loaded efficiently. Scrapers can cut a layer of material while moving forward and can travel at speeds up to about 30 miles per hour (mph) when fully loaded. At the dump area, they can slow down and dump the material by spreading it in layers at the specified thickness. *Figure 21* shows effective haul distances for different types of hauling equipment.

NOTE

The arrangement of the horizontal distance scale in *Figure 21* is called a logarithmic scale. Equal spaces on the scale represent powers (or logarithms) of 10. Logarithmic scales are used to display data that span a large range of values in a compact graph or diagram.

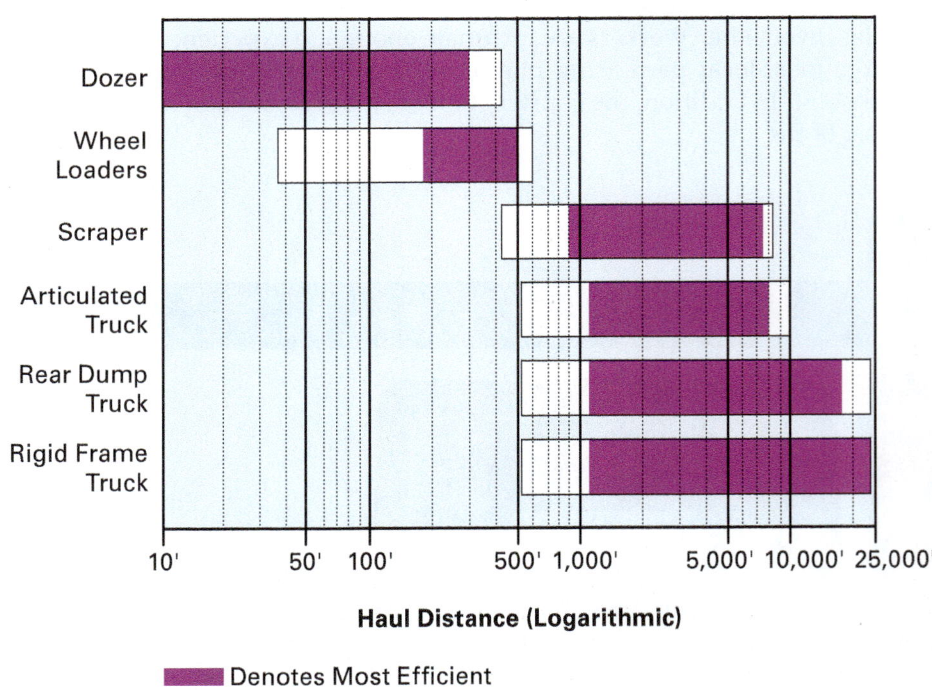

Figure 21 Effective haul distances for various equipment.

When this type of excavation is located far from the dump site, material is loaded onto trucks or other carriers and carried to the dump site. If this is the case, the loading area is set up to permit the smooth flow of traffic and quick loading of material.

2.1.6 Channel Excavation

Channel excavation is the excavation of loose, unconsolidated materials, of material lying under water (such as a drainage channel), or of saturated soils that prevent the equipment from traveling over the excavated area. This type of excavation is usually performed by equipment standing on solid ground at the same level or higher than the material being worked.

2.1.7 Limited-Area Vertical Excavation

Some excavation must be done by lifting the material vertically out of the pit because the sides of the excavation require shoring or shielding. This type of excavation is known as limited-area vertical excavation (*Figure 22*).

Figure 22 Limited-area vertical excavation.
Source: Kundel Industries

2.1.8 Trench Excavation

A trench is a temporary opening in which something, such as a pipe or box culvert, is placed and covered. The trench width, and usually the depth, are limited to 15'. One type of trench excavation is known as cut and cover. This sometimes includes excavating a large and perhaps deep trench, laying the specified pipe or culvert, and then covering everything over or building on top of it. This option is frequently used for highway and transit system tunneling when the excavation of the trench would cost less than boring a tunnel.

Trench excavation is different from pit or wide-area excavation. Pit or wide-area excavation focuses on moving large quantities of material. With trench excavation, the focus is on the rate at which the pipe or culvert can be placed. This factor affects the type and size of equipment selected for the

NOTE

Do not confuse a trench with a ditch. A ditch is a narrow slot cut in the ground and left open. A trench is an opening into which something is placed and then covered.

trenching job. The trench should not be excavated any wider than is necessary for bedding and setting the structure. Shoring can be used to reinforce the sides of the trench.

The three main classes of equipment used in trenching are backhoes, excavators, and trenching machines. The trenching machine is specifically designed for trench excavation. *Figure 23* shows different types of trenchers. *Figure 24* shows a skid steer digging a trench with a trenching attachment.

(A) Pedestrian Trencher

(B) Large Construction Trencher

Figure 23 Trenchers.
Source: (B) TESMEC USA M5 Trencher

Figure 24 Skid steer with trenching attachment.
Source: Image Provided as Courtesy of John Deere

2.2.0 Loading Excavated Soil

Regardless of the type of project, if the material is to be moved from one area to another, it must be loaded, hauled, and dumped. The equipment that transports the earth must return to the loading point for another load. Selection of the right type and size of equipment to do these basic operations is important for efficiency and cost. Usually, there is more than one way to perform the work, but the most efficient and cost-effective method should be selected.

Loading is the first step in the earthmoving cycle. Loading is performed as a basic function, such as loading material from a stockpile or spoils, or as a secondary function when some other work is being performed. This could be some type of excavation, such as trenching or cutting with a scraper for cut and fill operations.

The specific job and the type of material encountered has a great influence on both the selection of equipment and the method used.

2.2.1 Selection of Loading Equipment

Selection of loading equipment depends on factors such as cycle time, type of material, and equipment efficiency. For operations that move large amounts of material, the type of hauling equipment must also be considered when choosing the loading unit. Another important factor is the overall cost of the operation, which includes the hourly cost of using a certain type of equipment.

The options available for mobile loading equipment include the following:

- Front-end loaders (and variations)
- Excavators
- Scrapers
- Shovels

Front-end loaders can be used for loading from stockpiles, **windrows**, and spoils piles. They can also be used to excavate and load in the same operation. Configurations include wheel and track loaders. The size range and tight turning radiuses of front-end loaders make them well suited for loading operations.

Windrows: Long, straight piles of placed material for the purpose of mixing or scraping up.

Excavators come in many sizes and configurations. The term *excavator* includes backhoes and telescoping excavators. Although they are primarily used for excavation of material below the grade or into banks, they can also load material into dump trucks and other haul equipment. When the job requires only one operation, such as clearing or excavating ditches, an excavator can be used to do the excavation and load the truck.

Scrapers are special pieces of equipment that typically are used where large quantities of earth need to be moved a long distance. They are built to pick up the material through a scraping action, load it into the bowl, and transport it to the dump site. Scrapers also can be used as haul units, with other pieces of equipment loading material into the scraper bowl.

Shovels are mainly used to excavate material in pits and quarries. Because they are track mounted, they do not move very quickly. However, shovels can be manufactured in large sizes that can dig and load large quantities of material in one cycle.

Other types of loading equipment include conveyors, overhead buckets, and cribs.

2.2.2 Methods of Loading

Equipment unit size and maneuverability establish the method of loading. Some equipment can load as well as excavate. Loading consists of two basic methods: loading from piled material and loading from an excavation.

When loading from stockpiles and spoils, the front-end loader is the primary piece of equipment used. It is able to maneuver well and work with many different configurations of haul units. Loaders usually follow an I- or Y-pattern when loading from a bank of material. The configuration used depends on the amount of available work area. Generally, if the loader has to travel more than one and a half wheel revolutions, the operation is inefficient. *Figure 25* shows a front-end loader working against a bank while loading trucks using the Y-pattern.

Excavators are most efficient if they remain stationary while they work. Their method of loading is limited to loading directly from the excavation or first dumping the material in a pile next to their position and then loading from the pile into the haul unit. This method is not suitable if the area is too narrow to permit the excavator and haul unit access to the site at the same time.

2.3.0 Hauling Excavated Soil

Hauling is the movement of material from one point to another. This movement may cover a short distance within the construction site or a long haul to dispose of material, such as that from clearing and grubbing operations. Hauling is normally broken into the following two types:

- Over-the-road hauling is hauling over public highways using a haul unit such as a dump truck.

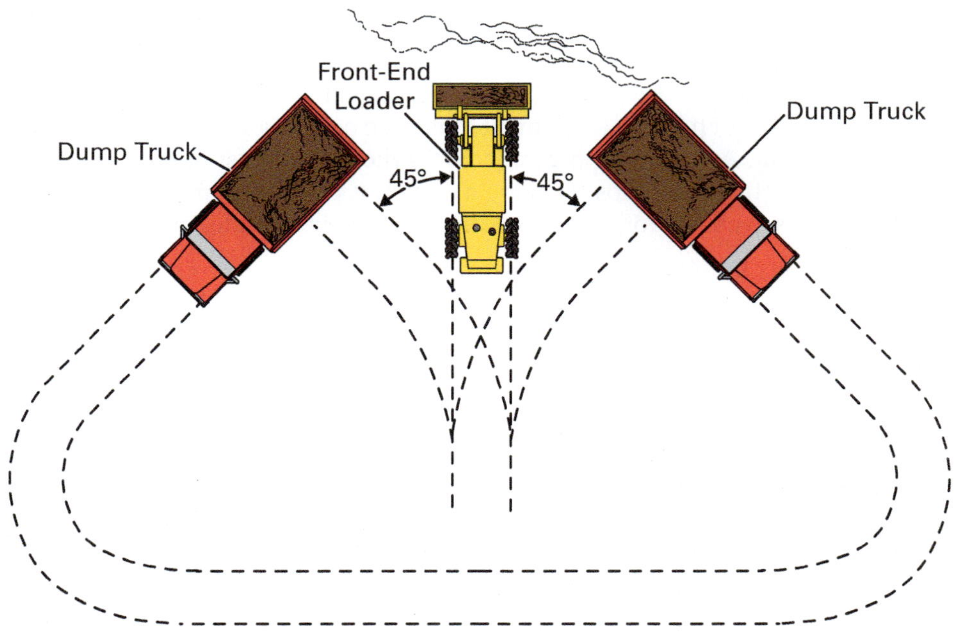

Figure 25 Loading dump trucks using the Y-pattern.

- Off-road hauling is done within a jobsite and not on public highways. Equipment used for off-road hauling includes large trucks that are not designed to travel on public highways, as well as scrapers, front-end loaders, and regular trucks.

2.3.1 Selection of Hauling Equipment

Equipment used for the two types of hauling is very different in design and construction. The specific operation determines which type to use.

Over-the-road haul units have distinct limitations because highways are designed for specific load conditions. Truck load limits are established according to their load-bearing capacity. These units are also restricted in width and height. State or federal laws enforce these limitations.

Off-road hauling units may include scrapers, front-end loaders, and trucks of various types. Off-road equipment is usually wider, higher, and therefore heavier than over-the-road dump trucks. This allows for an overall increase in the size and capacity of the equipment. For example, a tandem-axle over-the-road dump truck can carry a load of about 12 cubic yards of material; a large scraper can carry up to 44 cubic yards.

Selecting the most appropriate haul unit involves several considerations. The first is whether any portion of the haul will be over public highways. If so, equipment-use restrictions must be checked. The person operating the equipment is responsible. If there is a violation, the operator gets the citation. Also, the haul route needs to be checked for potential problems such as bridge height, width, and weight limitations. Hauling over narrow, curving roads with high crowns can be hazardous for larger trucks. High road crowns that can tilt the truck may limit the allowable heaping when the unit is loaded.

Individual job conditions dictate off-the-road types of hauling equipment. The following three configurations may be considered:

- *End-dump (rear-dump)* — These types of trucks should be used if material consists of rock fragments or masses of slate or shale that are too large for other units to unload, or if dumping is to be done over the edge of a waste bank or fill. Rear-dumps provide maximum flexibility for a variety of job conditions. They can be straight-body or articulated trucks.
- *Bottom- or side-dump* — These units are good for transporting free-flowing materials over reasonably level haul routes that permit a high travel speed. Bottom trucks can be used for dumping in windrows over a wide area. Side dump trucks are more suited for road-widening applications.

- *Scrapers* — Scrapers can haul a large amount of material at relatively high speeds for short distances. They have an added advantage of being able to load the material directly by cutting or scraping a thin layer of soil while moving forward at a slow speed. When the bowl is full, the scraper can travel at a higher speed to the dump site and spread the material in a thin layer. Because of their large size, scrapers are not normally used in small or confined areas. They are good for cut and fill work on highways and airports, as well as for general contouring of the ground.

2.3.2 Hauling Methods

Hauling is the largest single factor in cycle times for earthmoving operations. Assuming that the material is transported by some type of haul unit instead of a cable or conveyor system, synchronizing the hauling operation is the main concern. This includes the timing of the hauling as well as the traffic pattern. If the job requires a considerable number of haul units, a lot of time can be lost if work operations are not timed properly. Also, for jobs requiring the placement of large amounts of cut and fill or the removal of excavated material, it may be necessary to set up special haul roads and lay out traffic patterns.

A traffic control supervisor may be required for operations using a substantial number of haul units. The supervisor controls operating conditions at both the loading and dumping sites and keeps track of the haul units in transit.

2.4.0 Dumping Excavated Soil

Material may be dumped either to build up an area to make an embankment or to dispose of spoils. The dumping must be closely controlled when building up an embankment or when dumping at a spoils site.

2.4.1 Dump Sites

Spoils areas are used for the disposal of surplus excavation materials, materials that are unsuitable for fills, or for storage of materials that will be reused. The location of the spoils area may depend on the type of material and the haul distance involved. Soft materials going to a spoils area may require mixing with firmer soils. Organic debris may be burned or spread out in thin layers. Rocks and boulders may require placement in a specific way to reduce environmental concerns, such as erosion. Dumping of spoils outside the project site may be more convenient, but the cost of each option must be considered.

If the material to be dumped is for filling, backfilling, or making an embankment, these areas are shown on the plans and indicated with the survey and slope stakes. Material is sometimes dumped alongside the area to be filled and then moved into place with dozers or graders. The project engineer or activity supervisor marks these areas with stakes or paint, so the equipment operator knows where to dump each load.

Another type of dump site is a stockpile. Stockpiles are used to store material for later use in the project. A good example of this is the topsoil that is removed at the beginning of a project. Because good topsoil is an expensive **pay item**, it is saved for landscaping and contouring at the end of the project.

Pay item: A defined piece of material or work that the contractor is paid for. Pay items are usually expressed as unit costs.

2.4.2 Dump Patterns

Dump patterns depend on the material being dumped and the type of haul equipment being used. For spoils dumping, the material may be dumped into another excavation, such as an abandoned pit or over an embankment, or it may be dumped in a level, open area. Often, spoils are hauled in a rear-dump truck. Depending on the amount of spoils, dozers or front-end loaders may be needed at the dump site to move the material around and blend it in with the surrounding terrain.

The haul equipment for fill material usually includes bottom- or rear-dump trucks or scrapers. When backfilling, the material is dumped directly into the excavation or beside the excavation, and then another piece of equipment, such as a loader or dozer, is used to place the material.

Embankment construction fill material can be placed by either trucks or scrapers. *Figure 26* shows several different dumping patterns.

A dumped fill pattern for level ground is indicated in *Figure 26 (A)*. The material is placed in many individual piles, so it is easier to spread. This method reduces the amount of work required to move material over longer distances.

Figure 26 (B) and *Figure 26 (C)* show dumping alternating coarse and fine material along an embankment to provide for thorough mixing of the material. In *Figure 26 (B)*, the material is dumped as the truck is moving, in order to spread the material. The method shown in *Figure 26 (B)* is used for consolidated materials that will blend well. Note the safety berm behind the truck in *Figure 26 (C)*. Once the loads have been dumped, a dozer or motor grader is used to blend the coarse and fine material together. Another method used to place two different types of material prior to blending is to completely spread one type of material before the second type is dumped and spread on top. This method is used for

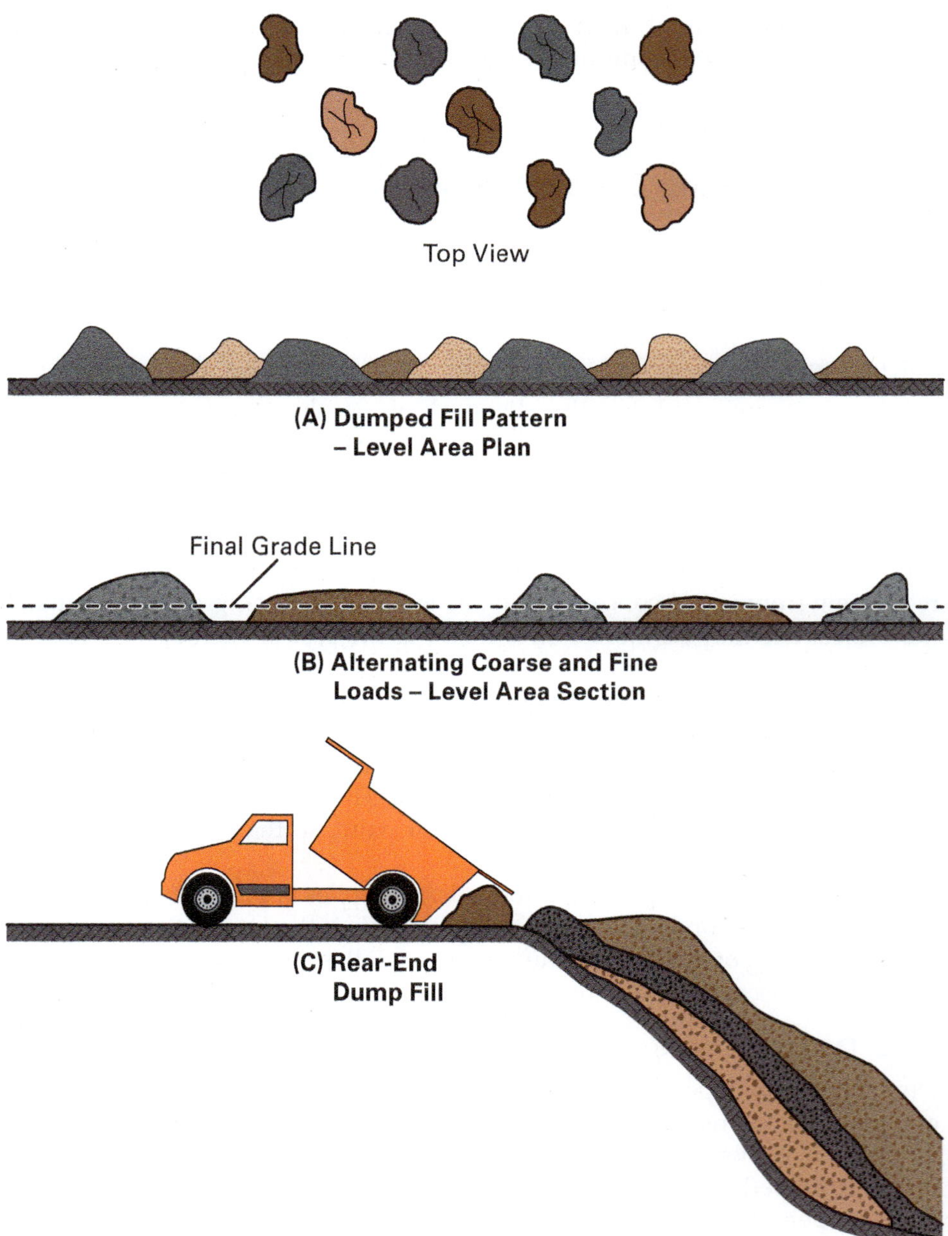

Figure 26 Dumping patterns.

materials such as clay and sand that do not blend easily. A scarifier is then used to mix the coarse and fine material before compacting.

In highway construction, material used to build an embankment is usually dumped in layers. This can be accomplished with scrapers, which can spread the material as they dump, or with trucks, which dump in piles or windrows. These piles and windrows are then leveled out with a dozer or motor grader.

2.4.3 Return Run

The success or failure of an earthmoving operation is based in part on how quickly a truck can safely return to the excavation site for another load, or **backhauling**. Equipment operators can help make the operation a success by the following behaviors:

Backhauling: The return trip of a piece of equipment after it has completed dumping its load.

- Driving cautiously at all times.
- Staying in line and following the directions of the traffic supervisor.
- Thinking ahead about how to maneuver the machine when approaching the loading site for the next cycle.

Always use the prescribed return route from the dump site. It is usually set based on the distance, traffic, and number of turns between the excavation area and the dump site. If haul roads are used, the pattern for the return trip is set and must be followed by all operators to avoid disrupting traffic flow. If there are no haul roads, the truck must be driven through the construction site. In this case, the project engineer should establish a pattern for hauling and backhauling. Operators follow markers laid down for this purpose. Ground personnel guide the equipment out of the area once the dump is made.

For scraper operations, the dozer is positioned where the scraper can come up to it (*Figure 27*). A good scraper operator, working with the dozer operator, can perform this operation smoothly and quickly. Otherwise, much time is lost in starting, stopping, and lining up the two machines.

Figure 27 Dozer pushing a scraper.
Source: Scott Colclough

Dump truck operators can help loader operators by getting into position quickly and keeping an eye on the loader. If a spotting marker has been placed, the operator should know where it is and how to position in relation to it.

2.5.0 Fill, Backfill, and Embankment Construction

Excavated materials can often be used as fill or backfill, or to build an embankment. However, the material may have to be subjected to one or more soil tests before it can be used as backfill or embankment. The number of tests required depends on the intended use of the material. Before filling, backfilling, or creating an embankment, always check federal and local regulations for the types of materials and the materials content allowed.

2.5.1 Fill Material

Some fill may be material stockpiled from excavations within the site. Other fill may be brought into the site from another area as needed. Fill material can be classified as three types: common borrow fill, waste fill, and select fill. In general, all fill should be kept free of organic matter if any future use of the filled area is planned.

Common borrow fill meets a pre-established project requirement. It is usually brought onto the construction site and dumped in piles. It is then spread uniformly over an area and compacted according to the job specifications.

Common borrow fill can be used to build up an area where little or no compaction is required. The fill is simply dumped into areas such as embankments, scraped areas, or around some types of buildings. When fill is dumped into a deep fill area and allowed to roll down the slope, it tends to separate because the larger fragments roll to the bottom and the fine material stays closer to the top. Dumped fills should be kept free of tree stumps, organic matter, trash, and sod if any future use of the filled area is planned. An exception to this process is the special case of sanitary landfills, where garbage and trash are spread several feet thick over the ground surface, covered with additional clean earth, and left to compost.

Waste fill is recycled, treated material approved for use in certain landfill areas. This could include embankment, landscaped areas, or backfill around some types of structures. Compaction requirements are based on where the material is being placed. When rock is used, it should be spread over the area rather than piled at a single point.

Select fill is material with specific properties that can be obtained only by selective excavation or manufacture. This category includes selected gravel or sand, selected rock used for stabilizing slopes, and dumped riprap. Again, the equipment used to place fill material includes the dump truck, scraper, dozer, and sometimes the front-end loader. Spreading of the material is accomplished by the scraper, dozer, and motor grader. Compaction can be done by wheel or tractor compaction or by a mechanical roller of some type. Select fill is usually more expensive than other types of fill and must be used, placed, and compacted as specified by the project plans.

2.5.2 Backfill

Consolidation: To become firm by compacting soil particles so they are closer together.

Backfill is the returning of excavated material to an excavation site. Before excavations, the ground tends to be stable because the soil has settled. Settling is also called **consolidation**. It is important to backfill and compact excavations so that a minimal amount of future settling occurs at the site. This may seem simple, but because of compaction issues, some backfills are poorly made and cause damage to the buried structure and the surrounding area.

The main concern with backfilling is choosing the best method for compacting the fill material. Although the specifications for some work require that all excavated material be disposed of and specifically selected materials be hauled in for backfilling, most contracts allow the use of excavated or dumped fill for backfill. If dumped material is used, compaction criteria must be varied to suit the soil encountered. While poorly compacted material does ultimately settle, over-compacted material does not settle at the same rate or under the same loads as adjacent natural soils, and this effect can be just as damaging.

For backfilling around small structures and buried utilities (*Figure 28*), a backhoe is the ideal piece of equipment. The backhoe operator needs to understand the process of compacting backfill material to place material at points where the workers need it to build up the backfill. Compacting too much in one area can cause pressure to build up against the structure, bending it out of alignment, or even breaking it.

2.5.3 Embankment Construction

Embankments are fills that are placed for specific purposes under carefully controlled conditions. Embankments usually consist of soils of a prescribed particle size placed with careful selection, compaction, moisture control, and mixing.

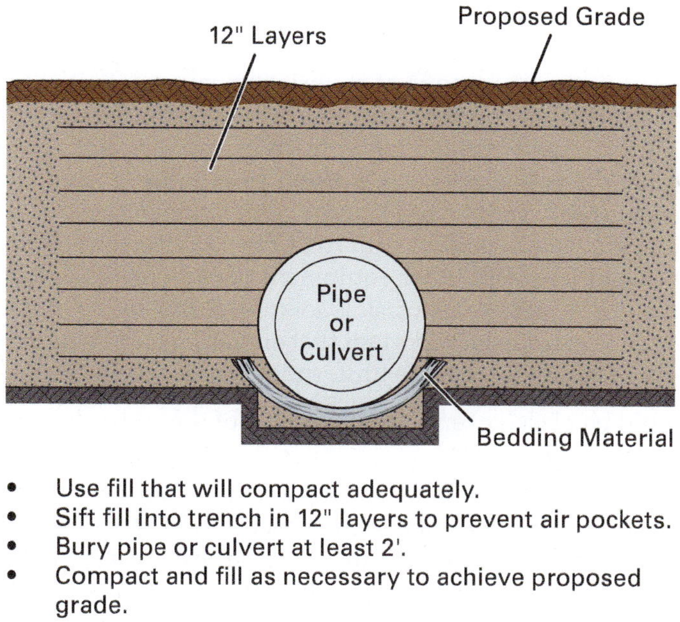

Figure 28 Backfill.

Embankments are used for highways and roads, earth dams, levees, canals, and runways and are classified as follows according to method of compaction:

- *Equipment-compacted fill* — This is generally a fill, often of select material, which is compacted by the wheels of the haul units. Moisture control may be required. This type of embankment construction normally does not have a compaction requirement.
- *Rolled-earth fill* — This type of compaction is used for roads, building foundations, dams, canals, and other sites where highly **impermeable** fills are required. The degree of compaction is normally specified for rolled-earth fills. The design is based on the properties of the soil to be used. *Figure 29* shows a rolled-earth fill.

Impermeable: The property of a material that does not allow movement of a fluid through it; for example, an impermeable soil will not allow liquid water to pass through it.

The embankment may be placed on the natural ground, or it may require excavation to another layer under the ground. Embankment material is spread and leveled in layers or **lifts**. The actual orientation and placement of lifts may vary from those shown in *Figure 29*, according to purpose, location, materials used, and the specifications of the project.

Lifts: When referring to soil compaction during construction, layers of soil that are placed in a fill area or on an embankment before compaction.

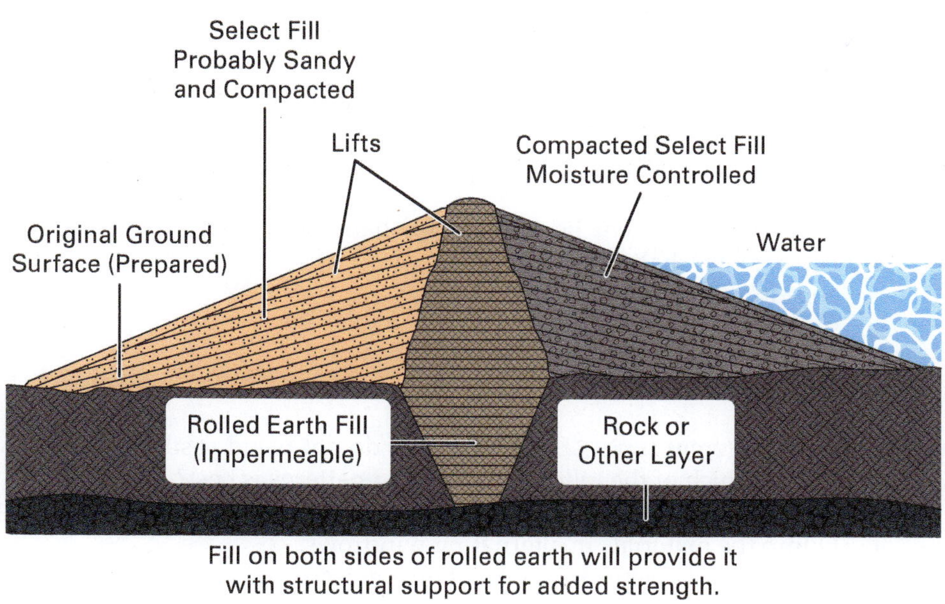

Figure 29 Rolled earthen dam.

The equipment and procedures involved in dumping and spreading the embankment material depend largely on the type of equipment the contractor has on the project and the type of material being used. Embankments are built in lifts, one on top of the other, for the full width unless otherwise specified. The contractor must have enough spreading, mixing, and compacting equipment and use the procedures necessary to meet moisture-density requirements and obtain uniform, well-mixed layers of embankment.

The scrapers usually follow patterns in picking up and depositing material that do not interfere with other equipment operations. The work should progress at a uniform rate without slowdowns caused by improper direction of equipment. It is extremely important that embankments be built uniformly. The following work activities ensure uniformity:

- Fill should be placed evenly from one side of the embankment to the other.
- Each lift should be spread to a uniform thickness that does not exceed the specified thickness before compaction. When excavated material consists mostly of rock fragments, the specification may allow the material to be placed in the embankment in layers not exceeding the thickness of the average size of the larger rocks. Each of these rock lifts must be leveled and smoothed with finer material, but the final rock lift must be several feet below the finish grade.
- Each lift must be compacted before the next lift is placed. Double dumping, or placing two lifts without compaction, is not allowed.
- Where large chunks of material are placed, extra compaction and blading are needed to spread the material uniformly.

Under embankment areas where fill is to be placed, the original ground must be broken up by scarifying it to a specified depth and then re-compacting it.

2.0.0 Section Review

1. Bulk wide-area excavations occur mainly for which of the following?
   a. Underground mining operations
   b. Strip mining
   c. Tunnel construction
   d. Building foundation preparation

2. The mobile equipment suitable for loading soil at excavations includes _____.
   a. mobile cranes
   b. front-end loaders
   c. stiff-leg derricks
   d. rigid-frame dump trucks

3. The type of equipment best suited for hauling large rock fragments or masses of slate/shale is the _____.
   a. end-dump truck
   b. scraper
   c. bottom-dump truck
   d. scarifier

4. To blend different types of materials that do not blend easily, such as clay and sand, which of the following dumping patterns is best?
   a. Dump the materials in alternating piles on a level area.
   b. Dump the materials in alternating windrows.
   c. Dump and then spread a layer of one material, and then do the same for the other.
   d. Dump the materials in alternating layers.

5. What kind of fill tends to separate by particle sizes when dumped down a steep slope into a deep depression?
   a. Select fill
   b. Common borrow fill
   c. Waste fill
   d. Consolidated fill

3.0.0 Stabilizing Soils

Objective

Describe the different methods used in stabilizing soils.

a. Identify the different types of binders.
b. Explain how to apply binders to soil.
c. Explain soil compacting.

Performance Task

3. Draw a plan for basic earthmoving operations:
   - Clearing and grubbing
   - Excavating the foundation
   - Constructing embankments
   - Backfilling
   - Compacting

Soil stabilization is the process of preparing subbase soils to provide a higher load-bearing capacity. The process involves pulverizing and mixing the soil thoroughly, often with binders, so that after compaction and curing, the soil is denser, more stable, and stronger. The general requirement is to stabilize the area well beyond the locations where heavy loads occur. For a road subgrade, this means stabilizing past the road boundaries. *Figure 30* shows the subgrade stabilization limits.

A soil's stability and load-bearing capacity depend on its cohesion and **shear strength**. The higher a soil's shear strength and cohesion, the greater its capacity to bear loads. Chemical binders are used to stabilize existing soils to increase shear strength and cohesion.

Shear strength: A property of solid materials that measures its resistance to deforming along planes, called shearing, like spreading a deck of cards. In soil mechanics, it is measured by the angle of internal friction, which relates to the maximum steepness of a soil embankment before it slides due to gravity.

Note: Shaping the subgrade in this manner is termed "boxing out."

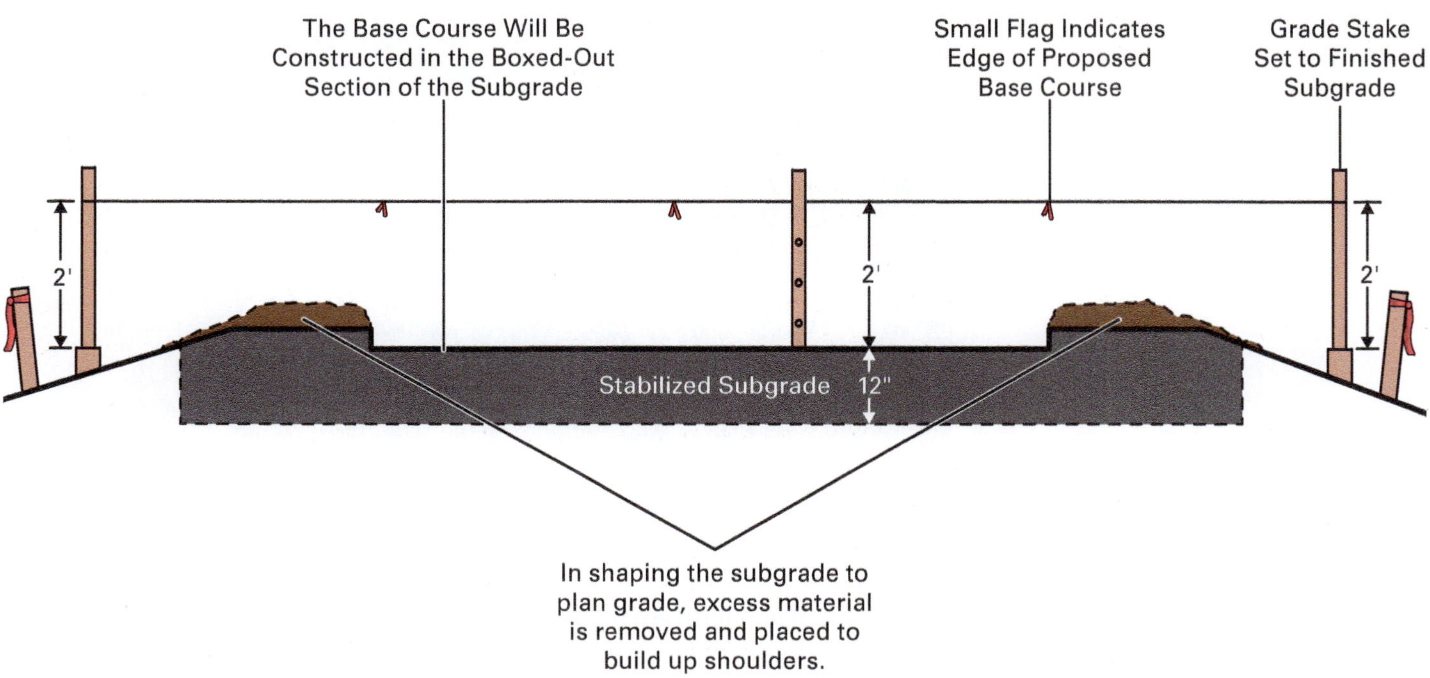

Figure 30 Subgrade stabilization limits.

The first three steps in stabilization are determining soil properties, calculating the amount and type of binder required, and determining the degree of compaction needed. These steps all result from analyzing the soil or material to be stabilized. Analysis includes factors such as gradation, moisture content, density, shear strength, and cohesion.

Stabilization of soils has many advantages, including the following:

- Allows use of in-place soil
- Eliminates need for excavating
- Eliminates the cost of replacing soil
- Reduces cost by mixing in place
- Reduces construction time

The stabilization process includes the following sequence of steps:

- Choosing a binder
- Preparing the subgrade
- Spreading the binder
- Determining the logistics of the stabilization job
- Stabilizing
- Compacting
- Checking quality

A soil stabilizer machine cuts, mixes, and pulverizes native, in-place soils or select material to modify and stabilize the soil for a strong base. Stabilizing agents may be mixed into the soil by the machine during this process. Stabilizers use blades on rotors or teeth on mandrels for cutting. Many have automatic depth control. Some models have a mixing chamber, which provides for consistent blending of materials. *Figure 31* shows a large stabilizer.

Figure 31 Large soil stabilizer.
Source: Jennyt/Shutterstock

To carry out stabilization by mechanical methods, the entire area must be mixed thoroughly to the designated limits. The simplest type of equipment is a traveling mixing plant with a single-rotor mixer design. This machine can move rapidly and can mix an area in several passes.

The cutting depth for stabilization is usually 12" to 15", and the cutting width is typically 8'. The production rate of the stabilizer can be calculated with these numbers and the travel speed, or it can be looked up on production charts.

Stabilizer features may include the following:

- Mechanical rotor drive
- Microprocessor control of major machine systems
- Four steering modes with automatic rear wheel alignment
- Choice of rotors
- Rollover protection
- Liquid additive and/or water spray systems in accordance with local environmental regulations

The machine should be able to cut up or down, because individual operations determine which type of cutting will result in more complete mixing. The ability of the machine to accomplish the specified mixing depth is critical to permit one-pass operation. Sometimes it is desirable to make an initial pass to establish proper gradation of the soil, followed by application of the binder, and then make a second pass with the soil stabilizer. Some stabilizer models offer the ability to reverse direction without turning around—a time-saving feature.

3.1.0 Choosing a Binder

Binders are selected to complement the particular soil being treated. The type and amount are dictated by the following:

- The soil type
- The size or gradation of soil particles
- Soil moisture content
- Load-bearing specifications of the project

Binders may be dry, such as lime or cement, or wet, such as lime slurry. Binders can be spread by dump trucks, spreader trucks, or tanker trucks ahead of a soil stabilizer. A soil stabilizer has a flat-bottomed mixing chamber attachment. The soil is picked up and can be mixed, blended, or pulverized in the chamber. The soil is then put back down and flattened on the roadbed or surface. A soil stabilizer can stabilize a road base to approximately a 12" depth in a single pass. If such machines are used, binders can also be added directly to the soil in the stabilizer mixing chamber.

The method of adding binders depends on the type of stabilizer equipment used, the size of the project, and the types of additives. The binder must be thoroughly mixed with the base material and the mixture pulverized for uniform color and size.

3.1.1 Cement

Cement is a fine, powdered hydraulic binder made from calcified lime compounds mixed with silica, alumina, and iron oxide. When mixed with water, cement sets and hardens in air or under water. Soil cement is made by mixing dry portland cement and water with the soil. The soil is compacted and paved.

Cement binder provides the highest load-bearing capacity of all binders. It can be used in gravel, sand, or reclaimed asphalt. Silt and clay may also be stabilized with cement.

3.1.2 Lime

Lime is especially effective as a binder in clay-bearing soils and aggregates because it reacts both chemically and physically with the soil. Lime used for stabilization includes quicklime (calcium oxide) and hydrated lime (calcium hydroxide). Both are forms of heated limestone.

Lime is applied as a dry powder, as a slurry, or in pellets. Lime reacts chemically with clay soils to produce a cementing action. It reacts favorably with soils with a higher **plasticity index**. Lime treatment is used for subbase stabilization, base stabilization, and lime modification.

Plasticity index: The range of percent water content in which a soil displays plastic or pliable properties. Soils with high plasticity indexes are usually clays; soils with lower indexes are usually silt, sand, or gravel.

3.1.3 Fly Ash

If lime is used on low plasticity-index soil, a second additive may be used to produce the required reaction to the lime. This is called fly-ash treatment. Fly ash is a residue recovered from smoke produced by coal-fired burners. Fly ash consists of minute spherical particles of glass, crystalline matter, and carbon. Its chemical analysis resembles portland cement without lime. It reacts chemically with lime to enhance the strength of the soil, giving it cement-like properties. Fly ash combined with lime also produces an excellent road base when mixed with aggregate.

Admixtures

Mineral admixtures are used primarily to reduce the amount of cement used in making concrete. These admixtures are either hydraulic or pozzolanic. Pozzolanic materials contain reactive silicates or aluminosilicates. When finely ground and used in a concrete mix, they react with other materials in portland cement to produce cement-like compounds. The Romans used a mixture of hydraulic lime and fine volcanic ash (a pozzolan) in the Appian Way, which connected Rome with southeastern Italy. The long road, portions of which are still in use today, was started in 312 BC. Fly ash, a byproduct of combustion of ground coal in coal-burning electric utilities, is a frequently used pozzolanic material.

3.1.4 Calcium Chloride

Calcium chloride is another material used as a binder. It increases soil density like other binders and is particularly valuable in colder climates. Calcium chloride lowers the freezing point of water. The roads that use it for stabilization of the base material are better protected against frost damage. The use of calcium chloride for stabilization may require less compaction. It can be applied dry or in a solution.

3.1.5 Polymers

Widespread use of **polymers** as soil stabilizers is a fairly recent occurrence. A polymer and one or more other materials are bonded together by chemical reaction. In soil stabilization, polymers bind with soil particles to create a durable surface that permits motorized traffic. Application requirements vary with each product, so it is important to follow the manufacturer's instructions. Most products are applied as a viscous liquid that dries quickly (in a matter of minutes to hours, depending on the weather). Some polymers can be used in a weaker solution to suppress dust for erosion control.

Polymers: Natural and synthetic materials made of long molecules containing many repeating chemical subunits. Familiar polymers include plastics and natural rubber. Soil stabilization polymers include acrylic materials.

3.2.0 Applying Binders to Soil

To prepare the subgrade, topsoil and large stones are removed from the proposed site and the subgrade is carefully shaped to grade. The desired site profile and the mix uniformity depend on accurate grading. The grading is done by a motor grader or by a fine grade trimmer. A trimmer creates an accurate profile and can accomplish the job in one pass.

When grading is complete, scarifying or disc harrowing to the depth of stabilization is sometimes used. However, a stabilizing machine capable of mixing the binder and pulverizing the soil in one operation can eliminate this step.

Dry binders must be evenly distributed over the area to be stabilized to assure that the specified soil-to-binder ratios are met. A box-type spreader attached to a dump truck or a rear-mounted spreader on a pneumatic truck is used for this purpose. The spreader meters out the material as the load is discharged ahead of the soil stabilizer.

Lime slurry or another liquid binder is best distributed by tanker truck, which spreads the product to pre-calculated depths as called for in the specifications.

3.2.1 Logistics

When planning the stabilization job, logistics play an important part. Planning considerations include the following:

- *Atmospheric conditions* — The temperature should be mild, with little or no precipitation expected. The initial mixing process should not be done in heavy rain. After the initial course is compacted and set, moderate rain is not a problem.
- *Binder material supply* — Binder material must be available as needed to avoid delays in application.
- *Water supply* — When cement is being used for a binder, it is critical to maintain the optimum moisture content. Equipment must be available to supply water for the mix as needed. Lime is usually added to soils with high moisture, so water is not an issue for mixing. However, enough water should be available to control dust when using lime.
- *Compaction equipment* — Equipment should be available for compaction to begin immediately after the stabilizer finishes the mixing and grading process.
- *Trimmer equipment* — After compaction, the surface should be trimmed to accurate grade and crown specifications.

3.3.0 Compacting the Soil

The stabilized area must be compacted after the stabilizer has finished. Only adequate compaction produces the required density. If cement is used, the compaction must be done immediately after mixing before the cement in the mixture begins to set. A cement mixture also requires that a vibratory roller compaction machine be used to obtain the needed results. *Figure 32* shows a vibratory roller.

When compacting a lime-stabilized area, the compaction must be done when the moisture content has dropped to a point within the specified range. This may occur immediately or after a short delay.

The number of passes made by the compactor depends on the amount of compaction to be achieved. This will, in turn, depend on the material moisture content, the layer thickness, the compactor type, and the degree of compaction called for in the specifications. With proper compaction, a final curing stage may be eliminated before the next course is applied.

Figure 32 Vibratory roller.

Compaction increases the density of soil to create a stable ground surface. It involves pressing soil particles together to force out air and water from the spaces between the particles. Compaction reduces the risk of future settling, so it provides a solid surface on which to build. After soil is disturbed, compaction occurs naturally over time as a result of wetting, drying, freezing, thawing, groundwater movement, and the weight of the top layer of soil.

In construction, it is necessary to speed the compaction process with equipment that imitates nature. Various types of equipment use pressure, kneading, vibration, and impaction, or a combination of these methods to compact the soil. The equipment provides the forces needed to quickly settle soil and allow building to begin.

Each layer of the roadbed must meet design specifications to ensure that the final product stands up under projected loads and meets smoothness requirements. Each stabilized layer must be checked for the following:

- Compaction or density
- Thickness
- Mix uniformity
- Gradation
- Load-bearing capacity
- Moisture content
- Binder content

3.0.0 Section Review

1. According to the definition of soil shear strength, soils with higher shear strengths _____.
   a. compact easier
   b. are less stable
   c. cannot form steep-sided slopes
   d. do not slide easily on an embankment

2. Which of the following statements about the logistics of soil compaction is true?
   a. Soil compaction works best with long, gentle precipitation.
   b. Soil compaction also shapes the grade to its specified profile and crown.
   c. Compaction occurs before soil stabilization to provide a firm base.
   d. Surface trimming shapes the compacted surface to accurate grade and crown specifications.

3. Which of the following is *not* a factor in determining the number of soil compaction passes required?
   a. Amount of compaction required
   b. Final layer thickness
   c. Initial soil density
   d. Moisture content

Module 22201 Review Questions

1. The solid earth layer that underlies the ground soil in the earth's crust is _____.
   a. the mantle
   b. common fill
   c. bedrock
   d. molten rock

2. The unique address given to a site that allows police, fire, and other emergency vehicles to find the location easily is called a(n) _____.
   a. fire road
   b. 911 address
   c. police locator
   d. emergency access

3. The soil type that aids cohesiveness in other soil types is _____.
   a. colloids
   b. silt
   c. sand
   d. gravel

4. During excavation and fill activities, slope control is handled with grade stakes that are placed by _____.
   a. the survey crew
   b. a competent person
   c. the inspector
   d. the contractor

5. When preparing a jobsite for excavation or construction, the removal of objects below the original ground is called _____.
   a. clearing
   b. stripping
   c. grubbing
   d. scalping

6. The EPA requires that operators of construction sites that are one acre or larger establish plans to reduce sedimentation, pollution, and _____.
   a. traffic
   b. riprap
   c. erosion
   d. dust

7. Open pits or holes constructed to collect groundwater that are then pumped out mechanically are called _____.
   a. sumps
   b. borings
   c. holding ponds
   d. drainage ditches

8. When working on a job, the first priority is to perform the job _____.
   a. efficiently
   b. quickly
   c. safely
   d. well

9. If a foul-smelling fluid seeping out of the ground is noticed while walking the new site before starting work, the operator should know that it _____.
   a. could be contamination
   b. is probably an underground spring
   c. needs to be filled before work begins
   d. needs to be shown on the site plans

10. What should an operator know or do if the site plans show that a particularly large tree on the building site needs to be preserved?
    a. Find the tree before starting any work.
    b. Try to look for the tree while working.
    c. Avoid working around any large tree.
    d. Ignore it.

11. A string of orange flags in a construction site indicates the presence of an underground _____.
    a. irrigation pipe
    b. electric power line
    c. fiber-optic communications cable
    d. natural gas pipe

12. Cycle time is defined as the length of time it takes to perform a repetitive operation, such as the time it takes for a piece of equipment to _____.
    a. load, haul, and dump its load and return for another load
    b. return to the excavation site after it has dumped its load
    c. haul its load to the dump site and return for another load
    d. load fill in all of the vehicles needed for the day's work

13. A temporary opening in which something, such as a pipe or box culvert, is placed and then covered is a _____.
    a. channel
    b. trench
    c. ditch
    d. pit

14. Backhoes, excavators, and trenching machines are the three main classes of equipment used in _____.
    a. trenching
    b. excavating
    c. slope control
    d. erosion control

15. When loading from stockpiles or spoils with a front-end loader, the most efficient movement pattern of those listed is the _____.
    a. M-pattern
    b. oval pattern
    c. Y-pattern
    d. triangular pattern

16. The type of dump site used to store material for later use is called a(n) _____.
    a. spoils area
    b. borrow fill
    c. stockpile
    d. landfill

17. Waste fill is recycled, treated material approved for use in certain _____.
    a. landscaping projects
    b. outdoor products
    c. playgrounds
    d. landfill areas

18. The binder that provides the highest load-bearing capacity is _____.
    a. lime
    b. fly ash
    c. cement
    d. crushed rock

19. Which type of equipment is often used to mix the binders and graded soil before compaction?
    a. Scarifiers
    b. Stabilizers
    c. Motor graders
    d. Trimmers

20. Compaction of a lime-stabilized area requires _____.
    a. that compaction be performed immediately
    b. that a soil stabilizer be used
    c. that the moisture content be reduced to within the specified range
    d. the addition of water to the soil

Answers to odd-numbered review questions are found in the Module Review Answer Key at the back of this book.

Answers to Section Review Questions

Answer	Section	Objective
Section One		
1. c	1.0.0	1a
2. b	1.2.0	1b
3. a	1.3.0	1c
4. d	1.4.0	1d
5. b	1.5.1	1e
6. c	1.6.1	1f
7. b	1.7.3	1g
Section Two		
1. b	2.1.5	2a
2. b	2.2.1	2b
3. a	2.3.1	2c
4. c	2.4.2	2d
5. b	2.5.1	2e
Section Three		
1. d	3.0.0	3a
2. d	3.2.1	3b
3. c	3.3.0	3c

MODULE 22109

Introduction to GPS/GNSS

Objectives

Successful completion of this module prepares you to do the following:

1. Describe the types of equipment used for finish grading.
   a. Describe how a Global Positioning System (GPS) and a robotic total station are used in finish grading.
   b. Identify equipment used in finish grading.
   c. Describe how laser instruments are used in finish grading.
2. Describe the process for site planning for use with GNSS.
   a. Identify equipment used in site layout for GNSS in 2D and 3D site layout.
   b. Describe how equipment uses GNSS for site contouring.
3. Describe how GPS/GNSS systems help autonomous equipment operate on a construction site.
   a. Identify GPS/GNSS support for autonomous equipment.
   b. Describe the benefits of operating equipment autonomously.

Performance Task

Under supervision, you should be able to do the following:

1. Establish a finish grade according to instructions.

Overview

The Global Positioning System (GPS) and the Global Navigation Satellite System (GNSS) are used extensively with earthmoving equipment. Using GPS/GNSS location services allows for precise excavation and placement of materials on today's construction sites. Special equipment-mounted receivers provide the information necessary to accurately locate heavy equipment on the site. This module will help you understand how GPS/GNSS systems are used on earthmoving equipment to mimic the design of ground-contouring plans.

Digital Resources for Heavy Equipment Operations

SCAN ME

Scan this code using the camera on your phone or mobile device to view the digital resources related to this craft.

1.0.0 Equipment Types

Performance Task

1. Establish a finish grade according to instructions.

Objective

Describe the types of equipment used for finish grading.

a. Describe how a Global Positioning System (GPS) and a robotic total station are used in finish grading.
b. Identify equipment used in finish grading.
c. Describe how laser instruments are used in finish grading.

Backslope: The area between the bottom of a ditch adjacent to a roadbed and the top of the cut where it blends into the existing, natural grade. If it is blending, it should not be a cut or fill.

Most construction sites use many different types of heavy equipment to perform final grading. Selection of the most appropriate piece of equipment depends on the work to be performed. A **backslope** can be finished with a motor grader up to a certain angle, but an excavator may be required if the motor grader is unable to get close enough to the backslope. An excavator would not be used to perform subgrade finishing on a road because the finish would be uneven, and the process would be inefficient. To perform finishing work effectively, an operator must know and understand the characteristics and limitations of each type of heavy equipment. Commonly used equipment includes the following:

- Motor grader
- Dozer
- Scraper
- Excavator

Ancient Roads

About 2,400 years ago, the Romans built a road known as the Appian Way. It was named after Appius Claudius, the official who undertook its construction. When it was completed in 244 BC, it was 362 miles (583 kilometers) long and averaged 20 feet (6.10 meters) wide along its length. The road had a stone foundation and was paved with lava blocks that fit tightly together. Like modern roads, it was crowned to allow water to run off. Parts of the road have been preserved and are used today by automobiles. Given that it was done entirely with human and animal labor, it is a marvelous feat of engineering.

1.1.0 Using GPS/GNSS and Robotic Total Stations

Robotic total station: A machine control system that uses infrared signals reflected by a target mounted on the grading equipment. This provides data to the machine's onboard computer about how to adjust the blade.

GLONASS: A Russian-owned Global Navigation Satellite System operated by the Ministry of Defense of the Russian Federation.

GNSS: An acronym for Global Navigation Satellite System. GNSS is the generic term used to describe a locating or grade control system that uses signals from GPS, GLONASS, BeiDou, or Japanese QZSS satellites.

In the past, site layout and grading were completed manually using setting stakes and strings. This method has been almost completely replaced—especially on larger projects—as technological advances make it possible for grading operations to be partially or fully automated. Today, entire projects can be completed without setting a single grade stake. Two common types of automatic grade control systems are satellite-based systems, and an optical positioning system called a **robotic total station**.

Historically, satellite-based systems have been referred to as GPS. In construction today, GPS is only one of the global satellite navigation systems used for machine control. Satellite systems in use or in development include the following:

- *GPS* — Developed and operated by the United States military and available for use by the public.
- **GLONASS** — A Russian-owned Global Navigation Satellite System operated by the Ministry of Defense of the Russian Federation.
- *Galileo* — A satellite system created by the European Union and the European Space Agency to provide time and location information to the receiver.
- **GNSS** — A satellite-based navigation system that provides geolocation and time information to users anywhere on Earth.

- BeiDou — A satellite system developed by the Peoples Republic of China (PRC), similar to GLONASS and GPS.

Signals from both GPS and GLONASS satellites are commonly used by grade control systems. For that reason, GPS and GLONASS are often referred to as GNSS, which is a more generic term. Some grade control system manufacturers use GNSS in the names of their equipment. For clarity and familiarity, this module uses the acronym *GPS* when describing a satellite-based system.

A robotic total station provides three-dimensional (3D) grade control capabilities. However, this type of system uses an electromechanical robot to locate and track the movement of a target mounted on a specific piece of equipment. The robot sends infrared signals to the target, which are reflected back to the robot. The robot then sends data via radio signals to the equipment's onboard computer, where it is used to automatically control the grading equipment.

Satellite-based systems and robotic total stations can complete a job faster and more accurately than conventional staking. The number of grade checkers on a jobsite can be greatly reduced. This saves labor costs. It also results in a safer work environment because fewer people are walking around operating equipment. The major disadvantage of these systems is the initial cost.

1.1.1 GPS/GNSS Constellation and Usage

The Global Navigation Satellite System (GNSS) comprises satellite configurations known as constellations. These constellations consist of satellites (*Figure 1*) that provide location and time based on the position of a piece of heavy equipment. The satellites generate signals detected by receivers mounted to a variety of earthmoving equipment. Using these signals, the receiver determines the precise location of the heavy equipment on the jobsite. The receiver needs a clear sky to receive a minimum of four signals from the satellites to calculate the receiver's exact location. The computer within the component can determine where it needs to move to get to its intended destination. When utilizing this technology on a construction site, the piece of equipment can accurately determine where the blade, equipment, or other component is on the construction site after it has been plotted. This technology will increase the rate at which the site is contoured properly, and it will be more accurate than human measurement. This increased efficiency will help to decrease costs, increase productivity, and satisfy all of the stakeholders in the project.

Figure 1 GNSS constellation.
Source: Monika Wisniewska/Shutterstock

1.1.2 GPS/GNSS-Based Automatic Grade Control Systems

GPS/GNSS-based automatic grade control systems can deliver extremely tight tolerances. They operate like laser-based systems, except the reference signal is from several orbiting satellites. These satellites broadcast radio signals containing information about satellite locations 24 hours a day. For proper GPS operation, there must be a clear overhead view. Other conditions may also affect GPS operation. For example, working underneath high-voltage power lines may create dead zones for satellite coverage and satellite signals may be weaker at certain times.

GPS-based systems are especially helpful on jobs requiring a lot of contouring. These systems compare the blade position to a three-dimensional digitized site plan to determine whether the blade position requires any adjustments.

GPS/GNSS technology may be used in all phases of the job. A surveyor using a GPS/GNSS receiver in a backpack can walk across an area in a set pattern to record position information. When the position information is fed into a computer, it produces a 3D surface model of the area. The surface characteristics information is then used to lay out a roadway using a computer-aided design (CAD) program. The end product of the design process shows lanes of the road, shoulders, drainage ditches, and other features. An accurate site design is a must when using GPS technology for grading. The CAD program also generates files that are downloaded into the GPS/GNSS grading system computer mounted in the cab of the equipment.

Before a site is graded, static GPS/GNSS observations are taken at the site to establish control points. For better grading accuracy, an on-site GPS/GNSS base station should be used. In this arrangement, satellites transmit radio signals that are picked up by receivers at a base station near the jobsite and on the grading equipment. The base station makes corrections to the satellite signals to determine its exact position. It then sends corrected data by radio to the grading equipment's onboard computer. The computer uses the satellite signals and the corrected signals from the base station to determine how to position the grading blade to meet the grade specifications in the site plan.

Virtual reference station (VRS): An imaginary reference station that is established by a computer based on data from numerous GPS receivers and the grading equipment that serves as the source of data for grading control.

Even tighter grading control can be achieved by using a **virtual reference station (VRS)**. In this type of system, data from numerous GPS/GNSS receivers is transmitted to a central computing station. The grading equipment's onboard GPS/GNSS system also signals the computing station of its approximate location. The computing station uses that data to establish an imaginary reference station in the immediate vicinity of the equipment. By doing this, the computing station is able to collect data from numerous receiver locations as if it were sent from a single receiver position (the virtual reference station). The computing station then sends the data of the virtual reference station to the control system of the grading equipment.

Global Positioning System (GPS)

GPS was developed by the US Department of Defense in the 1970s for military purposes. In military applications, it is vital to have accurate information about where you are and where the enemy is. A series of satellites in earth orbit constantly transmit signals to GPS receivers on the ground to calculate the latitude, longitude, elevation, and the time at the location of the GPS receiver. Signals from at least three satellites are used. Because the signals travel at a known speed (the speed of light), the time it takes the signal to reach the receiver helps to determine the position of the receiver. GPS technology also has many nonmilitary uses. In 1996, President Bill Clinton issued a directive that effectively made GPS technology available for nonmilitary use.

GPS/GNSS Modes

The following two modes are used when grading with GPS/GNSS: indicate-only and fully automatic:

- Indicate-only feeds GPS/GNSS guidance information to the equipment operator who controls the machine. It requires a minimum of one GPS/GNSS receiver on the grader and is the least complicated and less costly of the two modes. However, without a GPS/GNSS base station, indicate-only systems cannot deliver the best grading accuracy. There are several ways to notify the operator that the blade of the dozer or grader is in the correct position.
- Fully automatic systems require a GPS/GNSS base station and one or two GPS/GNSS receivers on the grader or a virtual reference station arrangement to deliver the best grading accuracy. The GPS/GNSS guidance system is fully integrated with the grader's hydraulic system for precise, three-dimensional blade control. Graders can be purchased with a fully automatic system installed, or machines can be retrofitted with aftermarket systems. An override feature allows the operator to always have control of the machine.

Figure 2 shows a typical GPS-based automatic control system used on an excavator's bucket. The information can be displayed on an in-cab display (*Figure 3*). The operator can make the required bucket adjustments, or a signal can be sent to the excavator's control system to automatically adjust the bucket.

Figure 2 GPS/GNSS grade-control system.

Sources: (A) Getty_dumy67/iStock/Getty Images; (B) Courtesy of CASE Construction Equipment; (C) Image Courtesy of Trimble Inc.

Figure 3 GPS-based control system cab display on an excavator.
Source: Image Courtesy of Trimble Inc.

1.1.3 Robotic Total Stations

A robotic total station is one of the most accurate 3D machine control systems available. In some respects, a robotic total station is similar to a machine control system based on GPS. However, a robotic total station uses infrared signals instead of satellite signals, and the robotic station must be in direct line of sight with a prism, or target, mounted on the grading equipment.

A typical robotic total station system is shown in *Figure 4*. The system consists of a robotic station, a target mounted on top of a mast attached to the grading equipment blade, and the onboard computer system that controls the blade movement.

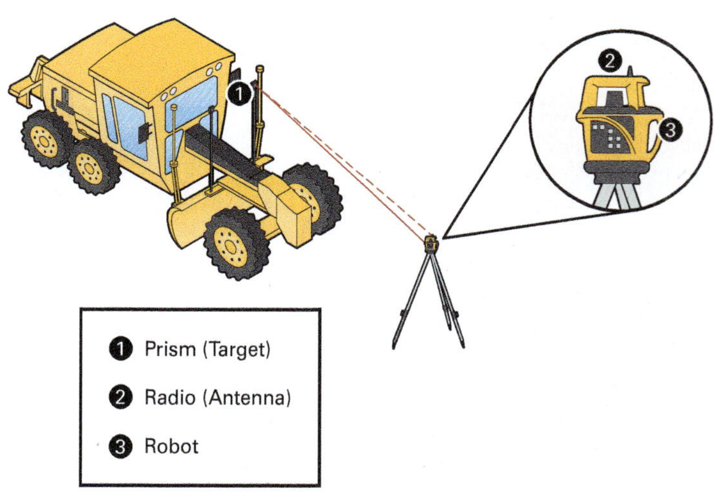

Figure 4 Robotic total station system.

During operation, the robotic station tracks and locates the target and calculates the vertical and horizontal angles of the target. It also sends an infrared signal to the target. The target reflects the signal back to the robot, which enables the robot to determine the distance to the target. The robot then sends the location information via radio signals to the onboard computer of the grading equipment. The onboard computer calculates in 3D how to adjust the blade to meet the site specifications.

Because robotic total stations do not use satellite signals, they are well suited for use in areas with heavy tree cover, inside buildings, underground, or any other situation where the sky is obstructed. Many companies find that the best method of machine control is to combine GPS and a robotic total station. This way, the advantages of both systems can be utilized.

1.2.0 Grading Equipment

During finish grading operations, the operator may work with a grade checker to ensure that the grading is at the correct elevation. The grade checker may use manual or laser instruments. The procedure is basically the same with either instrument.

In addition, technological advances now allow grading operations that are partially or fully automatic. These systems are computer controlled and may use either a laser signal, GPS signals, or optical signals as a reference. These types of systems are extremely accurate and help to save both time and money, but they are expensive to purchase. Contractors that routinely perform large commercial projects are most likely to own this type of equipment.

1.2.1 Motor Grader

A motor grader like the one shown in *Figure 5* is usually the first choice for finishing earthwork. It was originally designed for this type of work. It has a wide, sharp blade to cut bumps and high spots while moving the material in front of the blade to fill in holes and low spots. Accurate blade control allows the operator to perform very fine cutting and trimming. Because the blade can be moved outside the motor grader's wheel tracks, it can be positioned very close to curbs and other permanent structures. This eliminates the need for additional hand trimming.

Figure 5 Motor grader.
Source: Artmois/Shutterstock

When building subdivisions and industrial layouts, grading can be accelerated by adjusting the blade height to average cuts and fills. This speeds up the grading operation because the operator has longer runs without a grade change.

When fill is needed, the material is dumped and then spread with the dozer to the proper elevation before being compacted. After compaction, the motor grader will come back and trim the subgrade to the finished elevation.

Motor graders can also be used to finish fairly steep slopes. The wide blade and smooth operation make dressing a slope as easy as grading a subgrade or base course. For slopes greater than 3:1, some additional precautions may be needed to keep the motor grader from tipping. On steep slopes, the motor grader should be operated in first gear at a reduced engine speed, using the accelerator pedal, rather than the speed control, to control the speed of the machine.

When an articulated motor grader is used on a slope, crab steering should be used to keep the back of the motor grader on the downhill side of the cut. Using crab steering will place the frame in an offset position that provides maximum stability. Crab steering should be used any time the machine may tend to slip due to a heavy side load on the blade.

WARNING!

Evaluate each job carefully. Operating any heavy equipment on a steep slope places the operator at risk of tipping over. Tracked vehicles are more stable on slopes than wheeled vehicles but are not always available or desirable for use. If you feel that it is not safe to operate the equipment on a slope, you must tell the supervisor.

1.2.2 Dozer

A good dozer operator can perform grade work on many different surfaces. For slopes or embankments, a dozer is a good choice because it will spread and smooth the material, and the tracks can be used to compact the loose soil. A dozer can move across a slope, as well as up and down. Care should be taken to select the right size dozer for a specific job. Too large a machine can cause damage and tear up the surface if it makes a number of turns on a slope. Moreover, for fine finishing work, the tracks of the dozer will cause unwanted impressions in the surface.

WARNING!

Traveling across a slope creates a greater potential for tipping. If possible, travel up and down a slope. Do not exceed the manufacturer's recommendations for traveling on inclines.

Dozers can be equipped with automatic control devices that receive a signal from a laser level or GPS and manipulate the equipment's controls to keep the blade at a specified elevation. *Figure 6* shows a dozer equipped with laser receivers. A mast with the laser receiving unit is attached to the top of the dozer blade. After calibrating the height of the laser level receiving unit, the dozer is free to roam the area and perform grading. The main advantage of this setup is that the dozer can receive a continuous signal once the laser level is set and calibrated properly. This means that the grading operation is being checked for proper elevation continuously rather than only at the grade stakes.

Figure 6 Dozer with laser receivers.
Source: Spectra Precision

1.2.3 Scraper

Although a scraper (*Figure 7*) is one of the larger pieces of heavy equipment, it can also be used for some finishing to grade. Scrapers can be used to finish rough grade large horizontal areas where tolerances are not as close as those requiring the use of a motor grader. With its power, weight, and a very sharp blade, the scraper can neatly trim horizontal surfaces and pick up excess material without disturbing the compacted surface underneath.

An elevating scraper is used to pick up windrows of unneeded material made with a motor grader. Using a scraper to do this work saves time in cleaning up the area. It also saves the motor grader from having to push the material off the side of the embankment or to another location.

Scrapers can be used for grading gentle slopes that are long enough to allow for easy maneuvering. When performing this work, a dozer will usually cut a shelf for the scraper to use as a starting point. Because the blade angle on a scraper cannot be adjusted, it must make successive overlapping cuts to trim a slope. Slopes up to 4:1 can be cut and trimmed with the correct size scraper.

Figure 7 Scraper.
Source: Photosbyjim/iStock/Getty Images

Several operating procedures affect the scraper's grading performance. Good scraper operators always check their equipment before they go to work. A worn cutting edge on the blade will not cut properly. Unequal tire pressure will cause the scraper to lean to one side, causing it to cut unevenly. In addition, when the scraper bowl fills unevenly, it will cause the scraper to lean to one side. To ensure even filling of the bowl, be certain that the scraper is centered on the windrow.

1.2.4 Telescoping Excavator

Telescoping excavators are well-adapted for finish work. This equipment is designed to grade on both horizontal and vertical surfaces. Because the boom works back and forth in one plane, the motion is the same as the scraping motion made by a motor grader or a scraper. The operator can control the angles of both the boom and the bucket to produce a smooth, even finish on horizontal, vertical, and overhead surfaces.

In addition to a general excavation bucket, special attachments, such as a grading blade and a ditch cleaning bucket, can be used with a telescoping excavator for finish grading. Coupled with the excavator's long boom, these attachments make it possible to finish off areas that cannot be reached with other equipment.

The grading blade attachment is used in the same manner as a motor grader blade. Using the grading blade to finish or trim a slope is easily accomplished by extending the boom and pulling the blade toward the machine in a scraping motion. Tracked excavators provide a stable platform, even while working on a side slope. The track allows the machine to move up and down the slope and perform grading work over the entire work area.

The ditch cleaning bucket (*Figure 8*) is designed to provide a long, straight cutting edge that leaves behind a smooth, even surface in soft material. Because of its large size, it can collect large amounts of material as it smooths and shapes the ditch profile. This reduces the number of cycles required to do a job.

The excavator is also very effective for finishing sloped areas around drainage structures or near waterways. It can finish the slope, help lay any fabric matting for erosion control, and then place **riprap** on top of the mat.

Riprap: Broken stone, in pieces weighing from 15 to 150 lb (6.8 to 68 kilograms) each, placed on the ground for protection against the action of water.

Figure 8 Excavator ditch cleaning bucket.
Source: Gradall Industries LLC

1.3.0 Using Laser Instruments

Laser instruments are widely used in construction today for site layout and for checking rough and finish grades. Manual instruments rely on the skill of the operator to accurately read the instruments. Laser instruments have digital readouts, so when they are used properly, they can provide a greater degree of accuracy. The use of modern construction laser instruments provides many benefits, including the following:

- Ease of use
- Greater productivity
- Increased accuracy with a reduction in measurement errors

Figure 9 Handheld distance meter.
Source: Courtesy of Stanley Black & Decker, Inc.

Depending on the type of device being used, the laser beam produced may be visible or invisible to the human eye. An electronic receiver detects and uses these beams. Typically, a special receiver is used to check grades, because the visibility of the laser light is usually lost in daylight. Any obstacle or hindrance could block the beam, so the line of sight must also be clear.

WARNING!

Never look directly into a laser beam. The intensity of the beam could damage your eyes and your eyesight.

1.3.1 Laser Transmitter

Modern laser instruments used for construction layout work are typically battery-operated units that generate a fixed laser beam, a rotating beam, or both. Fixed laser beam instruments can generate either single or multiple beams.

One example of a fixed-beam device is a carpenter's level with a built-in visible beam laser. It can be used as an ordinary carpenter's level or switched on to emit a laser beam, extending the level's reference line, typically between 200 and 300 feet (about 61 to 91 meters). It can be mounted on a tripod or placed on a level surface.

Another example of a single-beam device is a handheld distance meter (*Figure 9*) used to measure distances up to about 100 feet (31 meters) without a target, and even longer distances with a reflective target. The auto-leveling laser alignment tool is a good example of a multiple-use laser instrument. It can be used to provide simultaneous plumb, level, and square reference points.

Rotating-beam laser level instruments (*Figure 10*) are widely used for establishing grades and leveling over long distances. Both visible-beam and invisible-beam instruments are made in manual, self-leveling, and self-plumbing models. Most self-leveling laser level instruments are designed to automatically turn off the laser beam whenever the instrument is not leveled. This feature is especially helpful if the instrument is accidentally bumped out of position. If this happens, the unit automatically stops rotating and shuts off the laser beam. The beam remains off until the instrument is reset by the operator and leveled again. The instrument head can be rotated rapidly or slowly, or it can be stopped and used as a single-beam device.

One advantage of using a rotating laser instrument is that it allows one person, instead of two, to perform many site layout operations. Setting up the laser will depend on the model of the tool. Most new tooling allows for automatic calibration once the stand is set up and leveled. Refer to the manufacturer's instructions to verify proper operational procedures. After the instrument has been set up and leveled, the laser head can be adjusted to rotate at various speeds, depending on the model. This creates a level, horizontal beam of light that sweeps over a large area so that it can be detected by multiple electronic laser detectors placed at different locations around the jobsite.

The procedures for determining and establishing elevations with laser levels are similar to those used with optical leveling instruments. To determine elevations, the base of the rod is held to the elevation to be determined. The laser detector is moved up or down on the rod until its display indicates it is centered on the beam. The reading on the rod is then recorded. Establishing a benchmark is key to operating this type of equipment.

An example of this procedure is shown in *Figure 11*, where a rotating laser is being used to check the elevations at the four corners of a simple building foundation. As shown, a rod with a detector attached is placed at each of the four corners where the elevation readings are desired. The rotating laser beam will hit the detector at each location when the detector is in line with the beam.

Similarly, to establish elevations, an electronic detector is attached to a grade rod. The grade rod is moved up or down until the beam indicates that the base of the rod is at grade.

Figure 10 Rotating-beam laser level.
Source: Spectra Precision

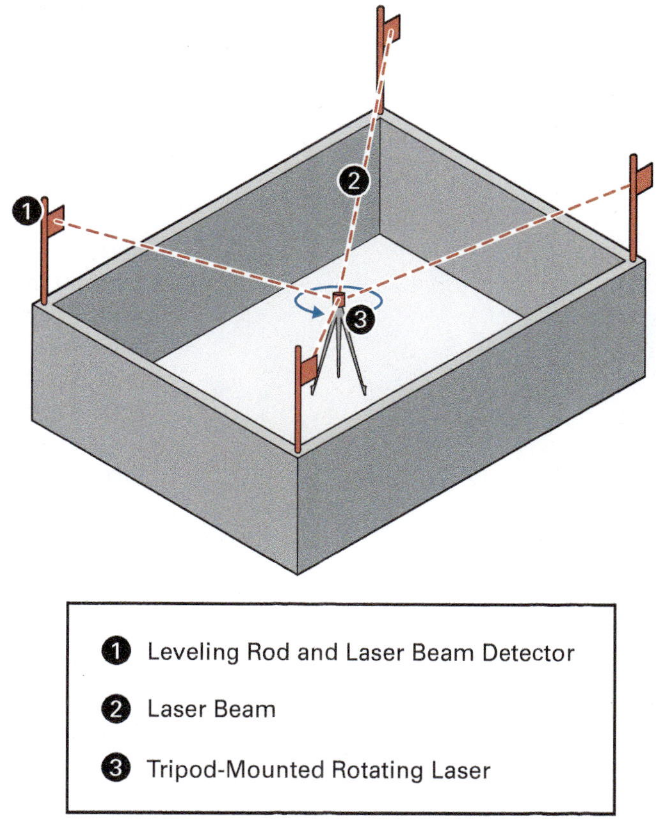

Figure 11 Checking foundation elevations.

1.3.2 Laser Receivers

A laser beam detector or receiver (*Figure 12*) is used with laser transmitters. It must be used with an invisible-beam (infrared) laser instrument such as an electronic level. In bright sunlight, it is also widely used to detect the beam of a visible-light laser instrument.

The laser beam detector is normally mounted on a leveling rod or similar support. However, handheld models are also available. During use, the detector is moved up or down on the rod to intersect the laser beam. A capture window on the beam detector receives the beam from the laser.

The operation of detectors made by different manufacturers is basically the same. Typically, they have an LED visual display that indicates the position of the detector (down, on center, or up) in relation to the laser beam. Most also have an audible device that gives an indication of beam position. For example, one manufacturer's detector uses a fast beeping tone to indicate that the detector should be moved down. A slow beeping tone means it should be moved up, and a steady continuous tone indicates that the beam is on center. Some detectors also have bandwidth and sensitivity adjustments used to adjust the detector for various conditions. These conditions include the degree of ambient light, laser wavelength, focus, power output, and bandwidth tightness required.

Figure 12 Laser receiver.
Source: Image Courtesy of Trimble Inc.

1.3.3 Use of Construction Lasers

General procedures for the setup and operation of laser levels to perform horizontal leveling and vertical plumb tasks differ with each manufacturer and model.

Setup and operation of the instrument must be done in accordance with the instructions given in the manufacturer's operator's manual for the specific instrument.

Using a laser level is fairly simple, but mistakes can still be made that will result in measurement errors. These include the following:

- Failure to follow good leveling practices
- Using instruments that are out of calibration
- Setting elevations incorrectly with the rod
- Performing laser measurements near other laser devices, which can cause unwanted signals to be received by the laser detector

1.3.4 Laser Instrument Safety

Under no circumstances should an unqualified worker operate a laser instrument. Workers must be properly trained, and the training must be documented before a worker can operate a laser beam instrument. Government regulations also require that laser manufacturers provide warnings and cautions on their instruments and in their literature regarding the hazards associated with the use of laser instruments. Qualified personnel must adhere to all such manufacturers' warnings and cautions. They must also follow the manufacturer's operating instructions exactly when using a laser instrument.

If a rotating laser level is operated in the rotational mode, the beam will usually not harm the eyes, because a lower power is used. Most lasers used on construction sites are low-power lasers. The laser equipment will have a label on it that indicates the maximum power output, usually less than five milliwatts, which is considered low power.

OSHA 29 CFR Part 1926 provides safety precautions that must be observed when a laser is in use. Manufacturers also provide various warnings and regulations to prevent hazardous situations. Some of these regulations are as follows:

- Prominent warning signs must be posted in the area where a laser is being used. Be alert to all such warning signs.
- Avoid going into areas marked off for laser operation unless it is necessary because of the job.
- Only those workers who have been properly trained are permitted to set up or operate a laser or make any adjustments to it.
- Qualified workers must carry an operator's card at all times when operating laser equipment.
- Always read and follow the manufacturer's operating instructions for the instrument being used.
- Avoid looking directly at the beam. Also avoid looking at any surfaces such as polished metal or a mirror, which can cause the beam to be reflected into the eyes.
- Special hazards are present when the laser is used in a fixed position, such as when it is used for tunnel or sewer work. Never look at the concentrated beam. Never look along the axis from the laser toward the point being sighted because the beam can be reflected from another surface.
- Wear approved safety goggles when the laser has a power output of 5 milliwatts or greater. The laser equipment must have a label indicating the maximum power output.
- Never point a laser beam at anyone. When possible, set the laser up so that it is above or below eye level.
- The laser beam should be turned off, shuttered, or capped when not in use.
- If you suspect you may have a vision problem caused by the laser, such as a persistent afterimage, seek immediate medical attention.

1.3.5 Calibration and Care of Laser Instruments

The calibration of a laser instrument can be altered by shock and vibration. Laser instruments should therefore be checked for proper calibration at regular intervals, particularly before starting on a new construction site. Calibration of the instrument should be performed by a qualified person and in accordance with the manufacturer's instructions. When an instrument is calibrated in the field, errors can be introduced because of temperature, humidity, and wind conditions. For this reason, it is recommended that the instrument be checked and calibrated in a controlled environment similar to that in which it will be used.

Although it is designed for construction site work, a laser is a precision instrument requiring the same care as any high-quality leveling device. Observing the following maintenance guidelines will help ensure trouble-free laser operation:

- Always follow the manufacturer's recommended maintenance procedures as directed in the operator's manual for the instrument.
- To prevent moisture and dirt from settling inside the unit, always make sure the laser, accessories, and carrying case are clean and dry before storage. To clean the laser and electronic detectors, wipe them off with a soft, nonabrasive cloth and clear, solvent-free water.
- When not in use, store the laser and its accessories in their original cases.
- Make sure the batteries in the unit and any spare batteries are fully charged to prevent work delays resulting from discharged batteries.
- Do not attempt internal repairs to a laser instrument. Return it to a dealer, manufacturer, or other qualified organization for repair.

A laser level's transmitter beam should be checked for proper calibration at regular intervals, particularly if the instrument is being used for the first time or if the unit has been handled roughly. Severe shock or vibration can cause the instrument to be out of calibration. The laser transmitter needs calibration when the laser beam emitted from one side of the unit is above true level, and the beam emitted from the opposite side is below true level.

When the instrument is correctly calibrated, the beam emitted is horizontal for the 360-degree rotation. If the unit is turned 180 degrees or 90 degrees from its original position, the reading is within the manufacturer's specifications, typically within $\pm^{3}/_{32}$" per 100' (±2.38 mm per 31 m) of the original position. Check the laser transmitter calibration as described in the manufacturer's instructions. Manufacturers of different makes and models of instruments recommend checking their instruments in different ways. A general procedure for one method is given here.

WARNING!

Only trained and authorized personnel can operate a laser instrument. Inform all workers within the operating range of the instrument that a laser instrument will be in use. Never look directly into a laser beam or point the beam into the eyes of others. Set up the laser transmitter so that the height of the emitted beam is above or below normal eye level.

Step 1 Set up and level the laser transmitter at a location that has a clear line of sight to an object at least 100' (31 m) away. Attach a laser detector (receiver) unit on a level rod or other calibrated rod and hold it plumb at the 100' (31 m) location.

Step 2 Turn on the laser transmitter and rotate it so that the +X side is aimed at the laser detector.

Step 3 Move the receiver into the beam to get an on-grade reading. Mark and record the elevation, noting that it is for the +X axis.

Step 4 Rotate the laser transmitter 180 degrees so that the –X side is aimed at the laser detector. Mark and record the elevation, noting that it is for the –X axis.

Step 5 Compare the +X and –X elevation readings. If the difference between the readings is less than $^3/_{32}$" (±2.38 mm), the X axis is within calibration. If within calibration, go to Step 9. If the difference in readings is more than $^3/_{32}$" (±2.38 mm), X-axis calibration is required, as described in Steps 6 through 8.

Step 6 To correct for a calibration error, locate a new mark midway between the +X and –X marks. Move the receiver on the rod until its center-marking notch is aligned with the new midpoint mark.

Step 7 Adjust the X-axis calibration screw to obtain an on-grade laser beam reading at the midpoint line. Most detectors have move-up and move-down indicators that show the direction the beams need to be moved up or down relative to the on-grade point.

Step 8 Rotate the laser transmitter 180 degrees back to the original +X face. The on-grade reading should be on or within $^3/_{64}$" (±1.19 mm) of the midpoint line. If not, repeat Steps 3 through 8.

Step 9 Rotate the laser 90 degrees. Repeat Steps 3 through 8 as required to check and adjust the beam emitted from the +Y and –Y sides of the unit.

1.3.6 Laser-Based Automatic Grader and Dozer Control Systems

Laser-based automatic grader and dozer control systems use complicated technology, but their operation is simple. *Figure 13* shows a typical automatic dozer control system. Signals from an off-board laser transmitter are used as the reference. The dozer is outfitted with electronic components that permit a computer processor to determine its exact location on a site and the position of the blade. As the equipment moves throughout the site, the computer determines its exact position and the elevation of the blade. This information is transmitted to an in-cab display so the operator can make necessary blade adjustments.

The complexity and degree of automation will vary among systems. Some systems use a three-dimensional digitized site plan so the computer processor can calculate the required position of the blade. In a fully automated system, a signal is sent to the grader or dozer and the blade position is adjusted automatically.

Figure 13 Automatic dozer control system.
Source: Image Courtesy of Trimble Inc.

1.0.0 Section Review

1. GPS-based grade control systems are helpful on jobs that require a lot of _____.
   a. contouring
   b. trenching
   c. soil compaction
   d. fill material

2. Automatically adjusting blade height to average cuts and fills results in _____.
   a. better variation
   b. faster grading
   c. lower engine speed
   d. greater stability

3. Grades are often established over long distances using a(n) _____.
   a. line of sight tool
   b. dynamic leveling piece of equipment
   c. LED beam projector
   d. rotating-beam laser

2.0.0 Site Planning with GNSS

Objective

Describe the process for site planning for use with GNSS.

a. Identify equipment used in site layout for GNSS in 2D and 3D site layout.
b. Describe how equipment uses GNSS for site contouring.

Performance Tasks

There are no Performance Tasks in this section.

Site planning for GNSS use is one of the most useful pieces of operational planning that can be employed on a modern worksite. Simple 2D projects like digging a basement or footers are great applications for a laser or **sonic tracer system** on a grader. Larger projects, such as airport runways, are ideal for 3D applications because controlling the slope is a more three-dimensional situation. Controlling the way the equipment grades the land is vital to hitting the required marks on the site plan every time. The key to using GNSS-controlled equipment is having the proper points plotted on the site plan and then having the equipment implement those points within their operational plan.

Sonic tracer system: Equipment that uses ultrasonic signals to maintain a set distance or elevation from an object, a design surface, or the ground.

2.1.0 Site Planning for GNSS Use

When surveying a site to set up a **universal total station (UTS)** for use with satellite-guided equipment, extensive staking and string lines are no longer needed. Using the UTS and GNSS navigation, the piece of equipment can easily be navigated to the proper locations and the blade can be angled exactly to create the proper grade specified by the site plans. The key to this type of guidance is setting up the site to be **geospatial information system (GIS)** compliant.

When surveying the site, the creation of X, Y, and Z coordinates (*Figure 14*) based on the GNSS location allows the surveyor to generate the needed information for the equipment to operate on a 3D axis. These coordinates are put into the surveying software to generate commands for the equipment that will operate on the site. As the equipment is operated on the construction site, antennas located on various components of the machine receive a position from the UTS setup on the site. The operator of the equipment downloads a 3D topographical map of the site into the machine that it uses to reference compared to the UTS on the site.

Universal total station (UTS): A combination of a digital theodolite and an electronic distance measuring device that can integrate with a GNSS module to provide a point of reference for 3D site modeling.

Geospatial information system (GIS): Computerized tools related to the physical mapping of data within a visual reference point.

Figure 14 Using X, Y, and Z coordinates with the GNSS location.
Source: Image Courtesy of Trimble Inc.

As the machine makes the calculations needed for the program, it adjusts the speed, blade angle, and bucket to the exact contours of the designed site (*Figure 15*). This system relies on the proper surveying being conducted before excavation occurs. **Civil information modeling (CIM)** provides the data-rich object information for the modeling software to generate a program for the equipment. When the surveyor comes to the site to lay it out, setting up the equipment to translate the site plans into a 3D model is vital to providing the equipment with the proper commands to follow when contouring the site.

Civil information modeling (CIM): The process of using data-rich objects for design of not only buildings but also civil structure and site components.

Figure 15 Blade angle is adjusted automatically by the program.
Source: Image Courtesy of Trimble Inc.

2.2.0 Benefits of Using GNSS-Operated Equipment

When using GNSS-guided equipment on the jobsite, the type of process determines the accuracy needed for the equipment. Something as simple as a basement for a structure that is only on a 2D plane requires less identification than a sloping site with complicated angles that must be precise. Equipment that is continually measuring from itself to the UTS can adjust in real time while it is operating. The operator of GNSS-guided equipment is able to spend more time in the cab and less time checking the worksite's survey lines and stakes.

With increased accuracy, the site becomes more efficient, and the chances of equipment-related injuries decrease. In addition, interpretation of contour or grade by the operator is taken out of the equation with GNSS-operated equipment. Everyone is working from the same set of plans and the computer ensures the contour is meeting specifications (*Figure 16*). This could also be an issue if the surveyors made a mistake in how they laid out the site plan beforehand. If the site plan is laid out incorrectly, the entire site could be in jeopardy.

Figure 16 GNSS increases jobsite efficiency.
Source: BanksPhotos/iStock/Getty Images

The precision that comes with GNSS equipment makes the operators' job much easier as they are tasked with just maintaining equipment operation and safely guiding the equipment on the worksite. These decreased operational requirements will free the operator to pay closer attention to safety on the jobsite. Keeping an eye on what is around the piece of equipment allows the operator to identify potential accidents before they happen. Lowering the requirements of the operator also frees them up to look for increased productivity.

2.0.0 Section Review

1. The universal total station is used for which of the following on the construction site?
   a. Better communication between the operator and the site supervisor
   b. A reference point for equipment to work from
   c. To provide location information for all jobsite vehicles
   d. To allow lower reference points and fewer workers on the jobsite

2. GNSS-operated equipment allows the jobsite to be more efficient and reduces the chances of _____.
   a. worker injury
   b. overspending
   c. environmental impact
   d. equipment damage

3.0.0 GPS/GNSS Guidance with Autonomous Equipment

Performance Tasks

There are no Performance Tasks in this section.

Objective

Describe how GPS/GNSS systems help autonomous equipment operate on a construction site.

a. Identify GPS/GNSS support for autonomous equipment.

b. Describe the benefits of operating equipment autonomously.

As technology progresses and autonomous technologies advance, they will be integrated into the operation of heavy equipment. While the need for trained operators will increase, these pieces of equipment will be operated with a smaller staff to maintain construction site operation. Autonomously driven equipment utilizes GPS/GNSS inputs to correlate its position with the coordinates of the site plan to make sure it is on the proper path and contours the site in the proper way.

3.1.0 How Does Autonomous Equipment Operate?

Radar: A type of sensor (Radio Detection And Ranging) that sends out radio waves and then measures how long it takes them to bounce back to the receiver to determine the distance of the objects in its viewpoint.

Lidar: A type of sensor (LIght Detection And Ranging) that sends out laser beam signals and then measures when they bounce back to the sensor. Lidar can generate a location image of those objects.

Ultrasonic sensors: A type of sensor that projects sound waves, then measures the time it takes for them to bounce off the target and return to determine the location of the target. Ultrasonic sensors work well over short distances (1'–5').

A piece of heavy equipment uses the GPS/GNSS for location tracking and for various equipment sensors to operate components present on the equipment. With all these inputs and a mapped site plan, the equipment can maneuver throughout the site and conduct operations necessary to meet the job requirements (*Figure 17*).

Having equipment talk to other pieces of equipment on the jobsite reduces their potential to strike each other. Using **Radio Detection and Ranging (radar)**, **Light Detection and Ranging (lidar)**, **ultrasonic sensors**, and GNSS, the equipment can calculate how close they can get to each other before an impact will occur. All these sensor inputs are sent to a central processing unit to process and control the various servos. As a result, operations of the equipment can be

Figure 17 Autonomous heavy equipment operates without human interaction.
Source: Image courtesy of Komatsu America Corp.

adjusted in real time as equipment is moving (*Figure 18*). As the equipment adjusts drive and implement positions based on input from the various sensors, it compares it to the site plan and correlates it to the needed adjustments. This system is constantly adjusting to make sure the equipment is running as planned and that the outcome is within the specification put into the system. As obstacles come into view, the equipment adjusts to the obstacle and then returns to course as directed by the site plan.

Figure 18 Various sensors enable autonomous operation.
Source: Scharfsinn/Shutterstock

3.2.0 Benefits of Operating Equipment Autonomously

Autonomously operated equipment provides many benefits that are not present with human-operated equipment. The ability of the equipment to operate independently of a human operator can allow it to operate in very dangerous environments. Highly hazardous environments can pose threats to workers who are not familiar with operating in those types of situations.

3.2.1 Repeatable Work

Autonomous features of equipment create a repeatable, predictable operation that can complete the job efficiently and correctly the first time. Along with hazardous environmental usage, establishing repeatable routes or operations over a small area is a great use of autonomous technologies. In mining or other large sites that have a set path, equipment must move in a repeatable pattern. Employing autonomous equipment frees up workers so they can be in situations that do not have set patterns. By utilizing your most valuable resource, human capital, the project can continue to increase efficiencies in a way to maximize production and minimize **lost-time injuries**.

Lost-time injuries: Injuries sustained by employees at work that result in absenteeism or delays in the employees' normal workload.

3.2.2 Improved Maintenance

When autonomous equipment is integrated with a construction site management program, the equipment can alert the supervisor when it will need to be repaired or refueled/recharged. The equipment can then be moved to the proper location on the site so service can be undertaken to keep the site up and producing. Any fleet of equipment should be scheduling for downtime to repair and refuel the equipment. In the autonomous site of the future, equipment will know when that is and will go to the proper location after the job is completed. It may also alert the area supervisor when it is time for service. These continuous operational efficiencies allow the site to maintain uptime, minimize time the equipment is not working, and reduce human error on the equipment.

3.2.3 Reduced Mechanical Failure

Autonomously operated equipment also helps keep the site functional by minimizing major mechanical failures. When humans operate the equipment, they may not see or hear the equipment when it is failing. They may ignore a mechanical failure and continue to operate it in an unsafe manner. A machine operated by a set of standards and requirements will not circumvent those standards and will ensure the equipment failure is addressed. This decrease in preventable mechanical failures can increase the bottom line and extend the life of the equipment. Maximizing the operational life of the equipment decreases the costs to operate as the piece of equipment will continually produce results when used properly and efficiently.

3.0.0 Section Review

1. When using autonomous technologies, which of the following will guide the job of an autonomous piece of equipment?
   a. The operator
   b. The weather
   c. The soil type
   d. The site plan

2. Autonomously operated equipment utilizes which type of sensors in long range operation by sending radio waves out and measuring their return?
   a. Lidar
   b. Radar
   c. Ultrasonic
   d. Camera

3. What is a key benefit of using equipment that operates autonomously?
   a. Increased torque
   b. Reduced engine sound
   c. Extended life of equipment
   d. Less refueling and recharging

Module 22109 Review Questions

1. When using an articulated motor grader to finish steep slopes, use _____.
   a. a ground anchor
   b. crab steering
   c. reverse blading
   d. larger tires

2. How often does a laser leveling system on a dozer check for proper elevation?
   a. Once a minute
   b. Once every 30 seconds
   c. Every 100 feet (30.5 meters)
   d. Continuously

3. What is a primary benefit of using GPS-controlled devices to regulate the blade height on a dozer?
   a. Blade height is being monitored continuously
   b. Blade height never changes
   c. The dozer can grade at steeper angles
   d. The dozer will require less maintenance

4. What is one of the primary items scraper operators should check before they start trimming?
   a. The drive belts of the engine
   b. Tire pressure
   c. Turn signals
   d. Proper equipment application

5. A telescoping excavator makes a good trimming machine because _____.
   a. its telescoping boom has a limited reach
   b. its bucket or grading blade can be worked back and forth in a scraping motion
   c. the bucket can turn 360 degrees
   d. several types of buckets are available

6. When laying out a site plan, which of the following is used by having a visual reference within the site?
   a. Virtual reference station
   b. Geospatial information system
   c. Lidar systems
   d. GNSS operation

7. Autonomous technologies are controlled on a jobsite by which of the following?
   a. The operator
   b. The site plan
   c. The site location
   d. Material on the site

Answers to odd-numbered review questions are found in the Module Review Answer Key at the back of this book.

Answers to Section Review Questions

Answer	Section	Objective
Section One		
1. a	1.1.2	1a
2. b	1.2.1	1b
3. d	1.3.1	1c
Section Two		
1. b	2.1.0	2a
2. a	2.2.0	2b
Section Three		
1. d	3.0.0	3
2. b	3.1.0	3a
3. c	3.2.3	3b

MODULE 22105

Utility Tractors

Source: Image Provided as Courtesy of John Deere

Objectives

Successful completion of this module prepares you to do the following:

1. Describe the different types of utility tractors and their equipment.
   a. Explain the drive system used on utility tractors.
   b. Identify and describe the instruments used on utility tractors.
   c. Identify and describe the controls used on utility tractors.
   d. Identify and describe the hitches and attachments used on utility tractors.
2. Describe the procedures for starting, operating, and shutting down a utility tractor.
   a. Describe the safety rules for operating a utility tractor.
   b. Identify the daily inspection and maintenance requirements for utility tractors.
   c. Explain how to properly start up, operate, and shut down a utility tractor.
   d. Explain how to connect attachments to a utility tractor.

Performance Tasks

Under supervision, you should be able to do the following:

1. Perform prestart inspection and maintenance procedures.
2. Properly start, warm up, and shut down gas-powered and diesel-powered engine utility tractors.
3. Perform basic maneuvering with a tractor.
4. Attach implements to a drawbar and three-point hitch.
5. Attach and detach implements to a power takeoff.

Overview

In addition to the larger, heavyweight types of construction equipment, there is a smaller category of machines called utility tractors. These machines can effectively and economically handle lighter construction tasks that would be a waste of a heavier equipment's capabilities. These tasks can involve pulling, towing, finish grading, and special operations such as hole drilling.

Digital Resources for Heavy Equipment Operations

SCAN ME

Scan this code using the camera on your phone or mobile device to view the digital resources related to this craft.

NCCER Industry-Recognized Credentials

If you are training through an NCCER-accredited sponsor, you may be eligible for credentials from NCCER. The ID number for this module is 22105. Note that this module may have been used in other NCCER curricula and may apply to other level completions. Contact NCCER at 1.888.622.3720 or go to **www.nccer.org** for more information.

You can also show off your industry-recognized credentials online with NCCER's digital credentials. Transform your knowledge, skills, and achievements into credentials that you can share across social media platforms, send to your network, and add to your resume. For more information, visit **www.nccer.org**.

1.0.0 Utility Tractors and Their Equipment

Performance Tasks

There are no Performance Tasks in this section.

Objective

Describe the different types of utility tractors and their equipment.

a. Explain the drive system used on utility tractors.
b. Identify and describe the instruments used on utility tractors.
c. Identify and describe the controls used on utility tractors.
d. Identify and describe the hitches and attachments used on utility tractors.

Many different companies manufacture tractors, and they come in various sizes and configurations. Many machines used on a construction site, such as equipment attachments, are pulled by tractors. This module describes a class of tractors called utility tractors, which range from less than 100 horsepower (hp) to well over 300 hp. Some have only rear-wheel drive, some are equipped with optional mechanical front-wheel drive (MFWD), and some have full-time front-wheel drive (FTFWD).

Utility tractors can be used on construction sites for light or medium work, including staging of materials and equipment, finish grading, fence installation, and moving construction trailers. John Deere 5 and 8 Series utility tractors are good examples of such tractors (*Figure 1*). The basic versions of these tractors are equipped with diesel engines, mechanical three-point hitches, drawbars, and an independent mechanical **power takeoff (PTO)**. They are also available with additional options that include a rollover protection structure (ROPS) operator cab, electrohydraulic (EH) three-point hitch, EH PTO, and EH selectable control valves (SCVs).

Power takeoff (PTO): Rotating mechanical output device on an engine or transmission to which cables, belts, or shafts can be connected to drive attachments or tools.

Figure 1 John Deere utility tractor.
Source: Image Provided as Courtesy of John Deere

Castered MFWD Axle

Some manufacturers of MFWD tractors use some form of a castered front axle to provide a tight turning angle. When the wheels are turned from a straight line-of-travel, the rearward tilt of the kingpins on both sides of the axle causes the top of both front wheels to tilt with a 12-degree caster toward the direction of the turn to reduce sideslip. The caster also allows the inside wheel to tuck under the frame slightly to achieve a tighter turning angle.

1.1.0 Drive Train Components

While the components, controls, and indicators of tractors vary with the manufacturer, model, and the options available, they are all similar in their basic functions and operations. *Figure 2* shows the general location of components on a typical John Deere utility tractor.

An internal combustion engine powered by diesel, gasoline, or propane propel most of the utility tractors on the market today. Battery and hybrid powered utility tractors are starting to appear in the industry as well. In a utility tractor powered by an internal combustion engine, air flows through the air cleaner into the cylinders. A pump pressurizes fuel from a fuel tank for injection. On some tractors, especially diesels, a **turbocharger** compresses air for better combustion and results in more power delivered to the crankshaft. The cooling system dissipates heat from the engine and, in some cases, from one or more hydraulic systems.

The crankshaft transmits torque to either a clutch that provides power to a mechanical transmission or to a hydraulic, or **hydrostatic**, drive unit. The clutch stops and starts power flow to a mechanical transmission. The rate, pressure, and sequence of fluid flow within a hydrostatic transmission controls its power flow. The differential transmits power from the transmission, through final drives, to the rear axles. The main power train drives the power takeoff (PTO) shaft and mechanical front-wheel drive axles through separate EH clutches.

Power from the engine is also coupled to a variable-displacement pump, providing hydraulic power for other systems. The variable-rate hydraulic system provides hydraulic fluid power through a filter for various operations. These include the EH clutches, the **rockshaft unit** and selectable control valve (SCV) controls to the rear SCVs, mid-mount SCVs if so equipped, and the rockshaft unit for the three-point hitch, which allows for movement of the attached component.

Turbocharger: Exhaust-driven centrifugal air compressor used to increase air flow to an engine, which improves combustion efficiency and increases power.

Hydrostatic: Relating to liquids at rest, or to the pressures they transmit, as in hydraulically powered controls or transmissions that stay in a neutral condition until the hydraulic fluid in them is put into motion or pressurized.

Rockshaft unit: A device that raises and lowers equipment mounted on a three-point hitch.

1 Engine	7 Transmission
2 Air Cleaner	8 Differential
3 Fuel Tank	9 Rear Axle
4 Turbocharger	10 PTO Shaft
5 Cooling System	11 MFWD Axle
6 Clutch	18 Three Point Hitch

Figure 2 Components of a John Deere utility tractor.
Source: Image Provided as Courtesy of John Deere

1.1.1 Transmissions

Tractors use a variety of transmissions that provide several forward speeds and one or more reverse speeds. The most common transmission in utility tractors is the synchromesh transmission. A transmission can also drive one or more PTO shafts to operate equipment attached to the tractor.

Mechanical shift transmissions, located after a clutch and before the differential, consist of two major types: sliding gear and constant mesh. Some should be equipped with a range selector as well as a gear selector. These allow the operator to select one of several speed ranges while the tractor is stopped, and then to select finer gear speeds within that range. Some tractors have a high-low shifter that is like a range selector but can change gears while moving. Mechanical transmissions include the following types:

- *Sliding gear* — Sliding gear transmissions have two or more internal parallel shafts. When moved by a shift lever, sliding spur gears are arranged on the shafts to mesh with each other to provide speed and direction changes. The tractor must be stationary to shift into different transmission power modes (high vs. low).
- *Constant mesh* — The two types of constant mesh transmissions are collar (or shuttle) shift and synchromesh transmissions. In the collar shift version, the selected free-running gears are locked to their shafts by a sliding collar or shuttle. These versions of the transmissions usually require that the tractor

be stopped to shift gears. Some give the sliding gear and collar or shuttle shift transmissions the term *standard transmissions*. In synchromesh transmissions, operators can select gears without clashing by synchronizing devices in the transmission that match the speeds of mating gears before they engage. With these types of transmissions, operators can change gears while moving with the use of the clutch.

- *Hydraulic assist* — A hydraulic assist transmission is a train of constantly meshed gears that operators can shift while moving, without a clutch, to prevent interruption of the power flow through the transmission. When the operator selects a range and gear speed, hydraulic fluid activates hydraulic disc clutches and/or brakes for the appropriate gear combinations required for the selected speed. This type of transmission, although built for heavier duty, operates in a manner similar to 4-, 5-, or 6-speed automatic transmissions in cars. The main difference is that the operator must manually select the gear.
- *Hydrostatic drive* — A hydrostatic drive (*Figure 3*) system allows infinite adjustment of tractor speeds within a range without shifting any gears. It is relatively expensive and inefficient compared to a geared transmission. This type of drive uses internal hydraulic fluid pistons instead of gear trains to transmit power to the final drive. The engine drives a pump and the fluid from the pump flows through a motor in a compact closed circuit. The output shaft of the motor connects to the final drive. The transmission fluid in a hydrostatic drive flows short distances back and forth between the motor and pump within the housing.

Figure 3 Hydrostatic drive transmission.

Source: Image Provided as Courtesy of John Deere

In use, the hydrostatic drive transmission has close-coupled piston blocks so that the ring of pump piston ports aligns with the ports supplying the pistons in the motor block. The incompressible fluid moving back and forth through the ports transfer energy from the pump to the motor. If the supply pressure is higher in one port compared to the other, the motor rotates in one direction—the greater the pressure difference, the higher the output shaft speed. If the pressure in the other port is higher, the motor rotates in the opposite direction. If the supply pressure is the same in both ports, the motor does not rotate.

The pump consists of a rotating housing inside the casing containing a ring of pistons in open-ended cylinders. The pistons attach to the rotating disk of a variable-angle swashplate. The swashplate disk, pistons, and cylinder block all rotate together in relation to the fixed-position motor block with two ports. The angle of the swashplate controls which pump pistons are applying pressure to the motor pistons and which motor pistons receive fluid from the motor pistons. In turn, the motion of the motor's pistons determines the direction and speed of the driven output shaft.

Controls change the angle of the pump's variable swashplate to increase or decrease the displacement of the pump pistons. This alters the supplied pressure to the motor pistons which controls the speed of the output shaft and the direction of rotation. In an actual hydrostatic transmission, the fluid also circulates through an external radiator for cooling, then through a reservoir and filter, and back to the transmission.

1.1.2 Clutches

Various types of disk, overrunning, and cone clutches are used on tractors to engage and disengage input power for transmissions and auxiliary power trains.

A tractor engine commonly uses a disk clutch to connect it to a mechanical transmission. Depressing a foot pedal on the operator platform disengages the engine power from a transmission. Releasing the pedal engages the power. Two types of disk clutches are possible: wet or dry. Both operate in a similar manner, except that a wet disk clutch is immersed in an oil bath or spray for cooling. The oil circulates through an external radiator to remove the heat.

Disk Clutch Components

Splines: Parallel grooves running lengthwise on a shaft that engage matching grooves on a part that slides onto the shaft.

A basic disk clutch consists of three main components: a drive plate, a clutch plate, and a pressure plate assembly. The drive plate is usually a part of the flywheel of the engine, which revolves all the time the engine is running. The clutch plate is a metal plate coated on both sides with a friction material. The clutch plate slides on **splines** located on the input shaft of the transmission. The clutch disk is what transfers power from the engine to the drivetrain. The spring-loaded pressure plate assembly, on the other side of the clutch disk, attaches to the flywheel with standoff bolts and rotates with the flywheel. As the operator engages the clutch, the springs of the pressure plate squeeze the friction surfaces of the clutch disk between the flywheel surface and the pressure plate surface. This gradually locks the clutch disk to the rotating flywheel and pressure plate, causing the clutch disk to turn and rotate the input shaft of the transmission drive unit.

Heavy-duty clutches contain more than one clutch disk splined to the input shaft and separated by steel plates that slide back and forth on the standoff bolts between the flywheel and pressure plate assembly. The standoff bolts cause the steel separator plates to rotate with the flywheel and pressure plate assembly. This provides multiple friction surfaces between the clutch disks to reduce heating and extend the friction surface life of the clutch disks under heavy loads. The pressure plate is pushed away from the clutch plate by a clutch release bearing fork (throw-out bearing) on the input shaft that is activated by mechanical linkages or a hydraulic system controlled by the clutch pedal. EH clutches used for accessories or MFWD operate similarly but are engaged and disengaged by electrical switch action instead of pedals.

Overrunning Clutch

An overrunning clutch automatically engages in one direction but freewheels in the opposite direction. This works using roller bearings and ramps on the input shaft inside the output shaft. When the input turns in the proper direction, the bearings roll up their ramps and jam between the ramps and output shaft. This locks the output shaft to the rotating input shaft. If the input shaft rotation slows or stops, springs force the bearings down the ramps, allowing the output shaft to freewheel until it begins to rotate slower than the input shaft. At that time, the bearings are again jammed against the output shaft by the ramps of the input shaft, causing the output shaft to rotate with the input shaft. These devices are most often used inside various transmissions.

Cone Clutch

Cone clutches are another form of friction clutch. They are simpler and more compact than an equivalent basic disk clutch because the friction surfaces are cone-shaped instead of flat. Cone clutches are composed of two pieces. The drive piece fastened to the input shaft has an internal cone with a machined surface. The driven piece is a cone that fits inside the drive piece cone. It is mounted, and slides back and forth, on the splines of an output shaft. The mating surface of the driven cone is covered with a friction material. When the driven cone moves into the drive cone by a throw-out bearing, the cones gradually lock up and the output shaft turns with the input shaft. Cone clutches commonly control gear trains that supply auxiliary power, including PTOs.

1.2.0 Instruments

An operator must understand the location and function of the various instruments for a tractor before operating the equipment. Study the operating manual for the tractor being used to understand the specific instruments, including engine operating temperatures, charging system readings, oil and hydraulic system pressures, and other operating characteristics. The following sections and figures describe the common instruments found on most tractors. Most of the figures in this book show John Deere or similar tractors. *Figure 4* is a view of a tractor operator's cab.

Operators must closely observe the tractor's instrument panel, located on the front control console, during operation to avoid serious damage to the tractor. Instrumentation varies among different makes and models of tractors. *Figure 5* shows a typical instrument panel inside an operator cab.

Figure 4 Overhead view of an operator cab.
Source: Image Provided as Courtesy of John Deere

Figure 5 Typical tractor instrument panel in an operator cab.
Source: Image Provided as Courtesy of John Deere

1.2.1 Charging System Warning Indicator

The charging system warning indicator comes on if the charging system is out of a normal range because the battery is discharging due to an excessive load or a malfunctioning alternator. Some tractors are equipped with an ammeter or voltmeter in addition to, or instead of, a warning indicator. All utility tractors provide a visual reference to systems based on colored lights or indicators. Red usually means stop, orange/yellow mean caution, and green means good. They also are equipped with meters or gauges that provide information about various systems on the tractor. These types of meters indirectly register the rate of electrical current charge or discharge by monitoring the electrical system voltage variations caused by loads and alternator input. After starting, the meter should show a moderate positive charging value, which tapers off to a lower value during operation.

NOTE

If the indicator light or ammeter shows a high negative discharge rate when the tractor is operating above one-third throttle, stop the tractor and inspect it to determine the cause.

WARNING!

System warning lights are color coded based on the severity of the system failure. Pay attention to any alarms so you know what system is failing.

1.2.2 Engine Oil Pressure Warning Indicator

The engine oil pressure warning indicator comes on if the engine oil pressure falls below an acceptable level. Some tractors are equipped with a warning indicator and/or an oil pressure gauge. The gauge shows the actual engine oil pressure and, usually, the normal operating range. Normal engine oil pressures vary widely due to the type of engine and manufacturer. Consult the tractor's operating manual for the normal oil pressure range.

CAUTION

To avoid serious engine damage, always stop the tractor engine immediately if the oil pressure falls below an acceptable value or an oil pressure warning light comes on during operation.

1.2.3 Fuel Level Gauge

The fuel level gauge indicates the amount of fuel in the tractor's fuel tank. On diesel-engine tractors, the gauge may contain a low fuel warning zone, or a low fuel warning light may be present. Avoid running out of fuel on diesel engine

CAUTION

Do not place a tractor under heavy load before attaining the correct operating temperature. Operating under heavy load with a cold engine increases fuel and oil consumption and shortens engine life. If the engine overheats under load with no visible loss of coolant, stop the tractor and take it out of gear, but do not stop the engine. Let the engine fast idle for several minutes until the temperature returns to the operating range. Next, shift to a lower gear or reduce the load to prevent overheating. Shutting off the engine while it is overheated can allow temperatures to rise even more and can result in burned valves, heat distortion, and lubrication failure. However, if the engine overheats from loss of coolant because of a broken hose or leaking radiator, shut the engine off immediately to prevent any more serious damage to the engine.

tractors because sometimes air must be bled from the fuel lines and injectors to remove air before restarting the engine. This task must be accomplished by a certified repair technician wherever the machine is sitting.

Weight Transfer Pressure Gauge

Some tractors equipped with a weight transfer hitch sometimes have a weight transfer and rockshaft pressure gauge. This gauge indicates the relative amount of weight transferred to the rear wheels of the tractor. The gauge normally has two zones. One zone indicates the weight transfer and the other indicates hydraulic pressure at the rockshaft. During operation, the operator should adjust the rockshaft control lever setting so that the gauge moves about slightly when the tractor is in motion.

1.2.4 Coolant Temperature Gauge

The coolant temperature gauge shows the temperature of the coolant flowing through the engine. Because operating temperatures vary between tractor makes and models, always refer to the tractor's operating manual to determine the correct operating range. Most gauges indicate the normal operating range and needle near the center of the gauge. Warning lights may also be displayed if the engine overheats.

1.2.5 Tachometer

The tachometer indicates engine speed in revolutions per minute (rpm). Most tachometers are marked in hundreds on the meter face. For instance, 15 on the meter face would represent 1,500 rpm. The normal operating range for tractors varies between makes and models. Check the operating manual for the correct range. Most tachometers also have special marks for operation of a PTO because some attachments are designed to operate at a specific input of 540 RPM. For tractors equipped with a 540/540E PTO, the marks, or the letter E, coincide with an economy engine speed (1,700 rpm) for 540 rpm PTO operations using light loads. For tractors equipped with a standard 540 rpm or EH 540 rpm PTO, with or without a 1,000 rpm stub shaft, you must use the normal full-power engine speed (2,400 rpm).

1.2.6 Speedometer/Hour Meter

The speedometer/hour meter indicates the forward speed of the tractor in miles per hour (mph) and the number of hours of operation. The speedometer is helpful when the effectiveness of an attachment depends on the speed of the tractor.

WARNING!

Never operate PTO-driven equipment above the rated speed listed in the tractor's operator manual. Doing so can cause catastrophic failure of the equipment connected to the PTO, which may result in injury or death to the tractor operator or nearby persons due to flying parts.

1.2.7 Additional Status and Warning Indicators

Many tractors have additional status indicators and system warning indicators, depending on the available tractor options. These lights call attention to the operating status of tractor systems such as the power train, PTO, or other hydraulic systems. Some may require immediate engine shutdown to prevent damage to the tractor. These indicators can include the following:

- Hazard warning/turn signal
- Headlight high beam

- PTO engaged
- MFWD engaged
- Hi/Lo: lights when the Hi position of Hi/Lo shift is engaged.
- Neutral: lights when gearshift and electrohydraulic reverser levers are in Neutral position; flashes when gearshift is in Neutral and reverser is in Forward position: goes out when gearshift is in Park position.
- Transmission (with reverser) information indicator: lights if malfunction occurs in transmission. Terminate operation when lit.
- Air cleaner restriction: indicates a clogged prefilter or air cleaner.
- EH hitch: indicates a malfunction in the hitch control system. Terminate operation when the light indicates a malfunction.

Other tractors may also have the following warning indicators that require terminating operation:

- Transmission wet-clutch low oil
- PTO clutch low oil

1.2.8 Performance Monitor

Some large tractors are equipped with a computerized performance monitoring system (*Figure 6*). These devices are capable of displaying, among other things, digital values for ground speed, productivity, area covered, wheel slip, PTO speed, engine speed, distance, and hours since last service. These monitors usually use a bar graph to continuously display **wheel slip**, along with an excessive wheel slip alarm.

Wheel slip: The moving of wheels without transferring the power to the ground and propelling the equipment forward.

Figure 6 Computerized performance monitoring system.
Source: Image Provided as Courtesy of John Deere

Other computerized systems are also available for agricultural use on large tractors. Among them are the following types of systems:

- *Global positioning system (GPS) guidance* — Programmable GPS guidance systems can perform hands-off steering of the tractor to maintain consistent, repetitive tracking and operation over long distances at night, in fog, or in other conditions.
- *Attachment management systems* — These systems allow two or more programmable sequences. Each sequence can perform 12 functions or more with a single operator command. The programming allows precise, repetitive control of the attachment and tractor for such things as attachment depth, raise/lower, float, SCV extend/retract, and speed shifts.

1.3.0 Controls

As with instrumentation, controls and their location can vary between makes and models of tractors. An operator must understand the location and function of the various controls for a tractor before operating the equipment. To aid in identifying controls and their functions, some manufacturers color-code them. Study the operating manual for the tractor being used to understand the specific controls and their markings. The following is one example of a color code:

- *Yellow* — PTO or auxiliary power
- *Red* — Gear shift levers or tractor movement
- *Black* — Hydraulics or miscellaneous functions

Controls for most tractors are located on the seat, under the front control console, on the side and rear control console, and on the overhead air conditioning control console.

1.3.1 Seat and Steering Wheel Adjustment

Proper seat and steering wheel adjustments are important to safe tractor operations. After entering the cab area using the steps on the side of the tractor, the operator should adjust the seat and steering wheel position and then fasten the seat belt before operating the tractor. *Figure 7* shows a typical seat used on utility tractors.

The adjustments are usually located left and right of the seat, but some may be located below the front edge of the seat bottom itself. Most seats can be moved up or down and forward or backward. Some provide an adjustable shock absorber function using springs or a hydraulic mechanism that can also be adjusted for your weight. You should adjust the seat so that your legs are almost straight when the clutch or brake pedals are fully depressed, and your back is flat against the back of the seat.

On some tractors, you can adjust the steering wheel (*Figure 8*) up or down and in or out, as desired, after positioning the seat. Because seat and steering wheel adjustment devices vary widely for various makes and models, refer to the operator's manual for specific instructions.

CAUTION

Never adjust the steering wheel or seat when the tractor is in operation.

Figure 7 Mechanical suspension seat adjustments.
Source: Image Provided as Courtesy of John Deere

Figure 8 Adjustable steering wheel in the tilt position.
Source: Image Provided as Courtesy of John Deere

1.3.2 Transmission Shift Lever Operation

Tractors equipped with standard, synchronized, or hydraulic assist mechanical transmissions have the following five basic transmission shift options:

- *Range selector* — Use the range selector lever, if present, to select a range of operating speeds for the transmission where the individual speed is then selected using the gear selector. Usually, the lever is separate from the gear selector, but on some tractors the same lever controls both functions. In some cases, a creeper gear selection for the range selector is an option that allows very slow tractor speeds in forward and reverse gears. For most tractors, you must stop the tractor and then depress the clutch before the range selector can be shifted. *Table 1* shows the typical range (letter) and gear selection (number) ground speeds for one series of John Deere tractor fitted with standard tires.

TABLE 1 Estimated Ground Speeds for Range and Gear Selections

Forward		Reverse	
Range-Gear	1,600/2,400 rpm km/h (mph)	Range-Gear	1,600/2,400 rpm km/h (mph)
A-1	1.16/1.73 (0.72/1.07)	A-1	1.26/1.89 (0.78/1.17)
A-2	1.57/2.36 (0.98/1.47)	A-2	1.72/2.58 (1.07/1.60)
A-3	2.15/3.22 (1.34/2.00)	A-3	2.34/3.52 (1.45/2.19)
A-4	2.88/4.32 (1.79/2.63)	A-4	3.14/4.71 (1.95/2.93)
B-1	3.33/5.00 (2.07/3.11)	B-1	3.64/5.45 (2.26/3.39)
B-2	4.54/6.80 (2.82/4.23)	B-2	4.95/7.42 (3.08/4.61)
B-3	6.19/9.29 (3.85/5.77)	B-3	6.76/10.13 (4.20/6.29)
B-4	8.30/12.44 (5.16/7.73)	B-4	9.05/13.57 (5.62/8.43)
C-1	9.64/14.45 (5.99/8.98)	C-1	10.51/15.77 (6.53/9.80)
C-2	13.12/19.67 (8.15/12.22)	C-2	14.31/21.45 (8.98/13.33)
C-3	17.90/26.86 (11.12/16.69)	C-3	19.53/29.30 (12.14/18.21)
C-4	23.98/35.97 (14.90/22.35)	C-4	26.16/39.24 (16.26/24.38)

- *Gear selector* — The gear selector on all tractors is used to shift into a selected forward speed or shift into reverse. On most tractors, the gear selector must be in the park position before it is possible to start the tractor. You can shift the gear selector using the clutch when the tractor is in motion.
- *High-Low shifter* — Some tractors have a high-low shifter that may be actuated electrically or mechanically. When used, it basically shifts the tractor speed a half-gear up or down from the currently selected gear without clutching while the tractor is in motion.
- *Reverser* — The reverser lever, if present, allows changing the direction of the tractor without clutching and changing gears. It may have two or three positions: Forward, Reverse, and on some tractors, Neutral. Engage reverse by pulling the lever downward. Some tractors are equipped with reverser gear ratios that are higher than the respective forward speeds. As a result, use the reverser at lower speeds or throttle down before engaging it. Operators often use reversers in tight places that require a number of back-and-forth motions. For most tractors, the reverser lever must be in the Neutral position to start the tractor.
- *Front-wheel drive selector* — On tractors with power-assist, front-wheel drive, use this lever to engage and disengage the front-wheel drive. For tractors with hydrostatic front-wheel drives, you can make the selection with the tractor in motion. On a tractor with an MFWD that is not equipped with a hydraulic device for engagement or disengagement, you must stop the tractor movement before engaging or disengaging the MFWD. If the MFWD has an EH device for engagement or disengagement like the one shown

Figure 9 An EH MFWD three-position selector switch.
Source: Image Provided as Courtesy of John Deere

in *Figure 9*, you can engage or disengage it while the tractor is in motion. Some of these switches are equipped with an automatic position or brake assist position that disengages the MFWD when the tractor is turned or if a brake is applied. The MFWD re-engages automatically under certain conditions.

WARNING!

MFWD greatly increases traction. Use extra caution on slopes because MFWD maintains more traction on steeper slopes than just two-wheel drive, increasing the possibility of the tractor tipping over backwards.

1.3.3 Clutch Pedal Operation

The clutch pedal (*Figure 10*) engages and disengages power flow from the engine to a mechanical transmission. The clutch is located on the left side of the operator platform and is depressed to disengage and released to engage. The twin pedals on the right side are brake pedals.

On most tractors, operators must cycle the clutch once at tractor startup to disengage an engagement override valve. Then depress the clutch again to shift the range selector lever to the desired range, if necessary, and to shift the gear-shift lever. To change the range selector lever during operation, always depress the clutch and stop all tractor movement before shifting the range selector lever.

1.3.4 Brake Pedal Operation

Tractors with two-wheel drive, including MFWD tractors, have two brake pedals on the right side of the operator platform (*Figure 10*). The left pedal controls the left rear-wheel brake and the right pedal controls the right rear-wheel brake. The brake pedals should be depressed simultaneously with the right foot to stop or park the tractor, or individually when making tight turns. For tight right-hand turns, you can depress the right-hand brake pedal while steering the tractor to the right. Use the left-hand brake pedal in the same manner when steering to the left. The operator must lock the pedals together for safe braking when driving the tractor at speeds over 15 mph (24 km/h). Full-time, four-wheel drive tractors have only one brake pedal.

CAUTION

Make sure to check that MFWD or FW assist is disengaged or disengages when a turn is attempted. Keeping them engaged will limit the turning radius.

WARNING!

Before operating a tractor on the road, lock the brake pedals together. Use of only one brake can cause the tractor to overturn or turn sharply in one direction or the other. Use reduced speed if a towed load weighs more than the tractor and is not equipped with brakes. Reduce speed and downshift before traveling downhill. Use both brakes.

Figure 10 Typical tractor clutch and brake pedals.

1.3.5 Differential Lock Operation

Some tractors have an electrohydraulic differential lock that automatically engages when the rear wheels begin to slip, and disengages when the traction equalizes, a brake is applied, or a turn is attempted. Operators must not turn tractors with mechanical locks while the lock is engaged. To use a mechanical lock, the tractor wheels must be stopped or turning at the same speed before engaging the lock. Engage the lock by pressing down on the differential lock foot pedal. The pedal remains down and engaged until the traction equalizes or a brake pedal is depressed. Then it disengages itself by spring action. If the wheels repeatedly slip, you can hold down the pedal to maintain traction.

CAUTION

Never operate a tractor at high speeds or attempt to turn the tractor with a mechanical differential lock engaged. To prevent drive train damage, do not engage a mechanical differential lock when one rear wheel is turning faster than the other wheel.

1.3.6 Tractor Light Operation

Tractors are required to be equipped with legally suitable lights (*Figure 11*) for any on-road use. In addition, many have front and rear work lights for off-road use.

Operators control the lights from a light switch on the front control console. Most light switches have several positions for various operational conditions, as shown in *Table 2*. The Warning position is for on-road use during daylight

TABLE 2 Light Activation in Various Switch Positions

Switch Position	Warning Lights Amber (A)	Taillights Red (A)	Work Lights (B) Rear Facing	Work Lights (C) Front Facing	Headlights Front Grille	Auxiliary Front Work Lights (Optional Cab) (D)
Off	Off	Off	Off	Off	Off	Off
Warning	On flashing	Off	Off	Off	Off	Off
Work light	Off	Off	On	On	On – high beams	On
Transport 1	On flashing	On steady	Off	Off	On – high beams	Off
Transport 2	On flashing	On steady	Off	Off	On – low beams	Off

Figure 11 Tractor work and road lights.
Source: Image Provided as Courtesy of John Deere

hours in clear conditions. The Work Light position is for off-road use only. The Transport positions are for on-road use at night or when visibility is limited. Switch between Transport 1 and Transport 2 to change the headlights back and forth between the high beams and low beams when meeting other vehicles on the road. When the high beams are on, the high-beam indicator on the instrument panel should also be on. When the Warning or Transport positions are used, both of the turn signal indicators on the instrument panel will flash, along with amber hazard lights. The turn signal overrides the hazard-flashing function.

WARNING!

Because tractors are slow-moving vehicles, always use the Warning or Transport switch positions when traveling on public roads. Never use the Work Light position because the very bright, clear lights can blind or confuse other drivers as they approach from the front or rear of the tractor.

1.3.7 Cab Heating and Air Conditioning Operation

Tractors with an optional enclosed operator's cab usually have a heating and air conditioning unit as standard equipment in the cab. The heating and air conditioning controls are usually on the operator's right side or overhead. Because of the wide variation of these units among tractor manufacturers and models, consult the operating manual for the tractor for specific instructions. Keeping the cab pressurized and the blower speed on high will keep dust out of the cab.

1.3.8 Power Takeoff Operation

A PTO supplies mechanical power to certain attachments connected to the rear of the tractor at the three-point hitch or drawbar hitch. The PTO is a splined shaft at the rear of the tractor above the drawbar hitch (*Figure 12*).

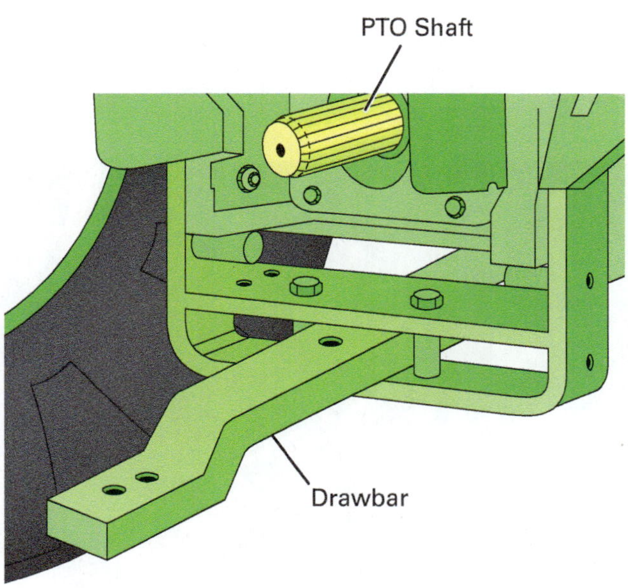

Figure 12 Tractor PTO and drawbar.

The attachment connects to the tractor engine via a PTO propeller shaft on the attachment that couples by a universal joint to the PTO splined shaft at the rear of the tractor. Tractors may have one of the following types of PTOs:

- *Transmission-driven* — When engaging a PTO selector lever, the tractor driveline transmission clutch operates the transmission and PTO at the same time.
- *Continuous-running* — A double-acting tractor driveline transmission clutch controls this type of PTO. With the clutch fully depressed, both the driveline and the PTO are disengaged. Only the PTO engages with the clutch released halfway; both the driveline and PTO engage with the clutch fully released.
- *Independent* — Most tractors use either a mechanical or an electrohydraulic clutch PTO that operates separately from the driveline transmission and clutch.

Changing the PTO Shaft

Tractors with the optional EH PTO are normally equipped to operate only at 540 rpm. However, an optional 540/1,000 rpm stub shaft conversion kit is available to allow 1,000 rpm operation. The exposed end of the inserted shaft has either 6 splines for 540 rpm operation or 21 splines for 1,000 rpm operation. Use the following steps to swap out a PTO conversion stub shaft:

Step 1 Stop the tractor engine, place the gear selector in Park, and disconnect any prior attachment from the tractor.

WARNING!

If an attachment powered by the PTO was in use, the shaft may be hot from operation. Wear gloves and allow the stub shaft to cool before changing it.

NOTE

Make sure any PTO-driven equipment attached to the tractor is rated for the power of the tractor and the speed at which the PTO will be operated. Always refer to the attachment manufacturer's specifications or operating manual for this information.

Step 2 At the PTO housing, align the snap ring with the flats on the stub shaft, remove the snap ring, and pull the shaft out of the PTO housing (*Figure 13*).

Step 3 After cleaning and lubricating with grease, ensure that the borehole in the stub shaft is clean.

Step 4 Insert the stub shaft into the PTO housing with the desired set of splines exposed.

Step 5 Align the exposed snap ring groove on the shaft with the snap ring groove in the PTO housing and reinstall the snap ring over the flats on the shaft into both snap ring grooves.

Figure 13 Swapping out a PTO conversion stub shaft.
Source: Image Provided as Courtesy of John Deere

Make sure the desired PTO-driven attachment is properly installed to the tractor drawbar or three-point hitch and the PTO shaft of the attachment is properly connected to the splined shaft on the tractor with all shields in place. *Figure 14* shows a tractor with a PTO-driven attachment connected to the tractor's three-point hitch and its PTO.

After properly installing the PTO-driven attachment, and while the tractor engine is stopped, follow the instructions in the tractor's operator manual for setting the PTO controls on the operator console of the tractor. After the controls are set, you can start the tractor and engage the PTO. Do not allow the tractor's engine speed to exceed the maximum specified for a given PTO operation. Therefore, adjust the tractor's speed using only the transmission range and gear selections.

Figure 14 PTO-driven spreader attachment on a tractor.

1.3.9 Hitch Control Operation

A three-point hitch can pull as well as raise or lower certain types of attachments attached to a tractor. Section 2.4.2 outlines the attachment procedures for a three-point hitch. *Figure 15* shows the components of a tractor with a three-point hitch.

In *Figure 15*, the attachment is connected to the **draft** links, which apply the pulling power of the tractor and establish the attachment's operating height above or below ground level (height/depth or draft). To accomplish this, the attachment's topmast must connect to the center link. The triangular attachment configuration, consisting of the two draft links and the center link, is normally adjusted so that the attachment is level behind the tractor.

Draft: In the heavy construction equipment trade, refers to the load-pulling capacity of a vehicle.

During tractor operation, the rockshaft lift arms, powered by hydraulic cylinders and connected to lift links, are used to raise or lower the draft links which, in turn, raise or lower the attachment. For some applications, the center link senses the load on the attachment and automatically adjusts the attachment's operating height for varying terrain conditions, depending on the rockshaft control settings. The sway bars provide lateral stability to the draft links. Other manufacturer's tractors may use different methods of providing lateral stability, including the use of chains in place of the bars.

WARNING!

A tractor hitch lifts and controls the attachments connected to the tractor. The movements of the lifting devices can create pinch points that can cause serious personal injury. Review all operational warnings listed in the operator manuals for each tractor and attachment.

NOTE

Because this training manual cannot cover the operational details of the hitch lifts on all tractors, operators must follow the instructions available in the specific tractor's operator manual.

Figure 15 Three-point hitch components.
Source: Mark Jones

For the hitch controls to work, the tractor's engine must be running. Hitch controls located near the operator's seat operate the rockshaft arms that move the hitch. The hitch control lever raises and lowers the hitch and sets the working height or depth of the attachment. Push the lever forward to lower the attachment and rear to raise the attachment. Most tractors also have a raise/lower switch that functions independently of the hitch control lever to raise or lower the hitch between the settings of the hitch control lever and height-limit control. This helps when turning at the end of an operation. Other hitch controls include the following:

- *External raise/lower switches* — With the transmission gear selector lever in Park, pressing and holding the raise or lower switches (usually mounted on the left fender of a tractor) allows for rockshaft control outside of the cab. The rockshaft raises or lowers the hitch at a very slow rate. These switches ignore any height limit and working height (depth) settings. The switches are not functional when the hitch control lever in the cab is in the transport lock position.

WARNING!

To prevent personal injury or damage caused by tractor movement, make sure the tractor transmission is in the Park position before using the external raise or lower switches. Stay clear of any pinch points when using the switches, and make sure that the attachment does not interfere with the tractor's fenders. Some machines that do not have EH controls and arms/hitches can lower without the engine running if the lever is moved.

- *Height limit control* — Turning the height limit control counterclockwise reduces maximum hitch height; clockwise increases maximum height. For the initial setting with an attachment connected, turn the control clockwise to the maximum position. You should raise the hitch with the hitch control lever to a desired maximum height setting without interference to the tractor. Then, turn the control counterclockwise until the hitch just begins to lower.

NOTE

Once any external raise/lower switch is pressed, the rockshaft is prevented from moving accidentally. To make the rockshaft ready for operation again, the operator must move the hitch control lever in the cab to a position that corresponds to the position of the draft links. Then, actuate the raise/lower switch in the cab.

- *Depth stop wheel* — Turning the depth stop wheel control sets a mechanical stop at the desired working height (depth) of the attachment. After setting the working height (depth) of the attachment with the hitch control lever, press and rotate the depth stop wheel to move the mechanical stop up to the forward edge of the hitch control lever. Then, if you move the hitch control lever rearward to raise the hitch for some reason, it can be moved forward to the stop, which will return the hitch to the preset working height (depth). If it becomes necessary to lower the hitch below the preset height (depth) for some reason, you can lift the hitch control lever and push it past the stop.
- *Rate-of-drop control* — Turning the rate-of-drop control counterclockwise slows the rate of drop of the hitch; clockwise increases the rate of drop. For anyone setting, heavy attachments will drop faster, and light attachments will drop slower. By varying the setting, you can achieve a satisfactory rate of drop of two seconds or more for the weight of any attachment rated for use with the size and power of the tractor.

WARNING!

Excessive drop speed may cause personal injury or damage to the tractor and attachment. Fully lowering the attachment should require at least two seconds.

- *Transport lock position* — Pulling the hitch control lever back to the highest setting and placing the lever in the lock slot raises the hitch to its maximum height limit control setting and disables all raise/lower switches. This prevents the hitch from lowering during transport. The hitch automatically rises when starting the tractor if the hitch control lever is in the transport lock position.

WARNING!

To prevent personal injury when starting the tractor, make sure no one is near the hitch or any hitch attachment. To prevent personal injury or equipment damage during transport, make sure the hitch control lever is in the transport lock position.

- *Load/height (depth) control* — The load/height (depth) control can be rotated to a number of positions for automatic adjustment of the working draft (height/depth) of the attachment due to load resistance and terrain contour variations. Turning the control counterclockwise reduces draft response; clockwise increases draft response. The higher the draft control setting number, the more the attachment raises and lowers due to load resistance variations. The draft settings of the control apply primarily to subsurface attachments, such as a ripper. They are also used to maintain tractor speed and load, and to maintain, as much as possible, the working depth of the attachment that was set by the height control lever. Rotation of the control fully counterclockwise to the position control detent holds an attachment, such as a scraper, at a previously set fixed-height position, regardless of the attachment loading. This setting, along with a minimum setting of the height control lever (fully forward), is also used for float control of a three-point mounted attachment with gauge wheels or skid pads so that the attachment freely follows any terrain contours.

WARNING!

When connecting or disconnecting an attachment to a tractor's hitch system, turn the load/depth control fully counterclockwise to the position control detent before attachment or detachment. Use only the hitch control lever or, if provided, the fender-mounted raise/lower switches when connecting or disconnecting an attachment. Do not use the raise/lower switch on the side control console hitch controls.

1.3.10 Selectable Control Valve (SCV) Operation

Most newer tractors are equipped with one or more SCVs to operate attachments with hydraulic motors and single- or double-action hydraulic cylinders. They can range from valves with basic hydraulic cylinder extend/retract functions to optional valves with enhanced operational capabilities. *Figure 16* shows a tractor equipped with a set of SCV couplers. Tractors may have multiple sets of SCV couplers, depending upon the number of functions needed on the attachments they are pulling. A set of SCV control levers are located near the operator's seat, but some tractors may also have an additional set of control levers mounted near the SCV couplers at the rear of the tractor.

Figure 16 Tractor with SCV couplers.

Most SCV control levers, including the enhanced SCV control levers, operate in the same way. Each control lever usually has four operating positions:

- *Neutral* — The control is spring-loaded to the Neutral position, which is in the middle of the movement range for the lever. In the Neutral position, hydraulic oil pressure equalizes for both couplers of the SCV, which hydraulically locks a cylinder in a fixed position.
- *Extend* — When the spring-loaded control lever is moved rearward from Neutral, hydraulic oil is directed out of the top extend coupler of the SCV to the hydraulic cylinder at a rate that is proportional to the distance the lever is moved. This allows for varying the rate of cylinder extension. For double-acting cylinders, the oil in the cylinder used for retracting is forced out of the cylinder through the depressurized bottom retract coupler. The top extend coupler is used for both single-action and double-action cylinders and as the fluid return line for hydraulic motors. When used for hydraulic cylinder operation, the lever should not be held in the extended position if the cylinder reaches its end of travel. Overheating of the hydraulic oil can occur.
- *Retract* — When the lever is moved forward from Neutral, hydraulic oil is directed out of the bottom retract coupler of the SCV to a double-acting hydraulic cylinder at a rate that is proportional to the distance the lever is moved. This allows varying the rate of cylinder retraction. For double-acting cylinders, the

oil in the extend cylinder is forced out of the cylinder through the depressurized top extend coupler. The bottom retract coupler is used only for double-acting cylinders and, along with an adjustable flow rate valve, as the pressure source for hydraulic motors. For single-acting cylinders, the bottom retract coupler is blocked by an internal check valve and the top extend coupler is depressurized when the lever is moved to a retract position. This allows the gravity of the load on the single-acting cylinder to push oil back through the top extend coupler in proportion to the distance the lever is moved forward. When used for double-acting cylinder operation, the lever should not be held in the retract position if the cylinder reaches its end of travel. Overheating of the hydraulic oil can occur.

- *Float detent* — When the lever is pushed all the way forward through retract to a detent position, both couplers are depressurized, and the hydraulic cylinder operates in a float mode that allows the cylinder to extend and retract freely. This mode allows an attachment, such as a loader, to follow ground contours or to stop operation of a hydraulic motor. Never place the control lever in the Float detent position when any attachment operation involves loads suspended in the air; doing so will cause the loads to immediately fall to the ground.

1.3.11 Ignition Switch Operation

The types and functions of ignition switches can vary widely between makes and models of tractors. Some only activate the starter and ignition system for gasoline engines. Others may activate fuel pumps, fuel valves, and starting aids for diesel engines as well as the starter. In some cases, the operator engages diesel fuel valves and starting aids by other manual controls. For specific instructions, refer to the operator manual for the specific tractor. The multipurpose ignition switch used on most tractors is located on the front control console. It normally has the following four positions:

- *Accessory* — Pushing the key in and turning it to the Accessory position applies power to an accessory power outlet located in the rear console or to devices such as radios that may be connected to accessory power. Other than a battery gauge or indicator, this does not activate any of the other electrical controls.
- *Stop* — Turning the key to the Stop position stops the low-pressure primer fuel pump that feeds the mechanically driven high-pressure injector pump and closes the electrical fuel valve that stops the engine. It also shuts off all other electrical power in the tractor, including accessory power.
- *Run* — Turning the key to the Run position starts the low-pressure primer fuel pump, opens the electrical fuel valve that enables starting and running the engine, and initiates a momentary instrument panel indicator bulb check. If the air temperature is below a given temperature, such as 5°C (41°F), the Run position activates the cold weather starting devices, such as glow plugs or intake air warmers. The key returns to the Run position after releasing at engine startup.
- *Start* — The key is turned to the Start position to crank the engine and high-pressure injector pump, which starts the engine. This position is spring-loaded to return to the Run position after releasing the key. To reduce battery load during starting, the ignition switch of some tractors may shut off power to accessories and lights when the key is in the Start position.

CAUTION

The average time allowed to crank most tractor engines is around 20 seconds. Attempting to crank longer than that can damage the starter. Before cranking the engine, always allow time for the gauges to boot up once the ignition is turned to Run. Always allow two minutes of cooling before trying to start the engine again.

1.4.0 Hitches and Attachments

Attachments for the tractor normally connect at the rear. The tractor can pull the attachment with either a drawbar or carry it using a three-point hitch. Different manufacturers have different setups for their hitches. Be sure to read the operator's manual before trying to connect an attachment. Sometimes special tools or connectors are required.

1.4.1 Types of Hitches and Attachments

Figure 17 shows various types of attachments used with tractors. Trailers or attachments with gauge wheels and their own manual or hydraulic adjustments are usually towed using a drawbar. Some construction attachments, such as grader blades, rotary cutters, and counterweights, are carried by a three-point hitch at the rear of the tractor.

(A) Tractor with Box Blade on Rear

(B) Basic Grader Blade for Tractor

(C) Rotary Cutter (Bush Hog)

(D) Front Bucket and Counterweight

Figure 17 Tractor attachments.
Source: Images Provided as Courtesy of John Deere

Finish grading can be accomplished with the adjustable rear grader blade. Such blades can be used with full hydraulic or manual adjustment for tilt, angle, and offset as well as an optional tail wheel to aid in leveling. Grader blades come in widths from 83" to 119" (210–300 cm).

Box scrapers can level, grade, remove material, and backfill. The model shown in *Figure 17* can be equipped with scarifiers that rip up and loosen soil ahead of the scraper. Box scrapers for utility tractors come in capacities of 15 ft^3 to 28 ft^3 (0.4 m^3 to 0.8 m^3) and widths varying from 60" to 84" (150–213 cm).

The rotary cutter, often called a bush hog, operates from the tractor's PTO. Hydraulic cylinders mounted on the tractor's lift arms push down some augers. The tractor's SCV controls actuate such as hydraulic cylinders. Augers up to 18" (46 cm) in diameter and 48" (122 cm) long are capable of drilling straight holes through solid rock, large rocks and dirt, or large tree roots and dirt/rocks.

When a tractor is equipped with a front-end loader bucket, a counterweight can be attached to the three-point hitch, as shown in *Figure 17.*

1.4.2 Drawbar Adjustments

When attachments are to be towed behind a tractor and powered by its PTO, the drawbar must be adjusted to the proper position. The following are the adjustments that can be made to the drawbar of a tractor:

- *Length* — The drawbar can be used in the short position or the extended position (*Figure 18*). The short position is for PTO equipment rotating at a lower rpm (around 540 rpm). Measure the positioning of the hitching point from the output end of the PTO shaft. For the lower rpm equipment, the hitching point is 14" (36 cm) from the end of the PTO shaft. The 16" extended position usually applies to PTO equipment rotating at 1,000 rpm. In both positions, center the drawbar with the PTO shaft. On some models, adjust the length from under the tractor body.

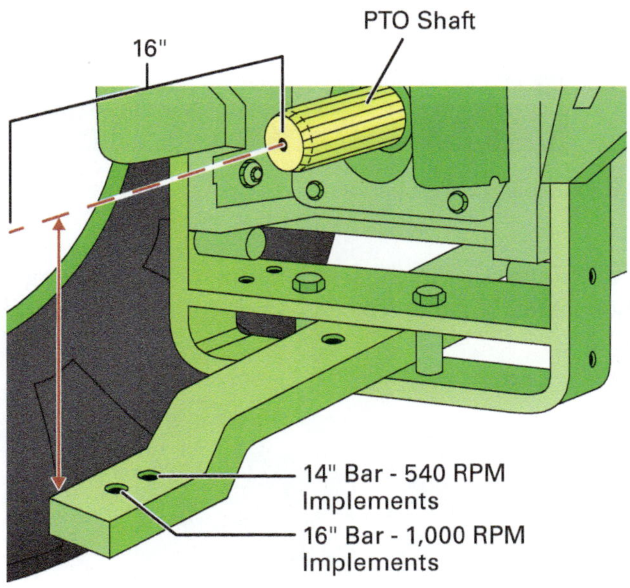

Figure 18 Drawbar length and alignment.

- *Height* — Maintain the proper drawbar height and length to prevent the tractor from upsetting and to protect the PTO shaft. Leave 13" to 20" (33–51 cm) ground clearance, measured from the top of the drawbar. Adjust the height by inverting the drawbar. *Figure 19* shows drawbars in the lowered and raised positions.

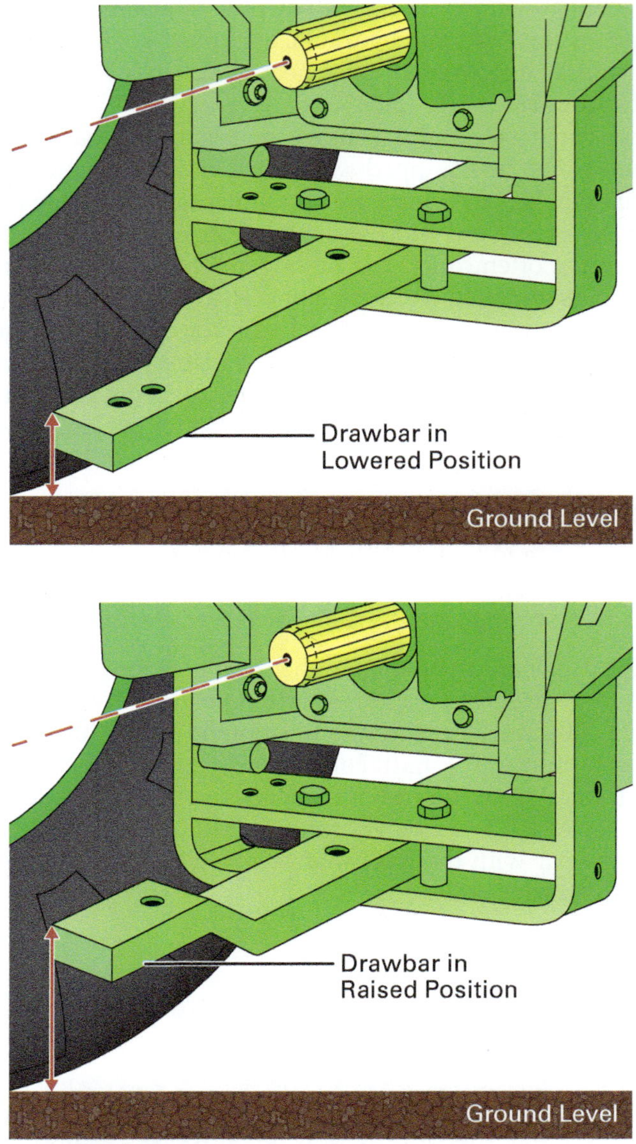

Figure 19 Drawbar position options.

- *Lateral adjustment* — Pin the hitch in one of the swinging positions for offset operations. Use the center fixed position for transporting (*Figure 20*).

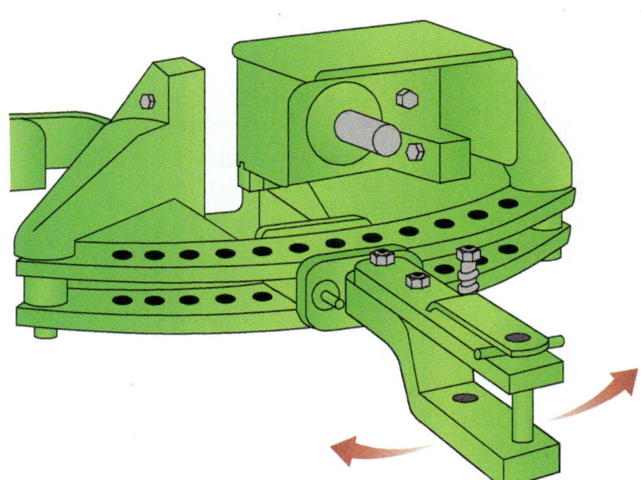

Figure 20 Lateral drawbar adjustment.

1.4.3 Connecting an Attachment to the Drawbar

Some attachments have a tongue for pulling, which must be connected to the tractor's drawbar (*Figure 21*). The drawbar height can be adjusted from 13" to 20" (33–51 cm) clearance between ground level and the drawbar hitch point.

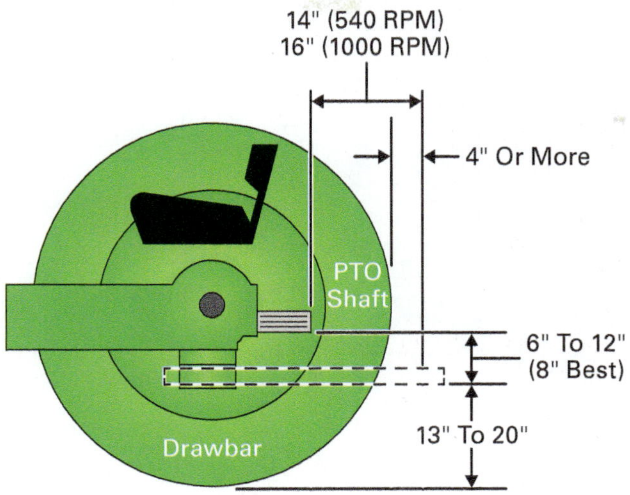

Figure 21 Connecting an attachment to the drawbar.

Be sure to match the attachment to the tractor power. Be very careful performing this operation. Accidents are common. Avoid injuries from the following possible hazards:

- *Lifting* — Know the proper way to lift heavy objects.
- *Cutting* — Wear gloves when handling sharp objects.
- *Pinching* — Use caution when connecting and disconnecting joints; keep hands and fingers out of joints.
- *Crushing* — Keep feet out from under equipment.

Read the operator's manual thoroughly and perform one or more of the following actions:

- Turn the drawbar over.
- Adjust the drawbar bracket.
- Adjust the drawbar frame assembly where it bolts to the tractor frame.

The final vertical distance between the drawbar hitch point and the PTO shaft should be about 8" (20 cm).

If the hitch will connect to an attachment that must be offset, extend the hitch. If necessary, move it laterally, and secure it into position.

To use the hitch for a PTO attachment, adjust the hitch as follows:

- For power takeoff of 540 rpm, the hitch point must be 14" (36 cm) from the end of the PTO shaft.
- For PTOs with a speed of 1,000 rpm, the hitch point must be 16" (41 cm) from the end of the PTO shaft. *Figure 22* shows the extended position hitch setup.

If the hitch connects to a PTO attachment, center the drawbar hitch point directly under the PTO shaft and fasten it securely. The vertical distance between the PTO shaft and the hitch point should be 6" to 12" (15–30 cm), ideally 8" (20 cm). Back the tractor into position so that the hole in the drawbar is in line with the hole in the attachment hitch. If the machine is on sloping ground, set and lock the tractor brakes. Then, set the hitch pin in place.

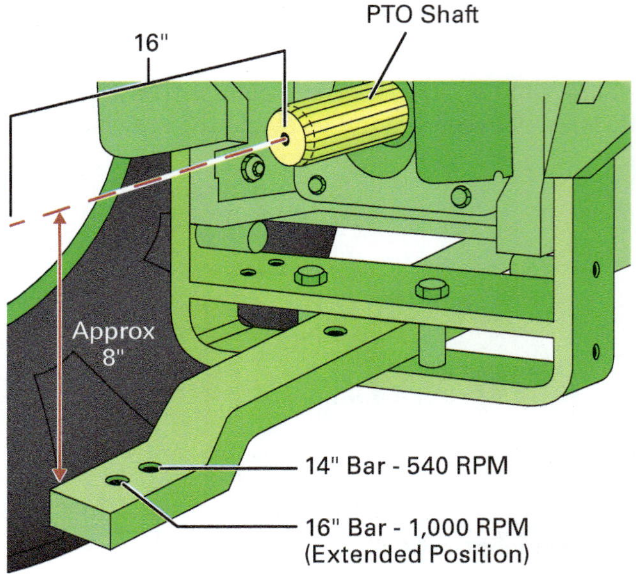

Figure 22 Extended position.

WARNING!

Do not try to position the tractor by operating the transmission controls from the ground. You may get pinned between the tractor and the attachment.

If you cannot insert the hitch pin by moving the attachment, get back on the tractor and realign the drawbar with the holes in the hitch. Be sure to use a cotter pin or some other means of holding the hitch pin in place so that it does not work itself out. Secure the cotter pin or other device. Make electrical or hydraulic connections if needed.

1.4.4 Ballasting a Tractor

Some tractors may require wheel weights to prevent wheel slip when towing heavy loads. MFWD tractors usually require front-end weights or front-wheel weights, as well as rear-tire weights to increase traction. For heavy attachments carried on a three-point hitch, you may have to add weights to the front of the tractor to act as counterweights for the attachment. This counterweight is necessary to prevent tractor tip-up when lifting the attachment and during transport, especially up hills. To see where and how to install ballasting weights on the tractor being used, refer to its operator manual. Make sure to securely fasten the weights.

1.0.0 Section Review

1. A type of transmission that permits infinite adjustment of speed within a given range without shifting is the _____.
   a. sliding gear
   b. constant mesh
   c. hydraulic assist
   d. hydrostatic drive

2. The instrument that should be closely monitored when operating a tractor under heavy load is the _____.
   a. fuel level gauge
   b. coolant temperature gauge
   c. tachometer
   d. charging system indicator/gauge

3. Which tractor control determines the desired ground speed of travel?
   a. Hitch control
   b. Global positioning system
   c. Transmission shift lever
   d. Differential lock

4. Which of the following statements about tractor hitches and attachments is true?
   a. Front buckets are usually attached to a drawbar.
   b. Bush hogs are powered by a PTO.
   c. When equipped with a front-end loader bucket, a counterweight usually is attached to the front bumper.
   d. All manufacturers use the same methods for attaching equipment.

2.0.0 Starting, Operating, and Shutting Down a Utility Tractor

Objective

Describe the procedures for starting, operating, and shutting down a utility tractor.

a. Describe the safety rules for operating a utility tractor.
b. Identify the daily inspection and maintenance requirements for utility tractors.
c. Explain how to properly start up, operate, and shut down a utility tractor.
d. Explain how to connect attachments to a utility tractor.

Performance Tasks

1. Perform prestart inspection and maintenance procedures.
2. Properly start, warm up, and shut down gas-powered and diesel-powered engine utility tractors.
3. Perform basic maneuvering with a tractor.
4. Attach implements to a drawbar and three-point hitch.
5. Attach and detach implements to a power takeoff.

Tractors have great potential for causing injury or death to the operator or bystanders if operated incorrectly. For these reasons, operators must have a thorough knowledge of the equipment they are operating, as well as accepted work procedures and safety principles.

2.1.0 Safety Guidelines

Safety is your number one priority as a tractor operator. Working safely involves knowing the proper operation of the machine you are using, understanding jobsite rules and regulations, and being aware of nearby workers.

2.1.1 Safety Rules for Operating Tractors

Keep the following safety rules in mind when operating tractors:

- Read the operator's manual thoroughly and know the machine. If you do not understand something, ask someone, such as a dealer or factory representative, to explain it.
- Study the company's safety manual.
- Do not wear loose fitting clothing.
- Wear ear protection if needed.
- Provide for ventilation.
- Become familiar with controls. Be sure to know how to stop the tractor and engine.

- Do not let another person on or near the tractor when it is in operation. With rotating attachments, shut down the tractor if a person or animal approaches.
- Remove or tie down loose tools and equipment.
- Keep your seat belt fastened.
- Stay seated during operation.
- Use all lights and safety warning devices.
- Check brakes.
- Disengage PTO when not operating an attachment.
- Do not smoke while working on the tractor, handling fuel or solvents, or checking or filling batteries. Hydrogen gas buildup from the battery may ignite.
- Never use attachments that are not matched to tractor size, weight, power, PTO speed, and drawbar or three-point hitch capacity.
- Make sure all safety devices, guards, shields, and safety decals or signs are properly installed as specified by the manufacturer's manual. Tractors should have ROPS equipment or a ROPS cab. PTO shields must be in place for the tractor and attachment shafts.
- Observe and follow all safety signs or decals and use any steps or handholds to mount the tractor.
- Always dismount the equipment the same way you mounted it. Do not jump down facing forward.
- Make sure front and rear frame counterweights, wheel weights, and wheel ballasts are used as recommended by the manufacturer for the attachment being used. Never add extra counterweights to compensate for an overload.

NOTE

Some tractor cabs have instructor seats, which must be equipped with a seat belt.

2.1.2 Safety Rules for Effective Tractor Operation

Always use the following tips to operate the tractor safely and efficiently:

- Perform prestart inspection and maintenance procedures each day before starting the tractor.
- Adjust the seat and steering wheel and fasten the seat belt before starting the tractor.
- When starting the tractor, follow the manufacturer's recommended starting procedure.
- Before moving a tractor, make sure that adequate clearance exists in all directions.
- After first starting a tractor, quickly check that the controls function properly before moving the tractor.
- When first moving the tractor, make sure that the steering works both left and right and the brakes function properly.
- Ensure that any attachments used do not exceed the tractor's load rating. Any attachments using a PTO must match the PTO speeds available on the tractor. Attachments using the SCV capabilities of a tractor must not exceed the tractor rating for this service.
- Keep all body parts inside the operator's compartment when operating the tractor.
- To prevent the tractor from tipping over backwards, do not lock both rear wheels with the brakes at the same time while the clutch is engaged, and the tractor is in a low range and gear with the engine at working speed. If both rear wheels become mired (stuck) or frozen and do not move and rotate, do not attempt to move forward. Always use reverse gear to free mired or frozen wheels. Also, make sure that any towed load is attached below the bottom of the tractor's rear axle.
- Keep speed down.
- Avoid sudden starts, stops, and turns.

- Always slow down before a turn. Some tractors will tip over at approximately 8 mph when turning on level ground.
- Use caution when entering the highway. Consider the condition of the road, the tractor, the traffic, and the weather. As speed doubles, the chances of a tractor tipping over becomes four times as great. Always use extreme caution.
- Signal all turns and stops.
- Travel completely on the shoulder whenever possible. Avoid driving half on the road and half on the shoulder.
- Depending on the soil conditions, stay at least 4' to 10' (1.2–3 m) from the edges of deep ditches, embankments, and excavations.
- Do not touch electrical wires with any part of the tractor.
- When traveling forward and turning with a towed attachment connected to the rear of the tractor, make sure that it does not come in contact with and ride up the tractor tire that is on the inside of the turn. In reverse, make sure that the tire does not ride up on and crush the attachment. If turning with an attachment connected to the three-point hitch, make sure that it will not strike anything on the outside of the turn.
- If possible, always park in a non-operating or designated area and on level ground. Check the ground to make sure it is firm. If parking on a grade, position the tractor at a right angle to the slope and block the wheels.
- Before dismounting a tractor, disengage the PTO, lower any attachments to the ground, and retract any hydraulic cylinders, especially single-acting cylinders, as much as possible. Then, place the tractor in Park (or Neutral) and/or set the brakes. Shut off the engine and cycle all the SCV control levers through all positions to relieve hydraulic pressure, then return the levers to the Neutral position. If the tractor is to be left unattended, remove the ignition key before dismounting.
- To help prevent water condensation, make sure the fuel tank is full if the tractor will remain unused overnight.
- Follow the company's preventive maintenance program.
- When transporting a tractor, make sure to properly secure it with the brakes locked. Cover the rain cap to protect the turbocharger.

2.1.3 Safety Rules for Operating a Tractor on Slopes or Hills

Follow these guidelines when operating on slopes or hills:

- Always reduce speed on slopes or hills.
- If forced to move forward out of a ditch, gully, or up a steep slope, use the lowest gear possible and engage the clutch very slowly. Be prepared to disengage the clutch immediately if the front wheels rise off the ground.
- Avoid driving forward out of deep ditches or gullies. Rearward tip-over can occur in less than a second. Always back out if possible.
- Go up steep slopes in reverse.
- If work on a slope is unavoidable, go up and down the slope, perpendicular to the toe of the slope.
- Do not go downhill with heavy loads.
- Keep the tractor in gear when going downhill. Use the same gear used for traveling uphill.
- Keep all raised equipment close to the ground when working on slopes.
- Make sure the heavy end of the tractor faces uphill.
- Do not cross slopes of more than 25 degrees. One rule of thumb is that most tractors will not tip over sideways on a slope during straight line travel if a vertical line from the side-to-side center of the top of a cab or ROPS falls inside the downhill rear tire.

- When crossing steep slopes, avoid turning uphill. If it is unavoidable, slow down and make a wide turn. Also, avoid any depressions with the downhill tires and any bumps, large rocks, stumps, ledges, or raised areas with the uphill tires.

2.1.4 Safety Rules for Operation of the Hydraulic System

Follow these guidelines when working and operating with hydraulic systems:

- Wear gloves and eye protection when connecting, disconnecting, or checking any high-pressure hoses. High-pressure fluid from pinhole leaks is usually not visible and can easily penetrate exposed skin. If an accident occurs, fluid injected into the skin must be surgically removed within a few hours or an infection may result. When checking for pinhole leaks in a pressurized hydraulic system, use a flat piece of cardboard as a search tool. Also, look for a stain below the suspected line, or a ring of dirt on the line.
- Before disconnecting high-pressure hydraulic hoses, always relieve pressure as much as possible by first lowering attachments and loads. Then retract any cylinders, especially single-acting cylinders, as much as possible. Shut off the tractor engine and cycle the SCV control levers through all positions to relieve remaining pressure, and then return them to the Neutral position. Before connecting hoses, make sure the tractor engine is stopped and the appropriate SCV control levers are in the Float detent position if connecting hydraulic motors, or in the Neutral position if connecting hydraulic cylinders.
- When adjusting hydraulic motor speeds, never use the SCV control lever; always use an adjustable flow control valve with the control lever in full retract position. Always adjust the motor to its rated speed or lower. Never place an SCV control lever in Neutral with a hydraulic motor connected to its couplers.
- Never place the SCV control levers in the Float detent with a load suspended from an attachment. The load will immediately fall to the ground. Always lower the load first.
- Use of single-acting cylinders lowers the hydraulic fluid level in the hydraulic system reservoir. Always check that the hydraulic fluid level does not fall below the minimum level with all single-acting cylinders in the fully extended position.
- Stand clear of equipment when working with hydraulic lines and tools.
- Keep the hydraulic system clean. Always use dust caps on quick-couple disconnects when the hydraulic system is not in use. Wipe dirt from hoses and connectors before attempting to connect the hoses.
- Use the proper fittings for hoses and attachments.
- Never depend on the hydraulic system to support machine parts. Block or jack up rotary wings; if they are raised to a vertical position, secure them with chains or fastening rods.
- Avoid prolonged contact with hydraulic fluid. Hydraulic fluid is a skin irritant.

2.1.5 Safety Rules for Operating the PTO

Tractors have 540 rpm PTO shafts, 1,000 rpm PTO shafts, or both. Make sure the attachment is compatible with the PTO. Some tractors have interchangeable stub shafts and are equipped with a selector lever to provide different speeds. With a selector lever, always be sure the selector is set for appropriate rpm.

WARNING!

Regardless of a PTO's rpm, remember that an uncovered PTO shaft is extremely dangerous, especially if it is engaged. Contact with a spinning shaft can result in serious injury or death. Always be sure PTO spinner (shaft) shields are installed and rotate freely. Disengage all power and shut off the tractor engine before checking spinner shields.

Always keep the following safety guidelines in mind when operating the PTO:

- Keep PTO guards and shields in place, even when the PTO is not running (*Figure 23*).

Figure 23 PTO warning decal.
Source: SB Johnny

- Always disconnect the PTO when not in use. Failure to disconnect the PTO when not in use accounts for many operator injuries.
- Never engage the PTO while the machine engine is shut off.
- Keep hands, feet, and clothing away from the PTO parts.
- Never step or cross over a PTO part.
- Do not wear loose clothing.
- Never operate the PTO shafts at extreme angles.
- Never ride, or let others ride, on the drawbar of the tractor.
- Always be sure the PTO stub shaft is properly secured to the machine's power shaft.
- Do not service, lubricate, or perform other work without first disengaging the PTO and shutting off the tractor engine.

PTO Safety

Every year, serious and often fatal accidents occur when a worker gets too close to an engaged PTO. Working around a moving PTO is very dangerous because it can easily catch loose clothing or hair. The PTO rotates very rapidly and has very high torque.

2.1.6 Safety Rules for Tractor Tires

An inflated tire can be very dangerous. Under pressure, tires can burst with explosive force. Always follow the manufacturer's instructions for tire inflation pressures and tire changing procedures along with the following rules:

- Check tires daily for damage or noticeably low pressure.
- Check inflation pressure with an accurate gauge to prevent overinflation. Overinflation reduces performance and increases the strain on both the tire and wheel rim, which can cause tire or rim failure.
- Inflation pressures of less than 12 psi (83 kPa) should be monitored regularly because the increased risk of tire **bead** leaks can result in tire damage.
- If the tires wrinkle, buckle, or slip on the rim under high-traction conditions, the tire pressure can be temporarily increased by 4 psi (28 kPa) to alleviate the problem, but traction will be reduced.
- Tire pressures of tractors operating on side slopes or hills should be increased by 4 psi (28 kPa) to compensate for weight transfer from side-to-side or front-to-back and back-to-front.
- When inflating a tire, never stand parallel to the rim in case it explodes. Always stand in line with the tread of the tire.

Bead: The rigid inner ring of an inflatable tire that fits against the rim and creates an air-tight seal when the tire is inflated.

- Stand clear of the tire and rim if it is being lifted with a chain or cable sling during a tire changing procedure.
- Never weld or perform torch work on wheels with inflated tires.

2.2.0 Daily Inspection and Maintenance

Preventive maintenance of the equipment is important. Each day before beginning operation of the tractor, perform a walk-around inspection and any prescribed daily maintenance checks.

2.2.1 Servicing a Tractor

Servicing a tractor involves checking the operating components for safe operation and providing any necessary service materials. The following equipment and materials may be needed to service a tractor:

- air supply
- cleaning solvents
- engine coolant
- engine oil
- fuel
- grease gun
- hydraulic fluid
- lubricants
- operator's manual
- rags
- tire pressure gauge
- tool kit
- transmission fluid
- water

2.2.2 Typical Daily Inspection and Maintenance Procedures

The following basic servicing actions are only examples. Follow the company's preventive maintenance program when servicing equipment. *Figure 24* shows an image from a portion of service interval guidelines for a tractor. In this case, the image shows various filters that need to be inspected and maintained on a regular basis.

In addition, attachments connected to a tractor usually have periodic service requirements that must be performed. Always refer to the tractor and attachment operator's manual for the lubrication chart and service for specific components. The following are typical inspection procedures:

- Visually inspect for leaks and broken, worn, or missing parts. Report major repairs and replacements to the proper authority; a trained mechanic must complete these maintenance actions.
- Check the engine oil level (*Figure 25*). Add or change engine oil if necessary.
- Check the engine coolant level. Add coolant if needed. If the engine has a radiator cap, do not remove it unless the engine is cool; open the cap slowly to release pressure.
- Clean the air filter and replace as needed.
- Check the fuel level and drain the water and sediment from the fuel tank and filter when applicable. Fill the fuel tank as needed.
- Check the hydraulic fluid level when hot and add fluid when necessary. Also, check the filters, hose, lines, and fittings and look for leaks.

NOTE

The location of the filters and points where you can check or replenish liquids vary from one tractor to another. Refer to the operator's manual for each machine to determine where these points are located, and at what levels the liquids should be. Also, check the operator's manual to see the permissible liquids for replenishing any oils or coolants.

CAUTION

Make sure to add only an approved oil. Refer to the operator manual or company guidelines to determine the oil needed.

Figure 24 Typical tractor filters that require servicing.
Source: Image Provided as Courtesy of John Deere

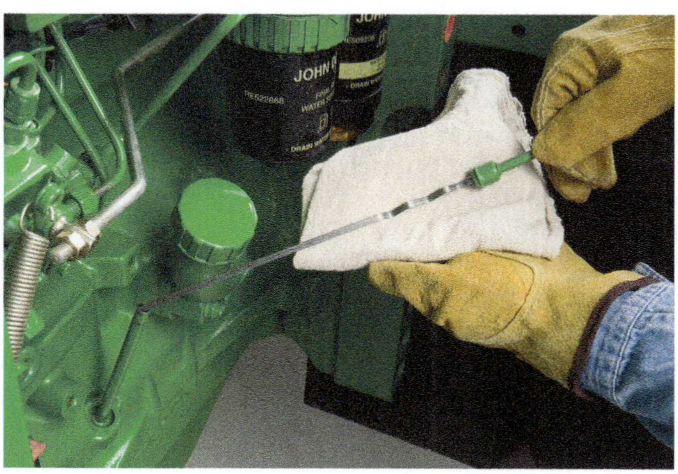

Figure 25 Checking tractor engine oil level.
Source: Image Provided as Courtesy of John Deere

- Check the transmission fluid level when hot. Add fluid if needed.
- Clean and lubricate grease fittings according to operator's manual.
- Check the tire pressure and inflate to recommended pressure.
- Check condition of tires.
- If operating in extremely dusty or trashy conditions, check the grille, radiator, and air conditioning condenser, if so equipped, for dirt or trash buildup. Clean as necessary.

Going Green

Fluid Disposal

To protect the environment, always contain and properly dispose of any leaking fluids or fluids removed from a tractor. When removing parts from a tractor, make sure to drain and contain any residual fluids in the affected parts.

2.3.0 Startup, Shutdown, and Operation of a Diesel or Gasoline Engine

Before attempting any prestart checks or startup, always read the operator's manual for the tractor to become familiar with specific controls and procedures.

2.3.1 Prestart Checks

Before starting or moving a tractor, perform any required daily inspection, maintenance, and safety checks. Also, make sure that the tractor's fuel tank is full. Check that all other people are clear of the tractor. Then, observe the following:

- *Pinch points* — Note if any attachments connected to the tractor will interfere with the rear tires when turning the tractor.
- *Obstructions* — Note any obstructions nearby or in the working area that could interfere with the movement of the tractor.

NOTE

Some tractors require that the clutch be depressed and released a second time while the engine is running to deactivate a hydraulic system lockout for power steering or other functions.

2.3.2 Startup

First, mount the tractor while maintaining three points of contact. Adjust the seat and then fasten the seat belt if the tractor is equipped with a ROPS. Make sure that the transmission is in Park, or the parking brake is set and the transmission is in Neutral. Check that the PTO, hitch, and any Hi/Lo shift and reverser controls are in the Off or Neutral position. If SCV controls are present, make sure they are in the Neutral position; however, if a hydraulic motor of an attachment is connected to any SCV, make sure that control is set to the Float position.

WARNING!

Always start a tractor from the operator's seat. It is dangerous to do otherwise.

CAUTION

If you crank the engine for more than 20 seconds, STOP and allow the starter to cool for two minutes. If you continue to crank the engine you could overheat the starter and cause it to fail.

When the controls are set, turn on the ignition and check that all warning/option indicator lights come on. Then set the throttle to the one-third or one-half position for gasoline powered tractors. For diesel in cold weather, activate any cold-weather starting aids. Next, press the clutch if required and activate the starter to crank the engine.

Release the starting device when the engine starts. If a choke is set, push it in slowly as the engine warms up. Check that the warning lights go off and that any gauges register the correct readings while the engine is warming up. After making sure the transmission is in Neutral, release the clutch while the engine warms up. If provided, observe the temperature gauge to determine when the engine has reached the proper operating temperature.

If the tractor has power steering, turn the wheels back and forth occasionally to determine if the hydraulic fluid is warm enough to allow easy steering. In very cold weather, consult the operator's manual for any procedures to allow rapid warming of the hydraulic fluid.

2.3.3 Safely Stopping a Tractor

First, depress the clutch (if present) or reduce forward speed and gradually apply both brakes. If the tractor will be stopped for a time with the engine running, shift the transmission to Park or set the parking brake and shift the transmission to Neutral. Then, set the throttle position to about one-third for gasoline engines or one-half for diesel engines for cooling purposes. If the tractor engine will be shut down, refer to the next section.

WARNING!

Start preparing to stop far in advance and look for possible hazards. Avoid sudden stops.

2.3.4 Shutdown

When ready to shut down a tractor, stop its movement. Shift the transmission to Park or set the parking brake and shift the transmission to Neutral. If attachments are connected, lower them to the ground and retract any hydraulic cylinders as much as possible to protect the cylinder rods from weather exposure. If the attachments will be detached, it is especially important to retract single-action cylinders as much as possible to return hydraulic fluid to the tractor's hydraulic system reservoir.

Set the PTO, hitch, and any Hi/Lo shift and reverser controls to the Off or Neutral position. Then, reduce the engine speed. Next, shut down the engine. Cycle all SCV controls through all positions to relieve hydraulic system pressure. Return them to Neutral and then turn off the ignition. Turn all other switches off and, if the tractor will be left unattended, remove the ignition key. If the tractor will be parked overnight, make sure that the fuel tank is full to help prevent water condensation. Then, dismount while maintaining three-point contact.

2.3.5 Basic Tractor Operations

Start and warm up the engine. Then depress the clutch (if present) and select the speed range (if present). Next, with the brakes applied, release the parking brake (if present) and shift the transmission as follows:

- *Standard (non-synchronized) transmission* — Select the gear to be used for operation and place the tractor in gear. Operate the tractor at medium to fast idle and slowly release the clutch pedal. Move the throttle to the desired operating speed after the clutch is fully released.
- *Synchronized transmission* — Set the engine at slow or moderate rpm and place the transmission in the lowest gear. Slowly release the clutch. To shift gears while the tractor is in motion, depress the clutch pedal and shift the gearshift lever, gear by gear, to the desired ground speed while changing the throttle to the desired engine rpm. The gears may take a few seconds to engage. Keep a steady pressure on the lever until it slides into position, then slowly release the clutch.
- *Power-assisted transmission* — In this transmission, the inching pedal replaces the clutch pedal. It is not used for shifting gears but for creeping speeds when connecting attachments, for making slow starts, and for emergency stops. It is possible to skip gears while shifting, but this results in a jerky motion. For safety, it is best to shift the tractor gear-by-gear. Place the gearshift lever into low gear with the engine running at a slow or moderate rpm. Shift the gearshift lever, gear-by-gear, to the desired ground speed while changing the throttle to the desired operating rpm.
- *Hydrostatic transmission* — Set the throttle to obtain the desired engine operating rpm. Then, move the speed control lever gradually forward or rearward until reaching the desired ground speed.

As the tractor moves forward or rearward, left or right turns are made by turning the steering wheel and, if necessary, applying a corresponding rear wheel brake to obtain a tighter turn. *Figure 26* shows how to make a tight right turn. Left turns are made in the opposite manner.

If equipped with a high-low shift, the speed of the tractor can be changed up or down by one-half gear without clutching while the tractor is moving. If equipped with a reverser, the tractor can be instantly moved backward while in a forward gear or forward while in a reverse gear without clutching, provided the selected tractor gear speed is for ground speeds under about 8 mph (13 km/h). If equipped, MFWD can be engaged for extra traction; however, it must be disengaged when making turns.

NOTE

Hand-shifted standard or synchronized transmissions may have a lever that works within various slots and notches of a forward or reverse quadrant, or levers that are positioned according to a chart that shows gear positions.

CAUTION

Do not make tight turns with a towed attachment. Doing so when moving forward can cause the attachment to catch on and then ride up the rear tire that is on the inside of the turn. In reverse, the attachment can jackknife, and a rear wheel could crush and ride up over the attachment. In both cases, injury and/or damage may result.

CAUTION

For tractors equipped with a speed range selector, the desired speed range can only be selected when the tractor is stopped.

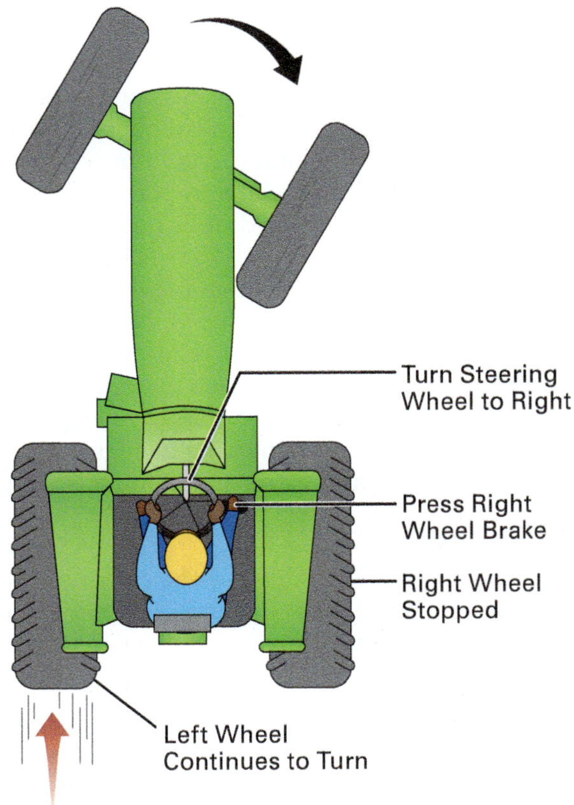

Figure 26 Turning to the right.

2.4.0 Connecting Attachments

A three-point hitch is the standard hydraulic linkage for tractor attachments. Three-point hitches consist of two lower links and an adjustable top link.

CAUTION

When connecting Category I attachments to a tractor, any sway bars may need adjustment to prevent binding and limiting full raise of the hitch.

2.4.1 Preparing the Attachment

A Category I three-point attachment hitch (*Figure 27A*) is narrower and is used for smaller attachments than Category II attachments (*Figure 27B*). Refer to *Table 3* to identify the attachment category. When connecting a Category I attachment to a Category II tractor hitch, install reducer bushings in the links to accommodate the smaller pins.

2.4.2 Connecting the Attachment

Follow these guidelines to connect an attachment to a three-point hitch (*Figure 28*):

Step 1 Check the operator's manual for attachment installation instructions.

Step 2 Set the rockshaft or hitch control lever to the lowest position.

WARNING!

On tractors equipped with EH hitch controls, make sure any load/depth control is set to Off or to a Float position before connecting or disconnecting a three-point attachment. If optional external raise/lower controls are used, the load/depth control is automatically disabled. This prevents the hitch from suddenly and unexpectedly moving due to weight changes caused by the addition or removal of an attachment.

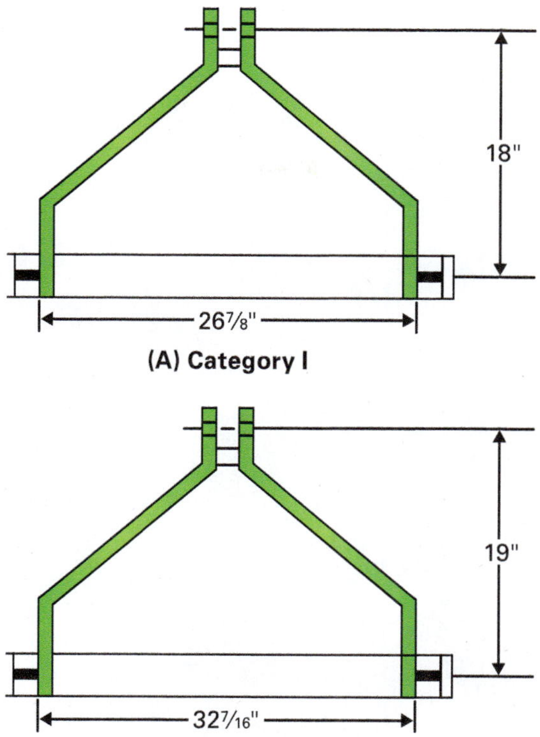

Figure 27 Category I and II attachment hitches.

TABLE 3 Hitch Category Sizes

Category	Mast Height	Width Between Lower Pins	Lower Pin Size	Upper Pin Size
I (A)	457 mm (18 in)	682.6 mm ($26^7/_8$ in)	22 mm ($^7/_8$ in)	19 mm ($^3/_4$ in)
II (B)	483 mm (19 in)	824 mm ($32^7/_{16}$ in)	28 mm ($1^1/_8$ in)	25.4 mm (1 in)

Source: Adapted from John Deere Sales Manuals, 2013, 2023

Step 3 Check the drawbar position to make certain it will not interfere with movement of the hitch. Some tractors have no separate drawbar. Others may require removal of the drawbar.

Step 4 Back the tractor so that draft links are in position to connect to attachment hitch pins.
- Make sure that the attachment is as laterally level with the tractor as possible. Use blocking if necessary.
- If the draft links are equipped with ball joints, back the tractor so that joints are over the hitch pins on the attachment.
- If the draft links are equipped with latches, align them so that latches are under the hitch pins of the attachment.

Step 5 Raise the draft links with the rockshaft unit or hitch control lever to the height needed to connect to attachment hitch pins.

Step 6 Place the transmission selector lever in Park or Neutral. When working on sloping ground, set the tractor wheel brakes, if possible, and block the wheels.

Step 7 Connect the left draft link to the left hitch pin of the attachment. Always attach the left draft link first. If the link connection is not in line with the hitch pin on the attachment, move the attachment ahead, if possible, or rock the tractor forward and backward if it is on level ground. On tractors with extendable or telescoping draft links, it is not necessary to rock the tractor or pull on the attachment.

Step 8 Insert the locking pin.

Step 9 Raise or lower the right draft link to the correct height by adjusting the lift link.

Figure 28 Connecting an attachment to a three-point hitch.

Step 10 Attach the right draft link and insert the pin.

Step 11 Unhook the center link (*Figure 29*) from the transport bracket. The center link enables the operator to tilt the attachment forward or backward to level it from front to rear.

Step 12 Adjust the center link to reach the mast by turning the outer housing or by using the adjusting handle.

Step 13 Adjust the center link so that both shafts inside the outer housing are extended equally.

Step 14 Drop the center link into the attachment mast and lock it into place by inserting a pin through holes in the mast and turnbuckle. Fasten the pin so that it cannot bounce out of the connecting point. If the center link is equipped with a latch assembly, depress the lever, fit the latch over the ball, and release the lever.

Step 15 Adjust sway bars or chains or install sway blocks in a lower position with the narrow, flat edge inward. This will eliminate sway during operation and transport. Be sure draft links do not bind with sway blocks.

Step 16 Install any front-end weights for balance, as specified for the attachment.

Step 17 Raise the attachment to the transport position, using the rockshaft or hitch control lever. Make certain that it clears at all points and that the connections are secure.

Figure 29 Location of center link.

2.4.3 Attaching a PTO

You should take precautions when connecting or disconnecting attachments from a PTO and adhere to the following list of steps and safety measures.

Step 1 Be sure to have the correct rpm PTO stub shaft for the attachment being connected.

Step 2 Set the wheel brakes.

Step 3 Disengage the power to the PTO shaft.

Step 4 Shut off the engine.

Step 5 Remove the PTO shaft guard or master shield if the tractor is so equipped.

Step 6 Check the tractor PTO coupling and the attachment PTO stub shaft to make sure they have the same number of splines. If not, attach the correct stub shaft and manually shift the shaft speed. See the manufacturer's instructions.

Step 7 Clean the inside of the coupling of all grease and debris.

Step 8 Oil the PTO shaft spline.

Step 9 Verify that the yokes on the shaft are aligned identically as shown in *Figure 30*.

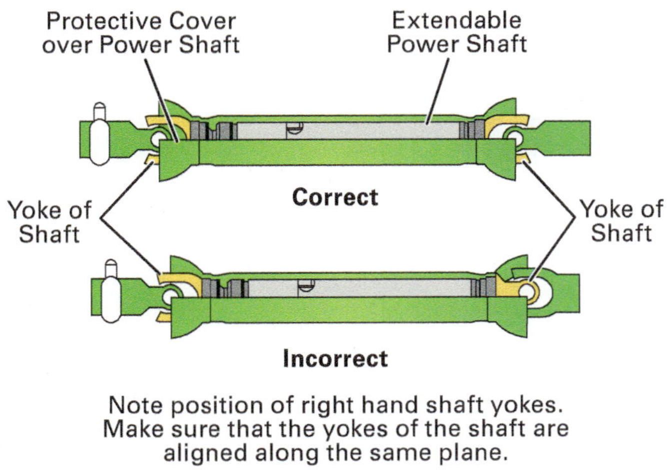

Figure 30 Positioning universal joints and shaft yokes.

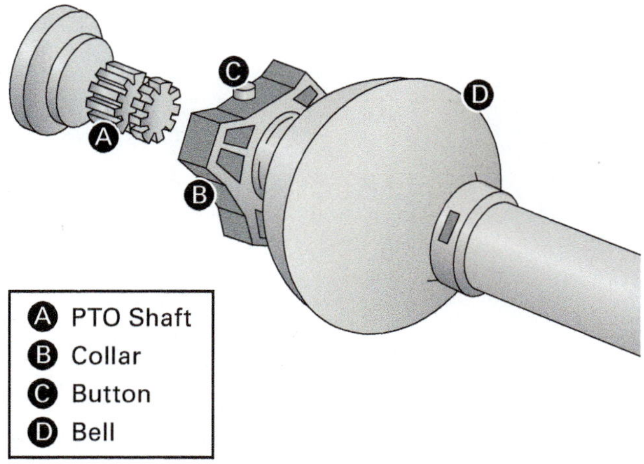

Figure 31 Power shaft with a spring-loaded lock.

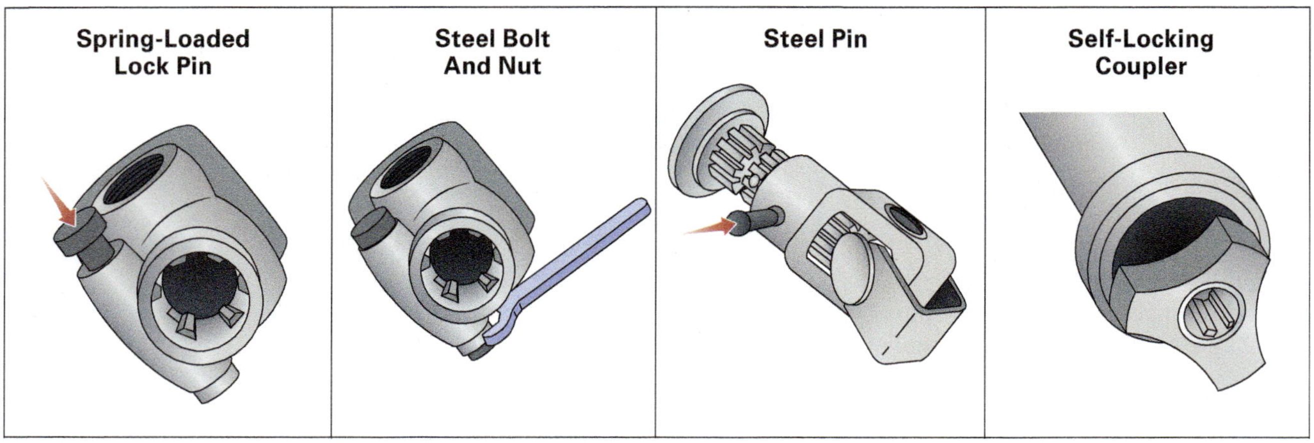

Figure 32 Locking devices.

Step 10 Attach the power shaft with a spring-loaded lock (*Figure 31*), if applicable. Other types of locking devices are common, as shown in *Figure 32*.

Step 11 Move the collar back against the stop on the yoke.

Step 12 Turn the collar to align the yoke with splines on the tractor PTO shaft. Some models require pushing the button down and turning the collar until the button stays down.

Step 13 Slide the power shaft onto the PTO shaft.

Step 14 Depress the pin or button and move the shaft forward until the collar snaps into the locked position.

Step 15 Check the locking device by pulling back on the bell or guard shaft. Do not pull on the collar; the latch will release.

Step 16 Slide the safety shield into place.

Step 17 Keep the alignment as straight as possible. Normally a PTO shaft should not be operated at more than a 15-degree angle from the opposite shaft of either U-joint.

Step 18 Maintain the proper drawbar height and length.

Step 19 Reinstall the PTO master shield.

NOTE

PTOs do not always require the use of the tractor clutch pedal; you can engage and disengage a PTO with the tractor stopped or moving.

2.4.4 Operating a PTO

The PTO provides mechanical power to the attachments. It should always be disengaged when not operating the attachment. Follow these guidelines to engage the PTO.

Start the tractor engine if it is not already running and, if required, depress the clutch pedal and both brakes all the way. If necessary, allow the tractor to come to a complete stop. Then, engage the PTO and, if applicable, release the brakes and engage the clutch slowly.

To shift gears and leave the PTO running, depress the clutch and, if required, both brakes. Change gears and, if used, release the brakes. Engage the clutch slowly.

To disengage the PTO and, if required, stop the tractor to do so, depress the clutch and both brakes. When the tractor stops, disengage the PTO.

2.4.5 Connecting and Disconnecting Hydraulic Hoses for Attachments

The following guidelines cover the connection and disconnection of attachment hydraulic cylinder or motor hoses to the SCV couplers at the rear or side of a tractor.

Use the following procedure to connect hoses:

WARNING!

Operators must wear gloves and eye protection when connecting, disconnecting, or checking any high-pressure hoses.

Step 1 Stop the tractor engine.

Step 2 Cycle the SCV control levers through all positions to relieve pressure at the couplings. Then, if hydraulic motors are to be connected, set the appropriate SCV control levers in the Float position. If hydraulic cylinders are to be connected, set the appropriate SCV control levers in the Neutral position.

Step 3 Remove and save any attachment hose nipple caps. Then wipe the nipples clean.

Step 4 Open the covers on the SCV couplers (*Figure 33*) and verify that they are clean.

Step 5 Select the proper coupler for a single-acting or double-acting cylinder extend and the corresponding hose nipple. Insert the nipple in the coupler and push in firmly to **seat** the nipple.

Step 6 Repeat Step 5 for the retract coupler and hose for double-acting cylinders.

Step 7 Pull all the hoses slightly to make sure they are seated in the couplers.

CAUTION

For hydraulic motors, the retract coupler must be used as the pressure source instead of the extend coupler, and a flow control valve must be present in the line either externally or internally to the valve.

Seat: To cause two surfaces to fit firmly together.

Figure 33 Typical SCV coupler covers.

WARNING!

Never put an SCV control lever in a partial retract or neutral position with the engine running if a hydraulic motor is connected to the SCV. The hoses may rupture, or the motor may be damaged.

NOTE

It is especially important to retract single-acting cylinders as much as possible to return hydraulic fluid to the tractor hydraulic system reservoir.

Step 8 Start the tractor engine and move the SCV control levers from Neutral to extend and retract the cylinders slightly; or from float to full retract and back to float to start and stop hydraulic motors. Then, check for signs of any visible leaks.

Step 9 For hydraulic motors, adjust the motor speed as described previously.

Use the following procedure to disconnect hoses:

WARNING!

Operators must wear gloves and eye protection when connecting, disconnecting, or checking any high-pressure hoses.

NOTE

Check attachment hoses and cylinders or motors for any visible signs of recent leakage.

Step 1 Lower all loads/attachments to the ground and retract all cylinders as much as possible to protect the cylinder rods from weather exposure.

Step 2 Stop the tractor engine, cycle the SCV control levers through all positions to relieve pressure at the couplings, and return the levers to the Neutral position.

Step 3 Pull the hoses from the SCV couplers.

Step 4 Replace the attachment hose nipple caps and close the SCV coupler caps.

2.0.0 Section Review

1. To keep a tractor from tipping backwards, _____.
   a. lock both rear wheels with the brakes at the same time when moving forward in low range
   b. do not attempt to move forward if the rear wheels are mired
   c. attach loads as high as possible with the hitch
   d. drive out of ditches by moving forward

2. Tire pressures of tractors operating on side slopes or hills should be _____.
   a. reduced by 4 psi
   b. increased by 10 psi
   c. increased only in wet conditions
   d. reduced by 10 psi only in wet conditions

3. When first starting a tractor, make sure that _____.
   a. the PTO is engaged
   b. all SCVs are energized
   c. the transmission is in Park
   d. the transmission is in Neutral with the parking brake off

4. When connecting an attachment, the draft links are raised with the _____.
   a. rockshaft unit
   b. throttle lever
   c. drawbar
   d. the hitch coupler

Module 22105 Review Questions

1. The most common type of transmission in utility tractors is the _____.
   a. sliding gear
   b. hydrostatic drive
   c. synchromesh
   d. hydraulic assist

2. The flywheel of a tractor's engine serves as what component of a basic disk clutch assembly?
   a. Drive plate
   b. Clutch plate
   c. Pressure plate
   d. Throw-out plate

3. If a tractor's charging system warning indicator shows a high negative discharge rate when the engine is at half throttle, the tractor _____.
   a. is operating properly
   b. should have its battery replaced
   c. should be run without lights or fans
   d. should be stopped and inspected to determine the cause

4. If a diesel engine has run out of fuel, it may need its _____.
   a. tank refilled
   b. tank refilled and engine primed
   c. fuel lines and carburetor drained
   d. fuel lines and injectors bled of air

5. What component shifts the tractor's transmission forward or reverse?
   a. High-low shifter
   b. Range selector
   c. Gear selector
   d. Reverser

6. At the back of most tractors, the splined shaft that provides mechanical power to connected attachments is the _____.
   a. power takeoff
   b. SCV coupler
   c. rockshaft
   d. draft link

7. Most selectable control valve (SCV) control levers have _____.
   a. two operating positions
   b. four operating positions
   c. six operating positions
   d. eight operating positions

8. When attempting to start a diesel engine when the air temperature is below 40°F (4°C), what ignition switch position activates the glow plugs or the intake air warmers?
   a. Accessory
   b. Warm up
   c. Start
   d. Run

9. What device(s) are added to prevent wheel slippage when towing heavy loads?
   a. A heavier three-point hitch
   b. A heavier drawbar
   c. Rear-end ballast
   d. Wheel weights

10. If a person or animal approaches an operating tractor with a rotating attachment, the operator should _____.
    a. stop and turn off the rotating attachment
    b. carefully drive around the person or animal
    c. stop moving until the person or animal leaves
    d. slow the engine speed until the person or animal passes

11. When first moving a tractor, the operator should _____.
    a. test the brake system and steering in both directions
    b. accelerate to full speed
    c. lean out of the cab to check for clearance
    d. lock both rear wheels with the brakes

12. Because of the increased risk of a tire bead leak on a tractor tire, inflation pressures of less than what level should be monitored regularly?
    a. 8 psi (55 kPa)
    b. 10 psi (69 kPa)
    c. 12 psi (83 kPa)
    d. 20 psi (138 kPa)

13. Before starting a tractor with an attachment that has a hydraulic motor connected to any selectable control valve (SCV), make sure that the control is set to the _____.
    a. sink position
    b. extend position
    c. retract position
    d. neutral position

14. When shutting down a tractor with an attachment that is to be disconnected, the hydraulic cylinders should be retracted as much as possible to _____.
    a. level the tractor
    b. lower the attachment
    c. relubricate the cylinder rods
    d. return hydraulic fluid to the reservoir

15. The tractor component that allows the operator to tilt an attachment forward or backward to level it from front to rear is the _____.
    a. draw bar
    b. center link
    c. draft link(s)
    d. sway bar(s)

Answers to odd-numbered review questions are found in the Module Review Answer Key at the back of this book.

Answers to Section Review Questions

Answer	Section	Objective
Section One		
1. d	1.1.1	1a
2. b	1.2.4	1b
3. c	1.3.2	1c
4. b	1.4.1	1d
Section Two		
1. b	2.1.2	2a
2. a	2.2.2; *Figure 24*	2b
3. c	2.3.2	2c
4. a	2.4.2	2d

MODULE 22212

Skid Steers

Source: Courtesy of CASE Construction Equipment

Objectives

Successful completion of this module prepares you to do the following:

1. Identify and describe the components of a skid steer.
   a. Identify and describe chassis components.
   b. Identify and describe skid steer controls.
   c. Identify and describe skid steer instrumentation.
   d. Identify and describe skid steer attachments.
2. Describe the prestart inspection requirements for a skid steer.
   a. Describe prestart inspection procedures.
   b. Describe preventive maintenance requirements.
3. Describe the start-up, shut-down, and operating procedures for a skid steer.
   a. State skid steer–related safety guidelines.
   b. Describe start-up, warm-up, and shut-down procedures.
   c. Describe basic maneuvers and operations.
   d. Describe related work activities.

Performance Tasks

Under supervision, you should be able to do the following:

1. Complete a proper prestart inspection, maintenance, and housekeeping for a skid steer.
2. Demonstrate proper entrance and exit of the operator cab on a skid steer.
3. Perform proper start-up, warm-up, and shut-down procedures on a skid steer.
4. Execute basic maneuvers with a skid steer, including the following:
   - Grading
   - Removing stumps and boulders
   - Steering
   - Changing attachments
   - Loading a dump truck
   - Maneuvering on slopes
   - Using a fork attachment for moving materials
5. Properly load and secure a skid steer for transport.

Digital Resources for Heavy Equipment Operations

Scan this code using the camera on your phone or mobile device to view the digital resources related to this craft.

SCAN ME

Overview

Skid steers are compact, highly maneuverable machines designed to accept a wide variety of attachments. Their standard configuration typically includes a bucket for moving soil, crushed stone, and related materials. With optional attachments, these machines provide a platform to accomplish an array of tasks. This module explores the basic care and proper operation of skid steer machines, as well as some of the many attachments that make it an indispensable tool in both construction and agriculture.

NCCER Industry-Recognized Credentials

If you are training through an NCCER-accredited sponsor, you may be eligible for credentials from NCCER. The ID number for this module is 22212. Note that this module may have been used in other NCCER curricula and may apply to other level completions. Contact NCCER at 1.888.622.3720 or go to **www.nccer.org** for more information.

You can also show off your industry-recognized credentials online with NCCER's digital credentials. Transform your knowledge, skills, and achievements into credentials that you can share across social media platforms, send to your network, and add to your resume. For more information, visit **www.nccer.org**.

1.0.0 Skid Steer Components

Performance Tasks

There are no Performance Tasks in this section.

Objective

Identify and describe the components of a skid steer.

a. Identify and describe chassis components.
b. Identify and describe skid steer controls.
c. Identify and describe skid steer instrumentation.
d. Identify and describe skid steer attachments.

The term *skid steer* is a very appropriate name, as it describes the way these units maneuver. Compact loaders are built using tracks or wheels (*Figure 1*), but the wheeled versions, known as skid steer loaders, are more common and are the focus of this module. Although the earliest designs had three wheels, skid steer loaders are typically four-wheel machines. However, unlike common loaders, skid steer loaders have no steerable wheels. Instead, the wheels on each side are driven together, always turning at the same speed and direction. (*Figure 2*) By powering the pair of wheels on one side forward or backward while the opposite side wheels remain idle, the unit can be steered. The idle wheels simply slide, or skid, in the direction they are forced by the driven wheels on the opposite side. Compact loaders with four-wheel steering capability are. available, but they do not skid to steer.

Figure 1 A skid steer loader rakes a yard prior to turf grass planting.
Source: Courtesy of Doosan Bobcat North America, Inc. All Rights Reserved.

Figure 2 A Bobcat compact tracked loader installing turf.
Source: Courtesy of Doosan Bobcat North America, Inc. All Rights Reserved.

The Melroe Bobcat

The Melroe Manufacturing Company was started by Edward G. Melroe in 1947. Earlier compact loader models with three wheels were introduced, but the first true skid steer, with four powered wheels, was introduced in 1960. Here is an advertisement for an early Melroe Bobcat machine, bearing the name chosen for it in 1962 by a Melroe associate.

Source: Courtesy of Doosan Bobcat North America, Inc. All Rights Reserved.

One factor that makes skid steers popular and versatile is that they can be turned in extremely tight spaces. Powering the wheels on one side and allowing the wheels on the opposite side to slide results in the ability to turn sharply, without creating a turning radius. However, these units can truly spin in a circle. By powering the wheels on the right backwards, while simultaneously powering the wheels on the left forward, the unit can literally spin as if it is on an axis. This special mobility is one of the primary features that make skid steers one of the most valuable tools on a jobsite.

Today's skid steers are far safer than their predecessors. Earlier models offered little or no protection for the operator in any direction. Required protection systems have reduced injuries remarkably. On the other hand, these units are so versatile, nimble, and responsive, they can make an operator feel invincible. This fact alone can lead to serious accidents.

This section introduces the basic components of a skid steer, along with the controls and instruments used in its operation. Several versatile attachments are also presented.

1.1.0 Main Chassis Components

A skid steer is built on an extremely rigid and strong frame known as the chassis. The extreme strength and rigid, compact construction of the chassis allows the machine to easily shrug off the stress imposed by skid steering. All other major components associated with the operation of the machine are mounted to the chassis.

Figure 3 provides an overview of the primary chassis components and structure. Skid steers are very compact units by design. They have fewer and less complex parts than their larger relatives. Of course, each brand and model of skid steer is slightly different in its construction, capacity, and operation. It is essential that operators become familiar with the specific model of skid steer in use. When beginning to use a new or different machine, always review the operator's manual.

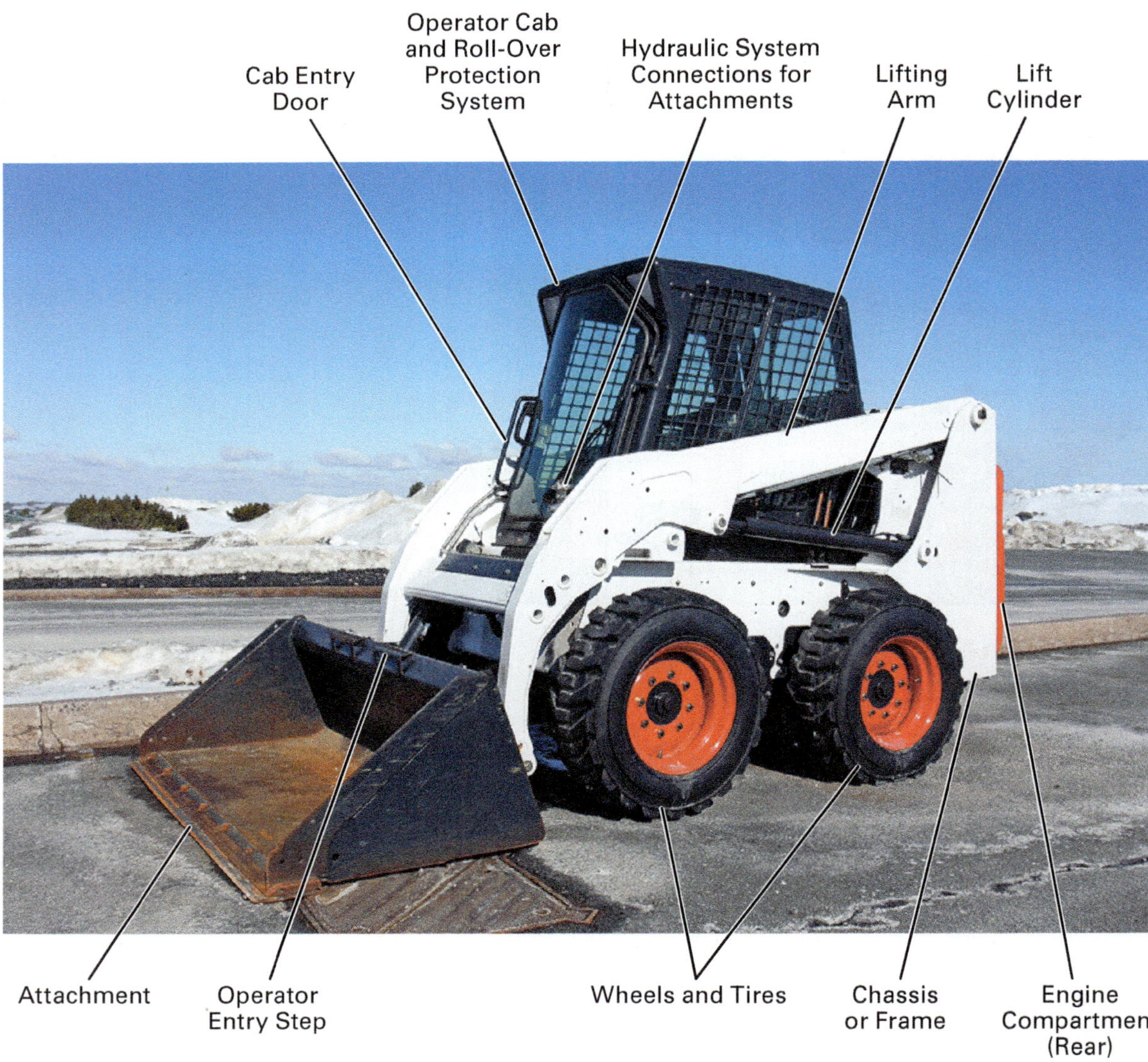

Figure 3 Skid steer chassis components.
Source: kozmoat98/iStock/Getty Images

One feature that allows their compact construction is the lack of a transmission. In fact, there is no direct relationship between the engine and the four wheels at all. The primary responsibility of the engine is to drive hydraulic pumps. All other functions, including movement of the machine, are then powered through hydraulic motors. Because the engine is not connected to the drive train, the differential is eliminated, along with axles that cross from one side of the frame to the other.

1.1.1 Engine and Engine Area

Although a few skid steers use gasoline engines, the vast majority of engines are diesel-powered. The horsepower generally ranges from around 20 hp to 100 hp. As shown in *Figure 4*, the engine and its accessories have been tightly packed into the compartment. However, serviceability is an important consideration. Although the compartment is quite crowded, most items requiring regular attention are positioned to allow easy access. The engine coolant radiator, as well as the fluid coolers, have been relocated in some models. This relocation is possible because the radiator is not cooled by an engine-driven fan blade. Instead, the fan blade is driven by a hydraulic motor. The fluid filters have been located near the access door to allow easy access to them as well.

Figure 4 Engine and engine compartment.

The engine on all active skid steer models is located behind the operator (*Figure 5*). The primary hydraulic pump is situated toward the center of the machine, beneath the operator. This keeps most of the weight at the rear of the machine to help balance the load.

Figure 5 Skid steer engines are located behind the operator.

1.1.2 Drive System

The drive system is one of the features that make skid steers unique. Instead of a traditional drivetrain with a gearbox and differential, the wheels are driven by powerful hydraulic motors. This type of drive system, one that relies on

Hydrostatic drive: A system that relies on the flow and pressure of a fluid to transfer energy. In skid steers, the transfer of energy is from the engine, driven by diesel fuel, to the four drive wheels through hydraulic motors.

hydraulic fluid flow and pressure to transfer energy, is called a **hydrostatic drive**. Because there is no direct steering capability, there are no linkages or rods associated with turning the machine. In fact, there is no steering wheel for the operator. As a result, the drive system does not have to accommodate any change of direction in the wheels. They simply rotate in one direction or the other, and no more. The simplicity of the design adds to the reliability and reduces the maintenance requirements.

Because each motor is connected to the wheels by chains, the two wheels on each side remain synchronized at all times. If one of the wheels is turning, the second is turning as well. The two hydraulic motors are completely independent of each other. The drive sprockets and chains are sealed inside an enclosure and are bathed in oil to provide ideal lubrication and help transfer heat from the drive components.

1.1.3 Operator Cab

The design and functionality of the operator cab has changed a great deal over the years. Early models had no cab at all (*Figure 6*). This situation led to injuries. The lack of any obstructions around the driver also encouraged unsafe behavior. Some injuries occurred as the result of falling loads that had been lifted overhead. Many others resulted from the up and down movement of the lifting arms on both sides of the operator in very close proximity. One could argue that visibility was excellent, which added to the skid steer's popularity. Cabs and other safety features were continually added by manufacturers, some as the result of regulations. Today, the operator cab can be considered a self-contained capsule devoted to an operator's protection and comfort.

Figure 6 Bobcat Model M440, introduced in 1962.

Source: Courtesy of Doosan Bobcat North America, Inc. All Rights Reserved.

To maintain a compact size, every area is used to mount operating components. The area beneath the operator is no exception. The cab can be raised as a unit to gain access to the components below. *Figure 3* shows an example of a cab that tilts forward, while *Figure 7* provides an example of a cab that tilts toward the rear of the machine. This model places the battery under the cab. The area under the operator is the typical location for the hydraulic pumps, driven directly from the engine drive shaft. The cab is hinged at one end and is secured to the chassis for operation by pins or mechanical linkage.

Note the location of the fuel tank. The area between the cab assembly and the engine compartment is a common location for the fuel tank.

Most of the structural features of the cab have resulted from requirements set forth by the US Occupational Safety and Health Administration (OSHA). The primary safety concerns for an occupied cab were related to objects falling from above, reaching or climbing in and out of the seat through the lifting arms, and rollovers. Standards for a rollover protective structure (ROPS) and a falling

Figure 7 Operator cab tilted back for service access.

object protective structure (FOPS) have answered these concerns effectively. The design of most ROPS and FOPS must be approved by a professional engineer, who often requires substantial testing for new designs or modifications to existing systems. ROPS and FOPS must be capable of withstanding a force equal to twice the weight of the machine at the point of impact.

To increase safety further, a control interlock system is also built into many machines. In some cases, the machine controls are not enabled unless the hinged armrests are down in front of the driver. Some are fitted with left and right armrests that fold down toward the center. Others, like the skid steer shown in *Figure 8*, have one-piece armrests that hinge straight up and over the operator's head for seat access. Some control systems may also be interlocked through the seat, where a certain amount of weight is required to enable the controls. Most people are familiar with this feature in passenger cars, where a seat switch disables the passenger air bag unless a significant amount of weight is present.

Figure 8 Skid steer with one-piece armrests.

Fully enclosed operator cabs usually have a heating and cooling system to maintain operator comfort. Even radios and CD players are options for those who find them valuable during many hours on the job. Fully enclosed cabs are usually pressurized by the ventilating system that blows filtered outside air into the operator area. Maintaining a higher pressure inside the cab than outside helps keep dust and dirt out.

1.1.4 Hydraulic Systems

In skid steers, the primary job of the engine is to drive one or more hydraulic pumps. An alternator to charge the battery and power the electrical components may be the only other device that is powered directly by the engine. All the hard work is done by the engine indirectly, using the hydraulic system as a means of transferring energy.

The engine supplies power to the pumps that build pressure in the hydraulic system. The pumps are generally driven directly from the crankshaft of the engine (*Figure 9*). In most vehicles, the crankshaft is connected directly to a transmission or transaxle. Manufacturers save space by stacking the hydraulic pumps on top of each other. The crankshaft is connected to the first pump, and power is transmitted through its shaft on to the next pump, and so on.

Figure 9 Hydraulic pump.

Two hydraulic pumps are generally used to power the drive wheels. One pump provides power to the hydraulic motor on one side, and the other pump handles the opposite side. Another pump provides the operating pressure and flow for work tools and attachments (*Figure 10*). The smaller pump on the end of the assembly provides pressure to devices that require lower flow rates and may also push hydraulic fluid for the entire system through one or more filters. There is only one reservoir for hydraulic fluid; the entire system shares the same fluid.

Figure 10 Hydraulic system attachment points.

Some hydraulic pumps are designed for fixed displacement performance, while others provide variable displacement. In a **fixed displacement pump**, the flow rate of the fluid is based on the speed of the engine driving it. As the engine speed increases, the fluid flow rate increases. This type of pump is often used to power the lifting arms and attachments. Additionally, as long as the pump shaft is turning, flow is being produced whether it is needed or not. When a component is not being moved by the fluid, a relief valve must allow it to remain in motion and return to the pump inlet. Without this feature, the hydraulic fluid would soon overheat inside the pump.

Fixed displacement pump: A pump that cannot be adjusted to deliver more or less fluid volume in a pumping cycle.

1.1.5 Tires

The tires of a skid steer must be able to withstand the punishment of constantly skidding to one side or the other, without rolling off the rim. The stress on the tires to skid is even greater when the machine is carrying a load. Highly abrasive surfaces will cause a higher degree of wear and offer a higher resistance to skidding.

Many types of tires are available for skid steer loaders. The least expensive type is the standard pneumatic (air-filled) tire (*Figure 11*). These tires need exceptionally heavy sidewalls. The tire tread is generally deep, allowing for a great deal of use before contact is made with the casing of the tire. Pneumatic tires offer a better ride and increased comfort for the operator in most cases. Several different categories are available, depending on the need and budget. The categories are regular duty, heavy duty, and severe duty. The duty level typically is related to the number of plies used in the construction of the tire.

Figure 11 Pneumatic skid steer loader tire.

Although they ride more comfortably and are less expensive, pneumatic tires are subject to punctures or other damage that may both ruin the tire and cause a work stoppage. The type of terrain and work dictate the tire choice. In areas with many tire hazards, such as a building demolition site, a tire without air offers a more functional choice. Such tires are sometimes called flat-proof tires (*Figure 12*). These tires are usually more expensive, but the savings over time from avoiding lost time on the job can be significant. The air cells in the tire help to cushion the ride somewhat. Note that this example has a completely smooth surface. This tire would be a good choice for hard surfaces such as concrete. However, with no tread, traction outdoors would likely be a serious problem. Flat-proof tires are available with tread as shown in *Figure 13*, and with more or fewer air cells for cushioning. With so many different tires to choose from, the best choice is made through consultation with tire specialists regarding the work environment. It is not unusual for an owner to have more than one type of tire on hand.

Figure 12 Flat-proof tires for a skid steer loader.

All compact loaders tend to damage the terrain in two ways. Damage occurs as the tire is skidded while the machine is turning, and damage occurs from the weight of the vehicle rolling across the ground. The wider the tire, the more weight of the skid steer loader is displaced. Maximum displacement of the vehicle's weight is achieved by using tracks. Of course, compact loader machines are available as tracked models, without any wheels. These are known as compact tracked loaders (*Figure 14*). The drive system works much the same, with independent hydraulic pumps driving the tracks on each side forward or backward. The tracks help to more evenly distribute the weight and improve the stability of the machine. The tracks handle muddy conditions better than tires. For the best of both worlds, a skid steer with tires can be fitted with over-the-tire (OTT) tracks made from rubber or steel (*Figure 15*). Not all tracks are solid in construction. Some are made of steel bars several inches apart. They maximize traction but do not distribute the machine's weight as well. Depending on the make and design of the machine, spacers may be needed between the wheels and their mounting hubs. This pushes the wheel farther away from the chassis, providing added clearance for the tracks.

Figure 13 Treaded flat-proof tire.

Figure 14 Compact tracked loader.
Source: Courtesy of Doosan Bobcat North America, Inc. All Rights Reserved.

Figure 15 OTT tracks.
Source: Edgar Lee Espe/Shutterstock

1.2.0 Controls

All models of skid steer are either lever- or joystick-controlled; no steering wheel exists. Most current models are controlled by left and right joysticks, which initiate the most important and most common control actions. One distinct advantage of the joysticks is that they allow multiple control inputs to be made at the same time. A few other controls are in various locations in the cab. Review the operator's manual to fully understand the controls available for the model in use.

1.2.1 Vehicle Movement Controls

The skid steer has no direct capability to steer. Turns occur when the drive wheels on one side are turning faster than the drive wheels on the opposite side. Because no input is required to perform steering, there is no need for a steering wheel.

Driving some units may require the use of two control levers. To engage and turn the drive wheels on one side, the lever on the corresponding side is pushed forward or backward only. The amount of travel in the lever controls the speed and force of the movement. If the right lever is very carefully and gently pulled toward the operator, the skid steer responds gently, and begins backing up to the left. This happens because the right drive wheels turn in reverse, while the left wheels skid across the floor or ground. Pulling one lever back while pushing the other forward causes the machine to spin in a very tight circle. The drive wheels on one side are driving in the opposite direction from those on the other side.

Figure 16 Left joystick on a skid steer loader.

Later models of skid steers are generally controlled with joysticks. For example, some machines are driven using the left joystick (*Figure 16*). Pushing the joystick straight forward engages the drive wheels on both sides, and the machine moves forward. Pushing it forward and to the left executes a left turn. The farther left or right the joystick is pushed, the slower the drive wheels on that side turn in relation to the opposite side. If the tightest possible turn is desired, the left joystick is not pushed forward or backward at all. Pushing it directly to the right or left results in the machine spinning in place. If the normal engine speed is insufficient, the operator can use the foot throttle to increase the engine speed. The engine speed can also be controlled with a hand-operated throttle (*Figure 17*). It is not uncommon for both a foot pedal and a hand throttle to be on a single machine. However, remember that the ground speed of the machine is not dependent upon the accelerator pedal like most heavy equipment.

Some machines with advanced control schemes offer two or more control configurations. For example, Bobcat units equipped with their advanced control system (ACS) can be switched from drive control exclusively through the left joystick (standard) to the older H-pattern of control. In the H-pattern, the left joystick controls the left wheels, and the right joystick controls the right wheels.

Figure 17 Hand-operated throttle.

Figure 18 Right joystick.

1.2.2 Attachment Controls

Note the buttons on the joystick in *Figure 16*. A placard on the cab wall helps familiarize an operator with the function of the buttons and controls. One button on the left side controls the horn (top) and the other initiates continuous hydraulic fluid flow to an attachment, such as an auger. On the right side of the left joystick, the two buttons are used to control the position of three-way hydraulic valves. The trigger on the joystick is hidden from view.

This skid steer has two speeds, allowing the operator to choose the top speed based on the work and terrain. The operator can enable high-speed operation for travel by pulling and holding the trigger.

Manufacturers try to place multiple control functions on the joystick. This allows the operator to control a variety of devices and systems without having to let go of the joystick. The operator should always try to maintain contact with both joysticks. When a joystick is released, it should always return to the neutral position.

The right joystick (*Figure 18*) of the same machine provides control over the attachment. Moving the joystick forward and backward lowers and raises the lifting arms, respectively. Moving it left and right tilts the bucket or other attachment up and down. On some machines, the bucket or other work tool can be placed in a **float mode** by first pushing the joystick fully forward. It may stop in a slight detent and remain there to place the tool in the float mode. On other machines, a momentary trigger-pull is required while the joystick is fully forward.

Float mode: Placing the bucket or other attachment into the float mode allows the attachment to automatically follow the contour of the ground, remaining at the same elevation above it.

The top of the joystick has two buttons and a slide switch. The function of the buttons or switches depends upon the skid steer model and the attachment being used. If the attachment, such as an auger, also needs to be controlled, these switches and buttons may provide the input. The right joystick also has a trigger hidden from view. The black lever just forward of the joystick in the photo is a hand-operated throttle control.

Figure 19 Snow blower accessory with manually mounted controls.
Source: Andrew Ciscel

Some older skid steers may not have many control options for attachments. For these situations, attachments may have their own controls that are routed into the operator cab and mounted in a convenient location. *Figure 19* shows an older skid steer using a snow blower with a separate control.

Note that this is only one example of how a skid steer may be operated. There are many different control schemes and layouts, sometimes within the same model family. Operators must take the time to read and understand the operator's manual for the skid steer in use and be familiar with all the features and controls available. If a simple attachment, such as a bucket, is in use, then many of the other attachment controls may not be needed.

1.2.3 Other Controls

Because compact loaders are entered from the front, there is no room for a traditional dashboard. Some models may provide a control or instrument on the armrest for convenient access. Most controls, other than those on the joystick, must be located either to the left or right (*Figure 20*) of the operator's head, or forward of the operator to the left and right (*Figure 21*).

Figure 20 Right overhead control console.

Figure 21 Left and right forward consoles.
Source: Courtesy of Doosan Bobcat North America, Inc. All Rights Reserved.

Figure 22 Left forward console.

Figure 20 shows the controls for the heating and cooling system on an overhead console. Many of today's skid steers have fully enclosed cabs that can be heated and cooled. Models that are not fully enclosed typically have no heating or cooling system. *Figure 21* also shows two other control switches. The work speed control allows the operator to choose how quickly the work tool moves. The operator seat ride control allows air-cushioned seats to provide shock-absorbing motion. Compressed air is used to adjust the seat height, but the ride remains firm unless this switch is turned On.

Figure 22 shows the ignition switch on the left console. The switch directly above the ignition switch operates the parking brake. Remember that the parking brake may automatically engage on some models when the armrest is raised. This ensures that it is not forgotten, and that it is engaged as the operator leaves the cab.

A number of control switches are shown on an overhead console in *Figure 23*. Note that not all the controls shown throughout this section are found on every skid steer. Like automobiles, some skid steers offer only the bare essentials, while others have many more features. The difference in cost between a simple unit and a feature-rich model, assuming they both have equal capacity, can easily be $20,000. *Figure 23* shows control switches on a typical overhead console. Some of these include HVAC, rear and front work lights, and auxiliary hydraulic controls.

Other common control switches include:

Hydraulic lockout and interlock override — The hydraulic lockout control enables or disables the hydraulic system. This switch would be used to disable the

Figure 23 Overhead console switches.

hydraulics when driving the skid steer on a roadway for a distance, disabling the hydraulics for the trip. The interlock override bypasses the safety devices that normally prevent the machine hydraulic systems from operating unless the armrests are down, or some similar interlock is satisfied. There are some attachments, such as some backhoe attachments, which require the operator to leave the normal seat and operate from an alternate position.

WARNING!

The operator override should never be used for any other purpose beyond allowing skid steer operation from an alternate seating position. Using the switch to intentionally bypass safety interlocks can result in serious injury or death.

- *Auxiliary electric control* — This switch enables electrical controls on attachments that require an electrical interlock as well as a hydraulic pressure source.
- *Level bucket control* — Keeping the bucket opening level prevents the contents from spilling out as the lifting arms are raised. Note that this switch only affects the level of the bucket as it is raised; it has no effect while the bucket is being lowered.
- *Attachment locking feature* — This control actuates an automatic locking mechanism for the attachment, locking it to the coupler.

One additional operator-accessible control is used in case of a shutdown or malfunction in the machine while the lifting arms are in the raised position. A manual control that opens a hydraulic circuit valve and allows the lifting arms to drop in a controlled manner is available to the operator. It is typically to the left or right of the seat and has a red handle (*Figure 24*).

A Skid Steer by Any Other Name is Still a Skid Steer

Skid steers are known by several names. They are commonly called skid-steer loaders or simply skid loaders. Although a number of other companies manufacture and distribute them, the name Bobcat® has also become a common name used to describe any skid steer unit, much as Kleenex® has become synonymous with tissue.

Figure 24 Dead engine valve control to lower lifting arms.

Although some people may not think of skid steers as heavy equipment, they are the most versatile piece of equipment found on many construction sites. Skid steers can be fitted with a vast array of attachments. They can break concrete, haul debris, and load it into a truck; saw and plane concrete or asphalt surfaces; mix fresh cement and pump it to an elevated form; auger holes in soil for fence posts and landscaping plants; and rake an area to remove debris and rocks before grass is planted. These are just a few of the tasks that can be done quickly and easily with a single piece of equipment.

A single hydraulic motor drives both wheels on one side of the unit. Each of the two motors drives a dual sprocket, and the sprocket drives two chains. Each drive chain is then connected to a sprocket on a wheel. The sprockets on the wheels are larger than the drive sprocket on the motor. This reduces the speed of the wheels compared to the motor, producing greater torque.

Variable displacement pump: A pump that can deliver a variable fluid volume in a pumping cycle to support the changing needs of a hydraulically operated device.

A **variable displacement pump** is typically used to power the drive wheels on each side. This pump can be controlled to output hydraulic fluid only on demand. If there is no demand, the pump just spins without creating any fluid flow. Because the engine provides all the energy to the pumps, too much hydraulic load can stall the engine. For example, when an attachment is in use and being operated, sending too much fluid to the drive wheels could cause this to happen. The control of the variable displacement pumps by the skid steer helps to prevent stalls from occurring. However, the operator must take an active role to avoid overloading and stalling the engine on many skid steers.

The operational controls open and close valves to the hydraulic lines connected to the loader's steering and lifting hydraulic cylinders and pistons. The lift arms mounted at the rear of the frame are connected to the attachment plate. When the hydraulic fluid is applied to the lifting cylinders, the pistons extend and raise the attachment. When the flow of hydraulic fluid is reversed, the cylinders retract their pistons and the attachment lowers. Another hydraulic cylinder connected to the attachment plate tilts the attachment forward and backward. Ensuring that the lift arms, and the hydraulic devices attached to them, are not damaged is critically important to the safe operation of the skid steer. The same goes for all the hydraulic lines connected to the hydraulic devices.

1.3.0 Instruments

Most skid steers offer very few instruments related to the machine's operation. This is true for two reasons. First, skid steers are relatively simple and are built with much smaller engines than their heavier cousins. Because the engines may be as small as 15 hp to 20 hp, the need for instrumentation to monitor their operation is limited. The second reason is space. Skid steer cabs are quite compact, and visibility outside of the machine is very important. Very basic machines have only a few common lights to alert the operator to a problem.

The service hour counter is one instrument common to all models. Although some models use a type with a liquid crystal diode (LCD) display (*Figure 25*), older models use a simple mechanical version. The counter is a valuable tool used to ensure that maintenance is done on a consistent schedule. The value on the counter should never be changed.

1.3.1 Gauges

The fuel gauge is another common instrument (*Figure 26*). It is extremely important that diesel engines are not allowed to run out of fuel. Running the engine completely out of fuel can cause engine damage or significantly slow down the job while the matter is resolved. Once the fuel is depleted, fuel must be added, and the fuel lines must be bled of air before the engine will restart. Other possible gauges on a compact loader are an engine oil temperature gauge and a hydraulic fluid temperature gauge. Beyond these three gauges, compact loaders typically use lights to alert the operator to a problem or warn of a developing failure. Some feature-rich models do provide an LCD display that shows a variety of settings (*Figure 27*).

Figure 25 Service hours counter.

Figure 26 Fuel gauge.

1.3.2 Information and Warning Lights

Some of the lights in the operator cab let the operator know that a particular switch is enabled. Other lights warn that problems are developing or an unsafe condition exists. *Figure 28* is an example of a warning light panel with a variety of systems represented. Once the ignition key is turned to its first position, battery power is routed to the machine systems and all the panel lights illuminate for about three seconds. This provides the operator an opportunity to ensure that all the trouble lights are working. Failed lamps or circuits should be repaired immediately. Without these lights operating, the operator may not be aware of a serious problem developing. Note that there are three different colors of lights. Green lights simply provide information, yellow lights indicate a situation that requires caution, and red lights indicate a serious problem that should be addressed immediately.

Beginning with the top row and moving from left to right, the lights shown in *Figure 28* provide the following indications:

Top Row

- *Yellow diamond with exclamation point* — Indicates a driver alert and lights up when any other warning light is on; also warns of low fuel.
- *Battery* — Warns that the alternator output to the electrical system is too low or too high.
- *Circle with a P* — Warns that the parking brake is engaged.
- *Person seated with seat belt* — Warns that the seat belt has not been fastened and/or the armrest has not been lowered.

Middle Row

- *Hydraulic vessel with a swipe* — Indicates that the hydraulic high-volume flow feature is enabled.
- *Hydraulic vessel with arrows around it* — Informs the operator that the continuous flow feature is in use. This feature may be used to operate a constantly rotating device like a broom.
- *Hydraulic vessel with dashed lines* — Warns of a problem with the work tool hydraulic system.
- *Hydraulic vessel with exclamation point* — Warns that the hydraulic fluid temperature is too high.
- *Engine crankcase with a coil* — Indicates that the engine is cold, and the glow plugs are warming up the cylinders for cold-start situations. Other machines may require that the glow plugs be operated manually.
- *Crankcase with an exclamation point* — Warns of an engine problem.

Figure 27 Back-lit LCD display.
Source: Courtesy of CASE Construction Equipment

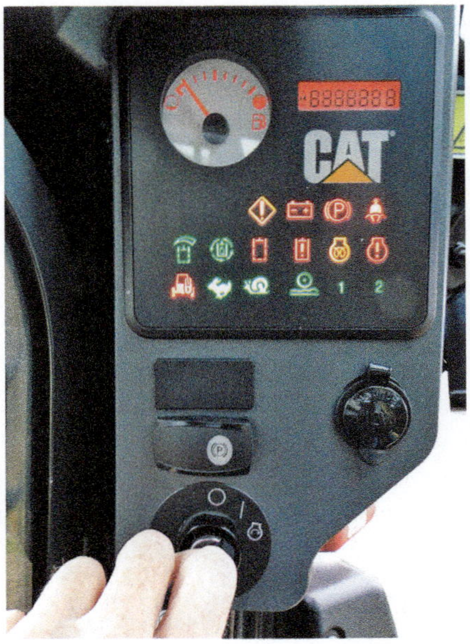

Figure 28 Information and warning light panel.

Bottom Row

- *Skid steer with a padlock* — Warns that the machine security system is active.
- *Rabbit* — Informs the operator that high speed has been selected.
- *Snail* — Informs the operator that a low work tool speed is in use.
- *Wheel rolling over rough terrain* — Informs the operator that the air-ride shock-absorbing feature of the seat has been enabled.
- *Numbers 1 and 2* — Informs the operator which of two joystick control patterns is in use.

At the top of the panel, the digital service hour counter is also testing its individual digit display and backlight.

1.4.0 Attachments

If there is one single feature that defines the skid steer among its users, it is the vast array of available attachments. There are many different types, and many different versions within each type. New or improved attachments are always being developed.

Skid steers have been developed around a simple yet strong and versatile system to connect attachments quickly and easily. Not only are attachments available through the skid steer machine manufacturer, but other companies also specialize in making attachments to fit any make of skid steer. Skid steer owners and users have also developed their own attachments for use in specific situations. Note, however, that using these attachments and the attachments of another manufacturer may affect an existing warranty on the machine. It is also important to note that most attachments go through a rigorous series of tests, and a great deal of data regarding their weight, capacity, and other specifications must be made available. None of this testing or data is available for home-built or site-built attachments. As a result, their safe use may be questionable.

A typical Cat skid steer attachment coupler is shown in *Figure 29*. Once the attachment is properly mounted on the coupler, the two locking levers are pushed toward the center, locking the attachment in place. Some manufacturers have developed controls that can automatically lock and unlock attachments from the coupler, without the operator having to exit the cab. However, these systems only allow for changing attachments that require no additional hydraulic connections. If hydraulic connections are required, the operator must shut down and exit the machine to complete the task.

The Ultimate Attachment?

Tired of sitting in the cab? Are dusty work conditions causing too much discomfort? Are hazardous materials possibly in the area? All of these may be reasons to try your hand at operating a skid steer without actually getting in.

The remote-control system receiver developed by Bobcat can be used to operate the piece of equipment safely without putting the operator in danger. Bobcat MaxControl allows operators to use an iPhone, or any device running the iOS operating system, to control the loader and attachments from outside the cab. The system allows a single operator to do jobs that normally would require both a spotter and an operator.

Source: Courtesy of Doosan Bobcat North America, Inc. All Rights Reserved.

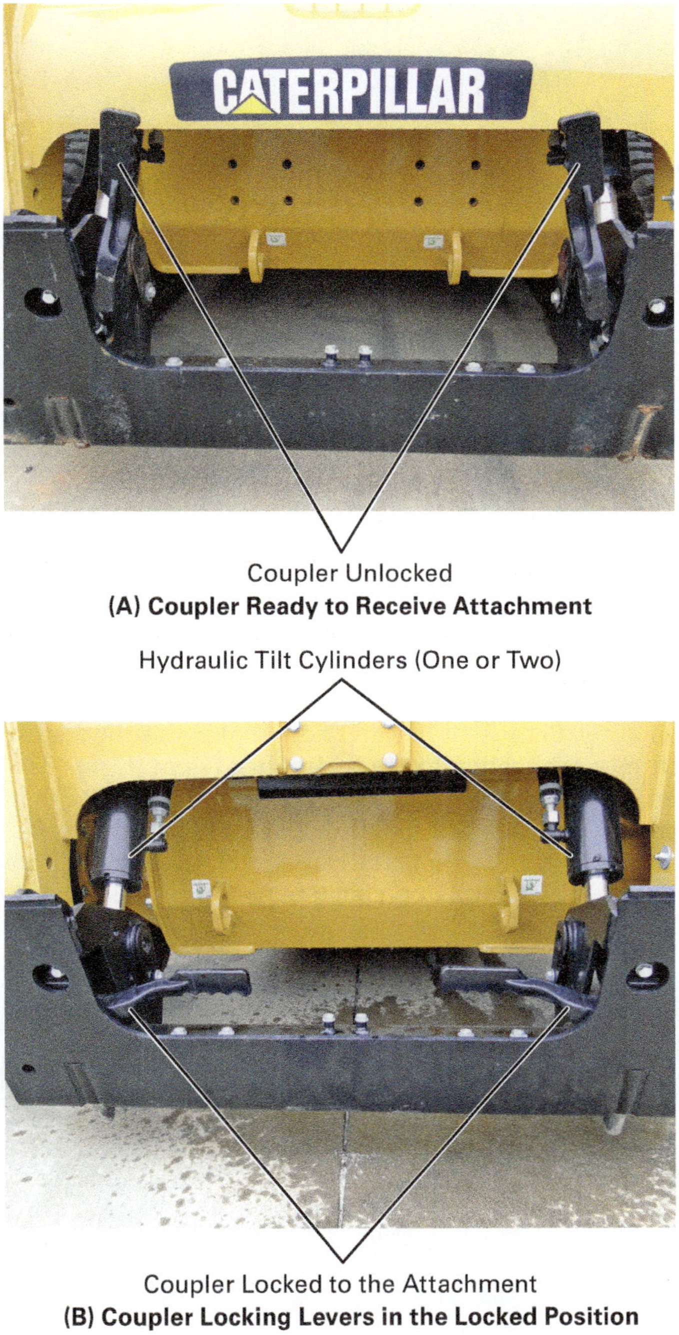

Figure 29 Skid steer attachment coupler.

The most common attachment is the standard bucket (*Figure 30*). The bucket can be used to move soil and rock, doze, level soil or gravel, dig, or load/unload materials. A number of different bucket configurations and sizes are available. Teeth can be added to the leading edge of most buckets. It is important that the bucket (or any other attachment) be properly matched to the capacity of the skid steer. Because the coupler specifications remain the same from the smallest to the largest skid steer models, an attachment that is beyond the capacity of the machine can be installed. It is the operator's responsibility to ensure that the attachments are appropriate for the machine in use.

Figure 30 shows two pieces of steel welded to the upper edge. These are step points provided to assist the operator in accessing the cab. Without any attachment on a machine, the first step up to the cab area is long and clumsy. It is quite common for attachments to provide this point for the operator's first step toward the cab.

Figure 30 Standard bucket attachment.

One modification to the standard bucket is the grappler (*Figure 31*). This style of grappler provides a relatively flat surface on the bottom, and a clamshell-like moving jaw above it. The lifting arms and tilt cylinders move the complete attachment up or down or tilt the forward edge up or down. The two cylinders on the attachment operate the jaw, closing it around material to be taken away. Because it has an additional hydraulic feature, this attachment requires a connection to the hydraulic system (*Figure 32*). The grappler jaw is then controlled by buttons on one of the joysticks that is in control over the chosen hydraulic circuit. This particular machine was prepared for duty in a recycling facility. Note the smooth tires, which indicate the unit will be driven almost exclusively on concrete or a similar surface. Grapplers are available in a number of styles.

Figure 31 Grappler attachment.

Figure 32 Hydraulic connections to support grappler jaw operation.

Another very popular attachment is a set of pallet forks (*Figure 33*), which allow the skid steer loader or compact track loader to function as a forklift. This set of pallet forks, and the related bale forks (*Figure 34*), do not require additional hydraulic connections. Bale forks are more slender than pallet forks, and the forks are well above the ground. Bales are picked up by simply driving into them with the forks, which are positioned near the center of the bale.

Figure 33 Common fork attachment (pallet forks).
Source: Courtesy of Doosan Bobcat North America, Inc. All Rights Reserved.

Figure 34 Bale forks.
Source: Courtesy of Doosan Bobcat North America, Inc. All Rights Reserved.

Figure 35 shows some examples of the many attachments available for a skid steer. While this module doesn't cover all of the attachments, a variety of them are presented here.

(A) Laser-Guided Grading Box

(B) Concrete Mixer

(C) Drop Hammer

(D) Concrete Pump

Figure 35 Common compact loader attachments.
Source: Courtesy of Doosan Bobcat North America, Inc. All Rights Reserved.

(E) Power Rake

(F) Enclosed Sweeper

(G) Snow Blower

(H) Concrete Breaker / Chipping Hammer

(I) Planer Attachment

(J) Excavator

(K) Earth Auger

(L) Brush Cutter

Figure 35 Common compact loader attachments (*continued*).

Source: Courtesy of Doosan Bobcat North America, Inc.

- *Bobcat box blade with laser measuring (A)* — This box levels loose soil, collecting it from high spots and spreading it in low spots, much like a grader. The model depicted interfaces with a laser system to provide elevation information to the operator. Laser guidance systems are available for both single-slopes and dual-slopes (slopes in two directions). When fully automated, the task requires less operator involvement, improved accuracy, and faster results. The Bobcat operator can also see the exact elevation of the box blade at any time.
- *Concrete mixer (B)* — Simply add water and the desired mix. When mounted on a Bobcat skid steer loader as shown, the mixer becomes very mobile and can be moved around at will.
- *Drop hammer (C)* — This attachment is primarily designed to break up large expanses of concrete or asphalt for demolition. The Bobcat model shown here delivers 3,600 foot-pounds of energy per blow, with a three-second delay between blows. This delay provides the operator enough time to slightly reposition the machine for the next blow. The drop hammer is significantly quieter than using a concrete breaker or chipping hammer.
- *Concrete pump (D)* — Cement is accepted into the hopper from a truck and is then pumped to a distant or elevated location.
- *Landscape rake (E)* — The landscape rake smooths soil surfaces and collects rocks and sticks, just as a common rake would. Rakes are mostly used on loose soil where new turf grass seed or sod is to be installed.
- *Sweeper (F)* — Sweepers are popular attachments. They do an excellent job of cleaning pavement after other work has been completed. Sweepers, which are enclosed, create less airborne dust and dirt. Sweepers like the one shown in *Figure 35* also collect the debris onboard also collect the debris onboard, so it can be disposed of elsewhere. Power broom attachments do not collect the debris—they are usually angled and simply sweep the debris off to one side of the surface.
- *Snowblower (G)* — The snowblower attachment provides a very effective means of deep snow removal, carried by a machine that performs well in adverse weather. The discharge chute rotates and adjusts so the snow can be placed as desired.
- *Hydraulic breaker (H)* — A variety of bit shapes and sizes are available for different surfaces and tasks. The model shown can be mounted just as easily on an excavator.
- *Planer attachment (I)* — The planer is capable of cutting or milling concrete and asphalt surfaces. The depth, width, and slope of the cutting device can be controlled. Milled material can be collected and recycled. These attachments are excellent for cleaning up areas where large, dedicated planing and milling equipment cannot reach, establishing drainage in parking areas, and leveling uneven pavement.
- *Backhoe loader attachment (J)* — The backhoe loader attachment turns the skid steer loader into a backhoe. Depending on the make and model, the attachment may require the operator to change seating positions and operate the backhoe from outside the normal cab, seated on the attachment itself.
- *Earth auger (K)* — Augers powered by a skid steer loader are excellent for boring post holes accurately and dependably. Because they are powered by hydraulic motors that are relatively small for their output, the head of the auger can follow the bit down into the hole for greater depth. This is another attachment that can also be attached to an excavator.
- *Brush rotary cutter (L)* — The brush cutter, shown in *Figure 35* mounted to a Bobcat unit, is as simple as it sounds. It creates a powerful hydraulically driven mower, driven by a very powerful and maneuverable skid steer loader.

Many other attachments are possible, including saws, trenchers and trench compactors, underground bores, tree spades, and tillers. If you can conceive it, it can probably be made to operate with a skid steer acting as a carrier. It is important to repeat that, even though an attachment may mount to a machine with no problem, it may not be an appropriate match for that machine. Before connecting and using any attachment, be sure that it is approved for use on the machine and does not exceed its capacity in any way. Attachments that perform the same function may have different operating methods and precautions associated with them. Always read the documentation for the attachment as well as that of the skid steer to protect the equipment investment and ensure the safety of nearby personnel.

1.0.0 Section Review

1. Skid steer engines are mounted behind the operator _____.
   a. for easier service access
   b. to help balance the load
   c. for improved engine cooling
   d. to keep the operator cooler

2. Most skid steers manufactured today are controlled by _____.
   a. levers
   b. a yoke
   c. joysticks
   d. a steering wheel

3. True or False: The service hour meter is reset after each significant maintenance inspection action is performed.
   a. True
   b. False

4. The drop hammer is primarily used to _____.
   a. drive pilings
   b. install fence posts
   c. pack soil in a trench
   d. break up concrete slabs

2.0.0 Prestart Inspections and Maintenance

Performance Task

1. Complete a proper prestart inspection, maintenance, and housekeeping for a skid steer.

Objective

Describe the prestart inspection requirements for a skid steer.

a. Describe prestart inspection procedures.
b. Describe preventive maintenance requirements.

Skid steers require very limited care because they are small and use smaller diesel engines than most heavy equipment. However, they still require a proper prestart inspection at the beginning of the workday. Consistently putting off common maintenance activities can lead to significant and expensive damage. Any time the machine is not operating smoothly on a jobsite, revenue is being lost. Operators have a responsibility to operate their machine as efficiently as possible, and that efficiency begins with timely inspections and maintenance.

2.1.0 Prestart Inspections

A thorough walk-around inspection should be conducted at the beginning of each shift or when a different operator is taking over. The skid steer should always be equipped with the appropriate operator's manual. Whenever there is any doubt about the tasks to be done, refer to the manual. Many machines are also equipped with a logbook so that inspections and other actions can be documented.

The prestart walk-around, also called the daily inspection, may occasionally require some additional actions. For example, greasing specific components is done on a schedule based on the operating hours. When greasing is due, it should be accomplished as part of the walk-around. If an area is to be greased every 50 hours, then the operator will need to do it once each week as part of the walk-around, assuming a normal 40- to 50-hour workweek.

Maintenance time intervals for most machines are established by the Society of Automotive Engineers (SAE) and adopted by most equipment manufacturers. Instructions for preventive maintenance are usually in the operator's manual for each piece of equipment. Typical time intervals are: 10 hours (daily), 50 hours (weekly), 100 hours, 250 hours, 500 hours, and 1,000 hours. The operator's manual also includes lists of inspections and servicing activities required for each time interval. For specific preventive maintenance actions, refer to the operator's manual associated with the machine being used.

2.1.1 Prestart Inspection Process

The daily inspection is also called a prestart inspection for a very good reason: it should be accomplished before the machine is started at the beginning of the workday. This helps to identify any potential problems that could cause a breakdown and indicate whether the machine can be operated. The items to be attended to on the daily inspection are often listed as 10-hour inspection items in the operator's manual. If the skid steer is in operation around the clock, with three different operators working normal shifts, then the daily inspection would be conducted three times each day.

The operator should walk completely around the machine checking various items. The machine should be parked on a stable, flat surface with the parking brake engaged and the lifting arms down. The key should be removed from the ignition.

Items to be checked and serviced on a daily inspection are identified in the following list. The operator can begin wherever it is convenient. In this review, the inspection begins at the right front wheel. It is always a good idea to carry a rag along to clean accumulated dust and dirt from grease points, hydraulic connections, and similar locations. Keep in mind that this walk-around example cannot apply to every machine and configuration. Always follow the procedures for the specific machine in use.

WARNING!

Do not check for hydraulic leaks with your bare hands. Use cardboard or another device. Pressurized fluids can cause severe injuries to unprotected skin.

- Check the right front wheel and tire assembly. Look for missing or loose lug nuts, damaged air valve stems, and cut or damaged tires. Check the tire pressure to ensure it conforms to the operator's manual.
- As the inspection continues down the right side, check the hydraulic hoses routed along the lifting arms for damage or leakage. Look for spots that may be chafing due to vibration or movement.
- Check the hydraulic fluid reservoir level (*Figure 36*). Some may be a little difficult to see. The reservoir level is typically checked by observing a vertical or round sight glass. The machine should already be parked with the lifting arms down. If the lifting arms are raised, a great deal of the hydraulic fluid will be in the operating cylinders, and the reservoir may appear to be low. Always check hydraulic fluid levels with the lifting arms down.
- Check the fuel tank cap for security (*Figure 37*). Note that the cap can usually accept a padlock for security if needed.
- Check to be sure that lifting arm locks (*Figure 38*), used to safely lock the lifting arms in the upright position, are properly secured to the side of the machine.

NOTE

It is best to refuel the skid steer at the end of the workday rather than waiting until the next day. As temperatures change overnight, water may condense inside the fuel tank. This problem is significantly reduced if the tank is left full of fuel.

Figure 37 Fuel tank cap.

Figure 36 Hydraulic fluid sight glass.

Figure 38 Lifting arm locks.

- Check the right rear wheel and tire condition and pressure in the same manner as the right front.
- Open the engine compartment and lock the door in the open position. Most machines have a retaining pin or other device used for this purpose.
- Make sure that all required guards are in place, such as the one covering the drive belt(s). Remember that the engine in a skid steer is typically backwards compared to its position in a rear-wheel drive auto.
- Inspect the cooling system for leaks and damaged hoses. The machine in *Figure 39* has the radiator positioned above the engine. The top guard is shown in its normal position, which keeps debris such as gravel from striking the radiator face. Remove any debris, then remove the retaining pin holding the guard down and lift it up (*Figure 40*).
- Check the face of the radiator for debris. Note that this radiator is split into two separate sections. The upper section is filled with engine coolant. The lower section contains the hydraulic fluid coolant. Some models also have a fill port on the hydraulic fluid cooler coil so that small amounts of fluid can be added easily (*Figure 41*). However, the hydraulic reservoir itself also has a fill port.
- This model also has a sight glass to make checking the engine coolant level easy (*Figure 42*). The coolant should be visible in the sight glass. Because this radiator hinges up for engine service (*Figure 43*), be sure to check this glass with the radiator in the operating position.

Figure 39 Radiator assembly with the guard in place.

Figure 40 Radiator assembly exposed (guard up).

Figure 41 Engine coolant and hydraulic fluid fill caps.

Figure 42 Engine coolant sight glass.

Figure 43 Radiator assembly raised for engine service.

WARNING!

The radiator may be under pressure, and the coolant may be at or near the boiling point. When the cap is removed, and the radiator returns to atmospheric pressure, steam and extremely hot water may be expelled. Use extreme caution and the proper PPE (including safety glasses and heavy rubber gloves) when removing the cap from a radiator.

- Check the oil using the provided dipstick (*Figure 44*). Add oil if necessary. Operating the machine with a low oil level threatens both lubrication and cooling, because the oil also carries heat away from moving parts.

Figure 44 Oil dipstick and air filter trap.

- Check the air filter. Most machines have a way to make a quick check without disassembling the air filter housing. The machine shown in *Figure 44* has a trap. This soft rubber piece has a slot at the bottom that is opened by simply twisting it slightly. If a great deal of debris falls out when it is disturbed, the housing should be opened, and the air filter element inspected and cleaned.
- Drain any accumulated water from the fuel/water separator (*Figure 45*). Note the instructions shown here to remove the entire bowl for cleaning if required. Drain fuel until all the water (which is clear in color, compared to the fuel) is gone. Dispose of the drained fluid in an approved container and according to policy. It should not be drained on the ground.
- Make sure the radiator and guard are returned to their normal position and secured. Make a final visual inspection of the engine compartment to look for obvious defects, leaks, or objects that do not belong there. Wipe down any surface that has grease or oil, to reduce the buildup of dust and dirt.

Figure 45 Fuel/water separator.

- Before closing and securing the door, check the lubrication schedule on the preventive maintenance placard (*Figure 46*), and compare the hours to the operator's log. Note the number of logged hours to determine when the machine should be serviced.
- Once the engine door is secure, continue the walk-around to the left side of the machine. Inspect the hydraulic lines along the lifting arms and other accessible or visible locations.
- Check both left side wheels and tires for damage and proper pressure, as was done for the right side.
- Examine and clean any visible fluid from the attachment hydraulic line connections (*Figure 47*). Make sure any protective caps provided are installed on unused lines.

Figure 46 Preventive maintenance placard.

NOTE

In cold weather, it may be preferable to grease pivot points at the end of a work shift when the machine is warm. Warm the grease gun before using it for better grease penetration.

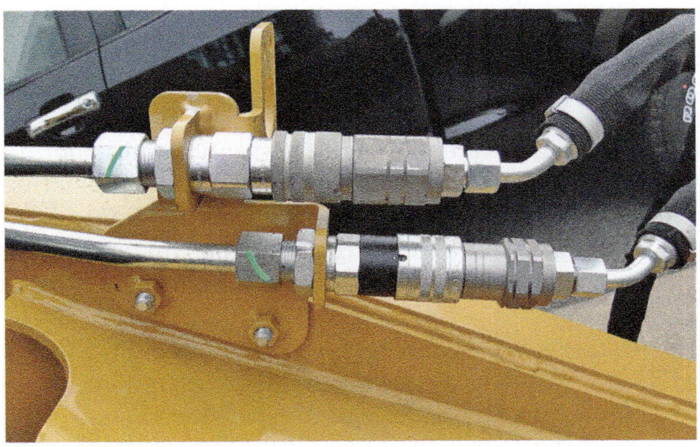

Figure 47 Hydraulic line connections.

- At the front of the machine, look for and remove any debris trapped between the coupler and the attachment, or the coupler and skid steer chassis. Examine the entire attachment or work tool for signs of damage or fluid leakage.
- Make a final visual check of light assemblies and windshield wipers to ensure they are serviceable. On this machine, the windshield washer reservoir can be checked as the cab is entered (*Figure 48*).

Figure 48 Windshield wiper and washer reservoir.

Before entering the cab to start the machine, redirect attention from the machine to the surrounding environment. Look for unusual obstructions or changes in the jobsite that may affect a heavy equipment operator. Once inside the cab, the operator does not have the same visibility. Look up as well as around the area.

2.2.0 Maintenance

Preventive maintenance is an organized effort to regularly perform periodic lubrication and other service work in order to avoid performance problems and breakdowns. By performing preventive maintenance on the skid steer, the operator keeps the equipment in a safe and reliable operating condition and reduces the possibility of breakdowns or expensive repairs.

Accurate, up-to-date maintenance records are essential for documenting the history of the equipment. Each machine should have a record that describes any inspection or service that was performed and the corresponding time intervals. Typically, an operator's manual and some sort of inspection log are kept with the equipment at all times.

Preventive maintenance of equipment is essential and is not difficult if the right tools and equipment are used. The leading cause of premature equipment failure is putting things off. Preventive maintenance should become a habit, performed on a regular basis. Refer to the operator's manual for the location of any specific component or lubrication point and perform the prescribed maintenance activities.

Figure 49 shows a maintenance scheduler as it appears on an engine compartment placard. Many skid steers require very limited maintenance. Note that this machine uses lines around the outline of the machine to show items that are to be accomplished on the same schedule. All the grease points shown are connected by the 10-hour interval line. Since inspection items identified as 10-hour requirements are considered daily or prestart inspection tasks, therefore, true preventive maintenance tasks are typically those that are conducted at longer intervals. The next line for this skid steer represents 500 hours. On this machine, there are no maintenance activities recommended between those intervals. For both owners and operators, this is a bonus. An operating interval of 500 hours is generally considered a quarterly (every three months) task, assuming normal, five-day work weeks with only one shift. However, a machine that is in operation around the clock and/or through the weekends would reach 500 hours much more quickly.

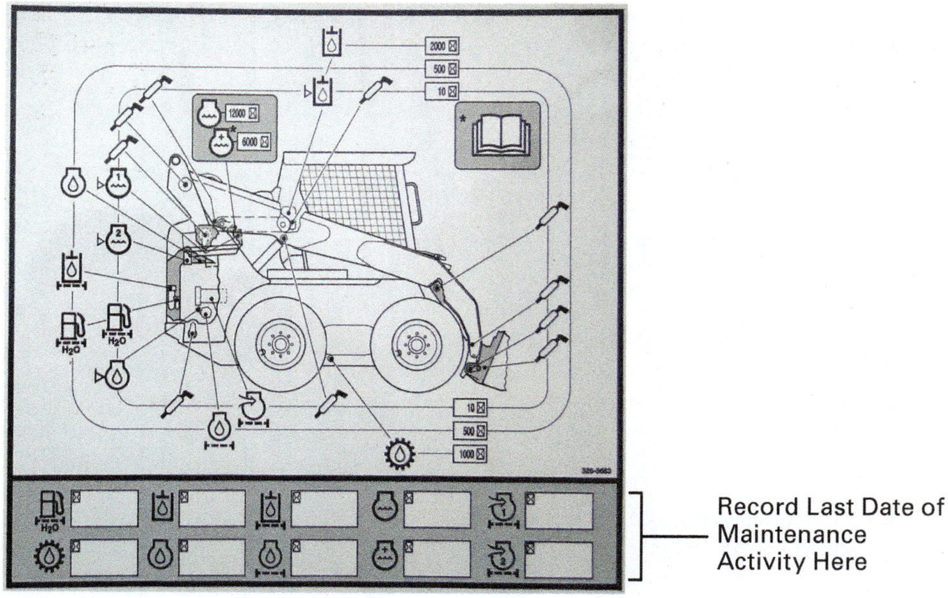

Figure 49 Maintenance scheduler placard.

It is important to note that not every maintenance requirement is shown on the placard. In this example, only common fluid and lubrication maintenance items are noted on the placard. For complete preventive maintenance requirements and guidance, always refer to the operator's manual.

2.2.1 Preparing for Maintenance

The first step in the maintenance process is to prepare the machine. The following steps should be taken:

- Park the skid steer on firm, level ground and apply the parking brake.
- Lower the attachment or coupler to the ground; the ground should support any attachment, rather than relying on the hydraulic system to hold it up. If the maintenance activity requires the lifting arms to be up, then the safety locks must be installed.

WARNING!

The operator should never exit the machine while the lifting arms are raised unless the safety locks are installed. A second person should install the guards while the operator raises the arms and waits inside the cab. Never raise the lifting arms and then exit the machine to install the locks.

- If the engine has cooled off, operate the engine at idle speed for three to five minutes.
- Shut the engine down and take the key out of the ignition.
- Firmly chock the wheels to prevent any movement.

2.2.2 Long Interval Maintenance

Although some machines do not require any specific maintenance beyond daily tasks for many operating hours, the environment and use of the machine may dictate otherwise. Even if it is not specifically called for, some activities should be carried out weekly, or as the operating conditions may indicate. The operator's manual likely states clearly that operating conditions can dramatically alter the maintenance schedule.

New skid steers may start out with a special maintenance schedule. For example, replacement of the engine oil is often required after the first 50 hours of operation. The new engine typically produces significantly more wear metals during the first 50 hours of operation than it will in the future. Replacing the lubricant in these components after this short interval removes these extra metal bits and flakes, preventing them from causing damage or plating other surfaces. Fluid filters may also be changed for the first time after a short interval. If a new skid steer is being driven, be sure to follow the required maintenance schedule for new machines. Otherwise, the warranty may not remain valid.

The battery is a common example of a component that may require inspection weekly or monthly. The case should be inspected for cracks, buckling, or any signs of damage from an external source. Check the electrolyte solution level if the battery is not sealed. There is typically a ring or other visual indicator around the bottom of the fill opening on each cell to indicate the proper level. Sealed batteries often have an eye showing the charge condition. Remember that ventilated batteries should be filled with distilled water only. Common tap water contains too many minerals that eventually form deposits on the battery plates, causing the battery to fail. Sealed batteries require only a visual inspection.

WARNING!

Proper PPE should be worn at all times when working near batteries. The electrolyte solution is a strong acid that can cause severe burns or blindness.

Some tasks are required at longer intervals. Generally, these tasks become more complex and are done in the maintenance shop environment. Tasks that may be required at the intervals shown include the following:

- 250 hours:
  - Replacement or cleaning of air filter elements
- 500 hours:
  - Replacement of the fuel filter
  - Replacement of the engine oil and oil filter (may be specified at 500 hours or 250 hours)
  - Replacement of the hydraulic fluid filter
- 1,000 hours:
  - Changing the hydraulic fluid and filter
  - Service and inspection of the drive sprockets and chains
- 1,500 to 2,000 hours:
  - Changing the hydraulic fluid and filter (if not earlier)
  - Changing the engine coolant

Remember that the preceding list is only a single example of a maintenance schedule. Make sure that the schedule for the specific skid steer model in use is followed. Service intervals can differ dramatically.

Hydraulic fluid maintenance is addressed in all maintenance schedules. However, the hydraulic fluid should be changed whenever it becomes dirty or breaks down due to overheating. Continuous operation of the hydraulic system in hot environments can heat the hydraulic fluid to the boiling point and cause it to break down. In dusty areas, the fluid cooler may become clogged or blocked, causing the fluid to remain hot. As a result, the hydraulic fluid condition should be observed consistently, and replaced if it shows signs of overheating and breakdown. Filters should also be replaced whenever the fluid is changed.

2.2.3 Preventive Maintenance Records

Accurate, up-to-date maintenance records are essential for knowing the history of your equipment. Each machine should have a record that describes any inspection or service that is to be performed and the corresponding time intervals. Typically, an operator's manual and some sort of inspection sheet are kept with the equipment at all times. Actions taken, along with the date, are recorded on the log. Various operators can then share information about fluids that have been added, for example, so that patterns in fluid loss are noted.

The operator's manual usually has detailed instructions for performing periodic maintenance. If you find any problems with your machine that you are not authorized to fix, inform the foreman or field mechanic before operating the machine.

2.0.0 Section Review

1. The hydraulic fluid level is typically checked by _____.
   a. removing a dipstick
   b. observing a sight glass
   c. observing a floating indicator arm
   d. removing the fill cap and observing the level

2. Where should an operator look for specific information regarding preventive maintenance on the skid steer?
   a. On the company's maintenance bulletin board
   b. On the service decals inside the operator cab
   c. In the operator's manual for the machine
   d. On the machine's maintenance record

3.0.0 Startup and Operating Procedures

Objective

Describe the start-up, shut-down, and operating procedures for a skid steer.

a. State skid steer–related safety guidelines.

b. Describe start-up, warm-up, and shut-down procedures.

c. Describe basic maneuvers and operations.

d. Describe related work activities.

Performance Tasks

2. Demonstrate proper entrance and exit of the operator cab on a skid steer.
3. Perform proper start-up, warm-up, and shut-down procedures on a skid steer.
4. Execute basic maneuvers with a skid steer, including the following:
   - Grading
   - Removing stumps or boulders
   - Steering
   - Changing attachments
   - Loading a dump truck
   - Maneuvering on slopes
   - Using a fork attachment for moving materials
5. Properly load and secure a skid steer for transport.

After operators learn all the basic skid steer components and perform inspections and preventive maintenance activities, the next step is to actually start and operate the machine. Before that can happen, operators must fully understand the safety issues associated with this versatile piece of equipment. The operator's manual for a skid steer contains safety information about that specific machine. All skid steers do have similarities in their operation. However, they also all have slight differences. Some may be a bit clumsier to operate than others. The type of steering modes available and the jobsite terrain also change how a machine handles.

3.1.0 Safety Guidelines

Safe skid steer operation is the responsibility of the operator. Operators must develop safe working habits and recognize hazardous conditions to protect themselves and others from injury or death. Always be aware of unsafe conditions to protect the load and the skid steer from damage. Become familiar with the operation and function of all controls and instruments before operating the equipment. Read and fully understand the operator's manual. Operators must be properly trained and certified. Take note of and read the many safety hazard placards on the machine.

3.1.1 Operator Training and Certification

Not everyone is allowed to operate a skid steer on a jobsite. However, skid steers are often rented by homeowners or other nonprofessionals who have no training in the machine's operation. This often leads to property damage or personal injury. Despite their size, skid steers are powerful and dangerous units in unskilled hands. Due to liability and insurance guidelines, only trained and qualified personnel are allowed to operate skid steers on a jobsite. Operators may receive a license certifying them to operate specific pieces of equipment. They also must be reevaluated at different times over the years. Because operators often move from one jobsite to another or from one company to another, they may be required to requalify at each new location.

3.1.2 Operator Safety

Nobody wants to have an accident or be hurt. Workers can take steps to protect themselves and those around them from getting hurt on the job. Be alert and avoid accidents.

The most frequent type of fatality with skid steers involves workers in the immediate area, not the operator. Operators are not killed nearly as often as are workers in the surrounding area, especially with ROPS and other safeguards now in place for operators. When operators are seriously injured or killed, the seat belt is usually not being used.

Know and follow your employer's safety rules. Your employer or supervisor will provide you with the requirements for proper dress and safety equipment. The following are recommended safety procedures for all occasions:

- Only operate the machine from the operator's cab.
- Wear the seat belt and tighten it firmly. The ROPS depends heavily on the operator remaining in the seat in the event of an accident. The ROPS should never be removed or modified in any way—no drilling, cutting, or welding should ever be done to a ROPS.
- Mount and dismount the equipment carefully using three points of contact. Never dismount a running machine, and never dismount when the lifting arms are up, unless the proper locks are in place.
- Wear a hard hat, safety glasses, and safety shoes when operating the equipment. Hearing protection should also be on hand and used with machines that are not enclosed.
- Do not wear loose clothing or jewelry that could catch on controls or moving parts.
- Keep the windshield, windows, and mirrors clean at all times.
- Never operate equipment under the influence of alcohol or drugs.
- Never smoke while refueling. Avoid any potential sources of static electricity when refueling.
- Do not use a cell phone or text while the machine is in operation.
- Never remove protective guards or panels.
- Never attempt to search for leaks with your bare hands. Hydraulic and cooling systems operate at high pressure. Fluids under high pressure can cause serious injury.
- Always lower the lifting arms and attachments to the ground before performing any service or when leaving the skid steer unattended.

3.1.3 Safety of Coworkers and the Public

As an operator, you are not only responsible for your own personal safety, but also for the safety of others who are working in the area. Sometimes, you may be working in areas that are very close to pedestrians or motor vehicles. In these areas, take time to be aware of what is going on around the site. Create a safe work zone using cones, tape, or other barriers. Remember, it is often difficult to hear when operating a machine. Use a spotter and a radio in crowded conditions.

The main safety points when working around other people include the following:

- Walk around the equipment to make sure that everyone is clear of the equipment before starting and moving it.
- Always pay attention to the direction of travel.
- Do not drive the skid steer up to anyone standing in front of an object or load.
- Make sure that personnel are clear of the rear area before turning.
- Know and understand the traffic rules for the area in which you are operating.
- Use a spotter when landing an elevated load or when you do not have a clear view of the landing area.
- Exercise particular care at blind spots, crossings, and other locations where there is other traffic or where pedestrians may step into the travel path.

- Do not swing loads over the heads of workers. Make sure you have a clear area to maneuver.
- Travel with the attachment or coupler no higher than 12 to 18 inches above the ground; always drop the coupler or attachment to the ground when parking the machine.
- Do not allow other workers to ride anywhere in or on the machine.

3.1.4 Equipment Safety

The skid steer has been designed with certain safety features to protect you as well as the equipment. For example, it has guards, canopies, shields, rollover protection, falling object protection, and seat belts. Know your equipment's safety devices and be sure they are in working order.

Skid steer rollovers are a significant cause of deadly accidents. In too many cases, the seat belt was not in use. However, the lifting arms tend to be involved in the majority of fatalities. Operators have tried to take too many shortcuts by leaving the machine running and exiting the cab. The lifting arms, due to their proximity to the machine and tight clearances, are the most dangerous part of the machine. Use the following guidelines to maintain a safe environment in and around the skid steer:

- Perform prestart inspection and lubrication daily.
- Look and listen to make sure the equipment is functioning normally. Stop if it is malfunctioning. Correct or report trouble immediately.
- Always travel with the coupler or load low to the ground.
- Never exceed the manufacturer's limits for speed, lifting, or operating on inclines.
- Know the weight of all significant loads before attempting to lift them. Review the appropriate load chart. Do not exceed the rated capacity of the skid steer.
- Always lower the coupler or work tool, engage the parking brake, turn off the engine, and secure the controls before leaving the equipment.
- Never park on an incline.
- Maintain a safe distance between your skid steer and other equipment that may be on the jobsite.
- Balance has everything to do with the operation of a skid steer. With no load, roughly two-thirds of the machine's weight is over the rear axle. The balancing point makes a dramatic shift forward with a significant load, changing the behavior of the machine.
- Skid steer stability decreases sharply during travel if the lifting arms are raised.
- Always move up and down inclines with the heavy end of the machine at the high end. If the machine is unloaded, the rear is the heaviest end. Therefore, the skid steer should be backed up a hill when it is unloaded. This may still hold true for lightweight attachments and work tools, such as an auger.
- Under no circumstances should an operator attempt to reach into the cab and operate the machine from the outside. Many serious injuries and fatalities on older, less-safe skid steers happened this way.

Know your equipment. Learn the purpose and use of all gauges and controls as well as your equipment's limitations. Never operate your machine if it is not in good working order.

A Tragic Month

In the period between mid-February and mid-March 2012, four skid steer accident fatalities occurred in the agricultural environment alone. One operator died after being crushed between the bucket and the machine. The other three fatalities resulted from rollovers. One was crossing uneven ground, while the other two resulted from loads being carried too high during travel. The seat belts were not in use.

All these losses could have been prevented easily. Don't be the next statistic.

3.1.5 Spill Containment and Cleanup

Skid steers are powered by combustible fuels. Fuels must always be contained during refueling activities. Most companies have designated refueling areas specifically to contain any fuel spills that may occur.

Skid steers also use hydraulic systems to operate all aspects of the machine. With so many of the operational controls relying on the hydraulic system, it is critical that hydraulic fluid be contained without spills or leaks. Anyone who has ever worked around any hydraulic system knows that it is almost impossible to stop all leaks. Leaks need to be cleaned up and repaired whenever reasonably possible. Hydraulic fluid spills, on the other hand, are much more of a problem. If a hydraulic hose ruptures under pressure, gallons of hydraulic fluid may be spilled. To protect the environment, companies maintain an oil spill kit easily accessible to the operator. A typical oil spill kit may include the following:

- A large (20-gallon or more) salvage drum/container used to store spill kit materials until needed. It is used for disposal after materials are used for cleanup.
- An emergency response guide for spills of petroleum-based products.
- Personal protective equipment (goggles and unlined rubber gloves).
- One or more 3" × 10' absorbent socks for containment.
- Half dozen 3" × 4' absorbent socks for containment.
- 20 or so absorbent pads for cleanup.
- 50 or so wiper towels or rags for cleanup.
- 3 to 5 pounds of powdery absorbent material.
- Several disposable bags with ties.

Although a flat-tipped shovel may not fit into a spill kit, having one handy is a good idea for spill cleanups.

3.2.0 Start-up, Warm-Up, and Shut-Down Procedures

Before starting skid steer operations, make sure that you are familiar with the task and the area of operations. Check the area for both vertical and horizontal clearances, and for obstructions that interfere with normal turns or pivots. Make sure that the path is clear of electrical power lines and other hazardous obstacles.

The following suggestions can help improve operating efficiency:

- Observe all safety rules and regulations.
- Analyze the tasks to be done and ensure the skid steer is capable of accomplishing the tasks safely. Visualize the work being done.
- Use a spotter if you cannot see the area where a load will be placed. A spotter may also be needed for work such as milling or cutting concrete.

3.2.1 Preparing to Work

Preparing to work involves getting organized in the cab, fastening the seat belt, and starting the machine. Mount your equipment using the grab rails and footrests. Getting in and out of equipment can be dangerous. Always face the machine and maintain three points of contact when mounting and dismounting a machine. That means you should have three out of four of your hands and feet on the equipment. That can be two hands and one foot or one hand and two feet. Skid steers have grab bars and secure, slip-resistant foot holds to help.

WARNING!

OSHA requires that approved seat belts and a ROPS be installed on virtually all heavy equipment. Old equipment must be retrofitted, and the seat belt must be used at all times. Do not use heavy equipment that is not equipped with these safety devices.

Adjust the seat to a comfortable operating position. Ensure that the throttle pedal on the floor is within comfortable reach. In a skid steer, the seat position is not quite as critical to operation because it has no steering wheel. Skid steers also do not have a service brake. Without these two concerns, positioning the seat is primarily related to maintaining a comfortable grip on the joysticks or levers, with sufficient legroom for comfort. The seat height is adjustable on most machines to provide comfort. The seat height will affect the shoulders and how the arms lie on the arm rests. Some models also offer the ability to adjust the position of the joystick controls forward or backward (*Figure 50*). Make sure you can see clearly and reach all the controls. Also adjust the mirrors before starting the engine, fasten the seat belt and ensure the armrests are down. Remember that late-model skid steers have an interlock in the armrests that will not allow operation unless they are in position.

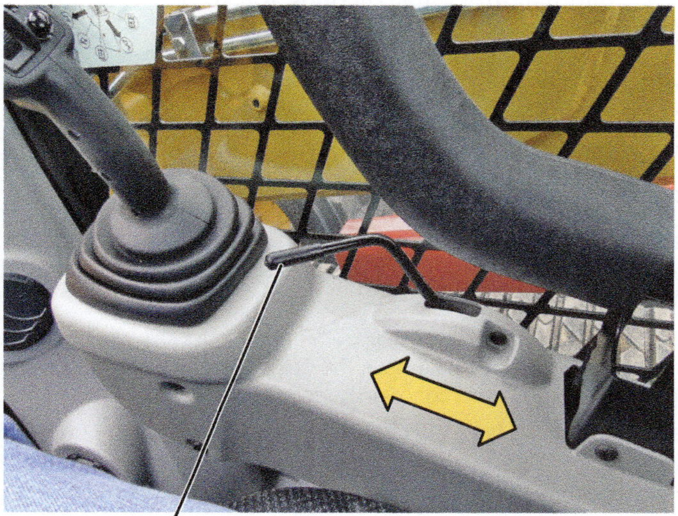

Figure 50 Joystick control module adjustment.

The startup and shutdown of an engine is very important. Proper startup lengthens the life of the engine and other components. A slow warm-up is essential for proper operation of the machine under load. Similarly, the machine must be shut down properly to cool the hot fluids circulating through the system. These fluids must cool so that they can help cool the metal parts of the engine before it is switched off.

3.2.2 Startup

There may be specific startup procedures for the piece of equipment you are operating, but in general, the startup procedure should follow this sequence. However, it is important to always follow the sequence offered in the operator's manual. Before beginning the startup sequence, the operator should be seated in the cab, with the seat belt fastened and the armrest(s) pulled down into position.

Step 1 Be sure the parking brake is engaged.

Step 2 Turn the ignition key to the first detent, or the On position. This should enable the electrical system, and the instrument lights should illuminate for several seconds. Watch the lights to ensure that they are all functional. If lights that alert the operator to serious problems (typically red lights) are not working, do not operate the machine until the problem is addressed. The parking brake lamp should remain On as the others turn Off.

Step 3 Set the throttle control to the middle of its travel, about half-throttle.

Step 4 Turn the ignition switch to the Start position. The engine should turn over. Never operate the starter for more than 30 seconds at a time. If the engine fails to start, wait two to five minutes before cranking again. The lesser amount of time typically applies in cold weather, which allows the starter to cool off more quickly.

Step 5 As soon as the engine starts, release the key; it should return to the On position on its own.

Maintain the engine speed at one-third to one-half throttle for the warm-up period. Always keep the engine speed low until the oil pressure light is extinguished. The oil pressure light should initially light and then go out. If the oil pressure light does not turn off within 10 seconds, stop the engine, investigate, and correct the problem.

There are special procedures for starting a diesel engine in cold temperatures. Many diesel engines have glow plugs that preheat the engine before starting. Follow the manufacturer's instructions for starting the machine in cold weather.

3.2.3 Warm Up

Warm up a cold engine for at least 5 minutes. In colder temperatures, warm up the machine for a longer period. While allowing time for the engine to warm up, complete the following tasks.

Step 1 Check all the gauges and instruments to make sure they are working properly. If the machine has a temperature gauge, watch it as the coolant temperature climbs.

Step 2 Determine what operating modes will be needed for the work ahead and set the controls accordingly. For example, the operator may or may not need to provide a high-flow hydraulic source for an attachment.

Step 3 Release the parking brake. This is often necessary to enable use of the controls. Ensure that, with the joysticks in their neutral position, the skid steer, lifting arms, and any attachments do not move.

Step 4 Once a sufficient warm-up time has been allowed, cycle the controls for the lifting arms. Lift them up and bring them down, then check the tilt controls for the proper response. Briefly operate the functions of the attachment in use to be sure it is fully operable.

Step 5 Position the lifting arms and attachment in the travel position and use the controls to move the skid steer slightly forward and backward. Then execute a turn in each direction.

Step 6 Reset the parking brake while the machine continues to warm up.

CAUTION

Do not pick up a load if the hydraulics are sluggish. Allow the machine to warm up until the hydraulic components function normally. The hydraulics can fail if not warmed up completely.

If there are any problems that have no obvious cause, shut down the machine and investigate or get a mechanic to look at the problem.

3.2.4 Shutdown

Shutdown should also follow a specific procedure. Proper shutdown reduces engine wear and possible damage to the machine.

Step 1 Find a dry, level spot to park the skid steer. Stop the machine by simply allowing the driving controls to return to a neutral position.

Step 2 Lower the coupler or attachment to the ground and engage the parking brakes.

Step 3 Place the speed control in low idle and let the engine run for approximately five minutes.

Step 4 Turn the engine start switch to the Off position.

Step 5 Release hydraulic pressures by moving the control levers until all movement stops.

Step 6 Remove the ignition key. If you must park on an incline, chock the wheels.

CAUTION

Failure to allow the machine to cool down can cause excessive temperatures in the engine or systems. The engine oil, hydraulic fluid, or transmission fluid may overheat or even boil.

Lock the cab door, and secure and lock the engine enclosure. Always engage any available security systems and devices when leaving the skid steer unattended.

3.3.0 Basic Maneuvers

To maneuver a skid steer, you must be able to move forward, backward, and turn. The following guidelines assume that the left joystick has full control over vehicle movement.

Remember that allowing the joystick to return to its neutral position stops all vehicle movement. There is no foot brake on today's models. With the skid steer's hydrostatic drive system, once the control returns to neutral, all hydraulic flow in and out of the hydraulic motors is stopped. The machine will not typically lock up the wheels as the joystick is released and it comes to a stop. However, it should stop rather quickly and surely. The machine will also come to a stop on an incline that is within its specifications, with only slightly more travel. It does not coast when the joystick is moved to the neutral position. The lack of a positive, dedicated brake in the skid steer is probably the most confusing feature to a new operator. However, the inherent stopping ability of the drive system has proven to be an effective and safe design.

3.3.1 Moving Forward

The first basic maneuver is learning to drive forward, as outlined in the following steps:

Step 1 Disengage the parking brake.

Step 2 Before starting to move, use the joystick to raise the coupler or attachment roughly 15 inches above the ground. This is the travel position.

Step 3 Push the left joystick straight forward, slightly and easily. All four drive wheels should engage and move the machine forward. To reach full speed at the current control settings, slowly push the joystick fully forward. The farther forward the joystick is pushed, the faster the machine will go, within its current or preset limits.

Step 4 To stop, allow the joystick to return to neutral.

CAUTION

Always travel with the coupler or attachment low to the ground (12 to 18 inches). Be aware of the forward edge of the attachment and avoid hitting things with it. On very rough terrain, a slightly higher travel position may be required.

3.3.2 Moving Backward

To back up or reverse direction, the only action required is to move the left joystick backwards. As the joystick is moved closer to the backward limit of its travel, the skid steer moves faster, up to its top speed for the current settings.

3.3.3 Turning

The concept of skid steering has been presented but not yet practiced. Remember that skid steer wheels are fixed on their axle and do not turn left or right. Turns happen as a result of the drive wheels on one side turning faster than the wheels on the opposite side. The skid steer can be made to pivot in place by driving the wheels on each side in opposite directions.

The sequence, using a single joystick to control machine movement, is as follows:

- With the joystick pushed forward and to the left, the machine executes a relatively gradual turn to the left. The drive wheels on the right rotate at a higher speed than those on the left.
- With the joystick pushed forward and to the right, the machine executes a relatively gradual turn to the right. The drive wheels on the left rotate at a higher speed than those on the right.
- Pushing the joystick from its neutral position directly to the left causes the skid steer to pivot in place to the left. It continues to spin in a tight circle in this position. The drive wheels on the left will be pushing forward, while the drive wheels on the right will be pushing backwards, thus causing the unit to pivot in place.

- Pushing the joystick from its neutral position directly to the right causes the skid steer to pivot in place to the right. It continues to spin in a tight circle in this position. The drive wheels on the left will be pushing forward, while the drive wheels on the right will be pushing backwards, thus causing the unit to pivot in place.
- With the joystick pushed backward and to the left, the machine gradually turns the front of the unit toward the left while backing up. The drive wheels on the left rotate at a higher speed than those on the right.
- With the joystick pushed backward and to the right, the machine gradually turns the front of the machine toward the right while backing up. The drive wheels on the right rotate at a higher speed than those on the left.

3.4.0 Work Processes

The operation of skid steers is straightforward, but it requires some practice. They can start and stop quickly, and new operators may feel a bit clumsy at first. The lack of a steering wheel and brake makes skid steers very different from vehicles that people are used to driving. Although many pieces of heavy equipment are quite different to drive and manipulate than a car, the vast majority retain a steering wheel and a service brake in common with standard motor vehicles.

With practice, good skills in operating the skid steer will be developed. Some basic work activities performed with a skid steer are described in this section.

NOTE

The controls on specific skid steer machines may be different from those described in the procedures. Check your operator's manual for information about the controls and limitations of your equipment.

3.4.1 Skid Steer Capacities and Limitations

Skid steer machines do not typically have load charts as extensive as a forklift may have. There are a few important specifications to consider about the machine itself before putting it to work. Much of the information an operator may need to know comes from the attachment or work tool documentation.

Skid steers are rated first by their **rated operating capacity**. This is generally the first specification noted. The rated operating capacity is considered the amount of weight the machine can safely lift and maneuver. However, the rated operating capacity is actually derived from another value that is based on testing.

Rated operating capacity: The amount of weight that a skid steer is projected to handle safely through common maneuvers. The rated operating capacity is typically derived from the tipping load test and is equal to 50 percent of the tipping load value.

Tipping load: The maximum load that a skid steer can hold in the air, with the coupler outstretched, before the rear wheels begin to leave the ground.

The **tipping load** of a skid steer machine is a specification that is based on testing. Full-size loaders may have a variety of tipping load values in their specifications. Straight static tipping load and full turn tipping load are two examples. Skid steers usually have only one tipping load value in their specifications. The tipping load is simply defined as the amount of weight that can be placed on the coupler with the lifting arms at their maximum forward reach, without causing the skid steer to lift the rear tires off the ground.

The rated operating capacity of a skid steer is based on the tipping load. It represents 50 percent of the tipping load. Therefore, a skid steer with a tipping load of 2,800 pounds (1,270 kg) would have a rated operating capacity of 1,400 pounds (635 kg). The rated operating capacity is considered a safe limit in comparison to the tipping load, indicating the machine can both lift and maneuver with this much weight. The tipping load test is done with the skid steer stationary. Note that skid steers can be fitted with counterweights (*Figure 51*) that increase their rated operating capacity somewhat. However, the increase in capacity is usually only 100 to 300 pounds (45 to 136 kg).

Figure 51 Counterweights at the rear of the skid steer.
Source: Dmitry Kalinovsky/Shutterstock

It is important to be fully aware of the machine's operating limitations. It is equally important that the weight of a load also be known. Knowing the machine's limitations alone is simply not enough. Pay attention to the weight of the load, and remember that a load that can be lifted off the ground without tipping the skid steer is not necessarily a load that can be safely maneuvered or carried to another location.

Lift and take control of all loads carefully. If the rear wheels of the skid steer begin to rise off the ground when trying to lift a load, stop and slowly lower the load back to its starting point.

3.4.2 Using a Bucket

The most common attachment for a skid steer is the bucket. It may be one of several different designs, but the bucket is the most popular overall. Unlike full-sized loaders, skid steers cannot typically reach over the sides of a large dump truck for loading. However, they are often used to load material, such as mulch or topsoil, into a smaller vehicle.

The procedure for carrying out a loading operation from a stockpile of material is as follows:

Step 1 Travel to the work area with the bucket in the travel position.

Step 2 Position the bottom of the bucket parallel to and just skimming the ground.

Step 3 Drive the bucket straight into the stockpile. Once the bucket stops penetrating, ease off the joystick to avoid spinning the wheels.

Step 4 Adjust the controls and raise the bucket to fill it, as shown in *Figure 52*.

Figure 52 Filling the bucket.
Source: Courtesy of Doosan Bobcat North America, Inc. All Rights Reserved.

Step 5 Tilt the bucket back as needed to keep the contents near the back of the bucket. Keeping the contents closer to the skid steer helps keep the machine stable for maneuvering.

Step 6 Use the joystick to back the skid steer loader away from the stockpile.

Step 7 Place the bucket in the travel position and move toward the destination.

Step 8 Raise the bucket high enough to clear the side of the truck or other obstruction.

Step 9 Move the bucket over the truck bed and tilt the bucket forward to dump the bucket. Moving the tilt control back and forth a few times may help move the contents out. *Figure 53* shows the proper position for dumping from a standard bucket.

Figure 53 Dumping the bucket.
Source: Courtesy of Doosan Bobcat North America, Inc. All Rights Reserved.

Step 10 Tilt the bucket back to level the bottom again and back the skid steer away from the truck.

Step 11 Return the bucket to the travel position and return to the stockpile.

Step 12 Repeat the cycle until the truck is loaded.

Loading material into a truck with a skid steer requires that the operator have good reflexes and distance judgment. The loader must be placed close to the side of the truck in order to get the bucket positioned to dump properly.

There are two main points where accidental contact is the most common. The first area of contact is between the bucket and the side of the truck. Either the operator has misjudged the height of the truck bed or has approached the truck too quickly, not allowing sufficient time to raise the bucket before getting to the truck. The second contact point is between the front wheels or tracks of the skid steer and the side of the truck body. Again, this is due to the operator misjudging the distance between the front of the loader and the side of the truck. Contact in these situations can cause severe damage to both pieces of equipment.

When loading from a stockpile or bank, the placement of both the truck and the skid steer are variable and must be adjusted to local operating conditions. Conditions to be considered are the weight of material, the gradient of the loading area, and available traction. If the skid steer works too close to the truck, it is necessary to pause during each cycle for the bucket to clear the side of the truck. If the machine works too far from the truck, the cycle is excessively long, with a resulting waste of time. The operator, by experience, must determine the most efficient arrangement for the particular operation and direct the trucks accordingly.

While there are many ways to maneuver a skid steer, the two most common patterns for a truck loading operation are the I-pattern and the Y-pattern.

For the I-pattern, both the skid steer and the dump truck move in only a straight line, backward and forward (*Figure 54*). This is a good method for small, cramped areas. The operator fills the bucket and backs approximately 20 feet away from the pile. The dump truck backs up between the skid steer and the pile. The operator dumps the bucket into the truck. The truck moves out of the way and the cycle repeats. This pattern requires constant truck movement, however, and is not generally needed with the quick and compact movements of the skid steer.

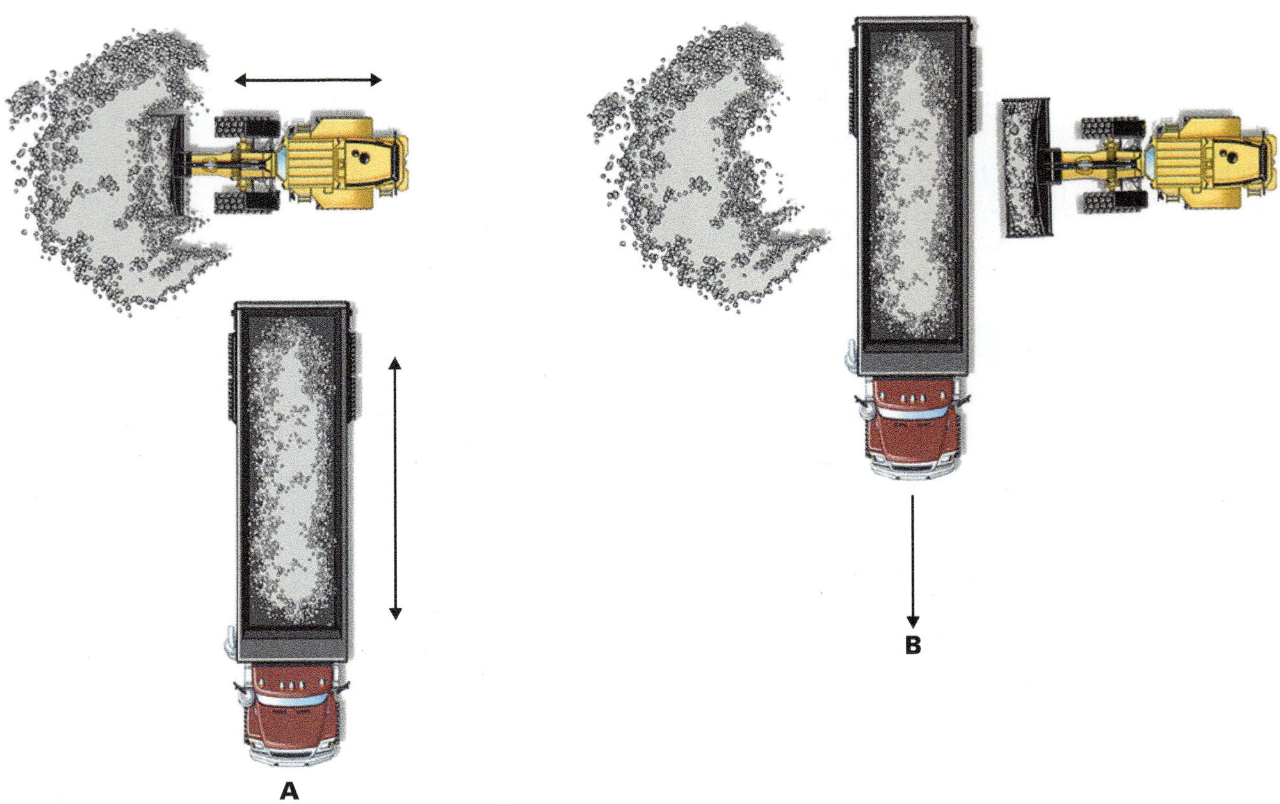

Figure 54 The I-pattern for loading.

To perform the I-pattern loading maneuver, position the skid steer so that it is on the truck driver's side of the truck. That way, eye contact can be made with the driver. Fill the bucket and back far enough away from the pile to allow the truck to back into position. Signal the truck driver with the horn. The truck backs up to a predetermined position, as shown in *Figure 54* (B). Move the machine forward and center it on the truck bed. Raise the bucket to clear the side of the truck and place it over the truck bed. Dump the bucket. When the bucket is empty, move the tilt control to shake out the last of the material. Back the skid steer away from the truck and signal the truck driver to move. When the truck is out of the way, lower the bucket and position the machine to return for another bucket of material.

The other loading pattern is the Y-pattern (*Figure 55*). This method is used when larger open areas are available. The dump truck remains stationary and as close as possible to the pile. The skid steer does all the moving in a Y-shaped pattern.

Figure 55 The Y-pattern for loading.

Position the truck so that eye contact can be made with the driver. Fill the bucket with material. While backing up, turn the skid steer to the right or left, depending on the position of the truck. Shift forward and turn the machine while approaching the truck slowly. Stop when the skid steer is lined up with the truck bed. Dump the bucket in the same way done for the I-pattern. When the bucket is empty, back away from the truck while turning toward the pile. Drive forward into the pile to repeat the pattern. Repeat the cycle until the truck is full.

3.4.3 Leveling and Grading

Minor leveling and grading operations can be performed with the skid steer under most conditions. A multipurpose bucket is well suited for grading. Leveling can be done by tilting the bucket down and placing the cutting edge on the ground surface. Backing up with the bucket in this position will smooth out loose material.

For grading operations, the bottom of the bucket should be parallel to the ground surface (*Figure 56*). While maintaining this position, material from high spots will be scraped into the bucket.

Figure 56 Grading with a skid steer loader.
Source: Courtesy of Doosan Bobcat North America, Inc. All Rights Reserved.

To perform a grading operation with the skid steer, follow these steps:

Step 1 Line up the bucket and skid steer on the area to be graded.

Step 2 Move the joystick to position the bucket at the desired height, with the bottom level.

Step 3 Begin moving forward through the area.

Step 4 Maintain a low steady speed and keep the bucket at a constant height. As high spots are encountered, they will be trimmed by the blade and loaded into the bucket.

Step 5 When low spots are encountered, tilt the bucket forward and dump some of the material. Back up over the dumped material and smooth it out with the back of the blade, then return to grading while moving forward.

Step 6 Reposition the machine and make another pass to the left or right of the area just graded.

3.4.4 Fork Attachments

Use a fork attachment with pallet-style forks (*Figure 57*) to move palletized materials. Forks come in several different sizes depending on the lifting weight and reach required. Operating the compact loader as a forklift is basically the same procedure for all models.

Figure 57 Fork attachment.
Source: Courtesy of Doosan Bobcat North America, Inc. All Rights Reserved.

The following steps are used to unload material from a flatbed:

Step 1 Position the compact loader at either side of the truck bed.

Step 2 Position the forks in order to obtain the appropriate fork height and angle.

Step 3 Drive the forks into the opening of the pallet or under the loose material. Use care not to damage any material.

Step 4 Move the controls to lift the material slightly off the bed.

Step 5 Tilt the forks back slightly to keep the pallet or other material from sliding off the front of the forks (*Figure 58*).

Figure 58 Lifting the load.
Source: Courtesy of Doosan Bobcat North America, Inc. All Rights Reserved.

Step 6 Use the joystick to back the machine away from the truck.

Step 7 Lower the forks to the travel position and travel to the stockpile area.

Step 8 Position the skid steer so that the material can be placed in the desired area.

Step 9 Lower the lifting arm slowly until the material is set on the required surface.

Step 10 Adjust the forks to relieve the pressure under the pallet.

Step 11 Back the machine away from the pallet.

Step 12 Repeat the cycle until the truck is unloaded.

Safe operation of the skid steer as a forklift requires extreme caution. Make sure to follow these safety requirements:

- Do not swing loads over the heads of workers. Make sure that there is enough clear area to maneuver.
- Do not allow workers to ride on the forks.
- The forks should always be lowered to the ground when the skid steer is parked.
- Using the fork attachment usually requires operation in a confined area. Always know what clearance is available for maneuvering.

3.4.5 Other Attachments

The skid steer can accommodate many different attachments. Each one, even those that do the same job but are different in brand or model, has different operating characteristics. For many attachments, the skid steer is merely a carrier. As far as the skid steer is concerned, the operator need only worry about the capacity of the skid steer to carry and maneuver the attachment into position. Operating the attachment, however, may be more complex. It is essential that the operator reads and understands the operating requirements and safety guidelines of the attachment before beginning work with one. Test the controls in a safe setting before beginning, to ensure that the attachment can be operated safely and efficiently.

3.4.6 Transporting a Skid Steer

Transporting a skid steer is a very common task. It is not unusual for a skid steer to be on a different jobsite each day.

The skid steer should be transported on a properly equipped trailer or other transport vehicle. Before beginning to load the equipment for transport, make sure the following tasks have been completed:

- Check the operator's manual of the transportation vehicle or trailer to determine if the loaded equipment complies with height, width, and weight limitations for over-the-road hauling.
- Check the operator's manual to identify the correct tie-down points on the equipment.
- Plan the loading operation so that the loading angle is at a minimum.

Once these tasks are complete and the loading plan has been determined, carry out the following procedures:

Step 1 Position the trailer or transporting vehicle. Always block the wheels of the transporter after it is in position, before loading is started.

Step 2 Place the bucket in the travel position and carefully back the skid steer onto the transporter. Remember that the heavy end of the skid steer should always be uphill. With the skid steer unloaded, the heavy end is always at the rear.

Step 3 Lower the bucket to the floor of the transporter.

Step 4 Engage the parking brake, shut down the engine, and remove the ignition key.

Step 5 Lock the door to the cab as well as any access covers. Attach any vandalism protection.

Step 6 Secure the machine with the proper tie-down equipment as specified by the manufacturer (*Figure 59*). Chocks are recommended at the front and back of all four tires and may be required. If so, make sure they are secure and cannot leave the trailer while the trip is underway. Use all four tie-down points for the machine as documented by the manufacturer.

Figure 59 Front tie-down points.

Step 7 Cover the exhaust and air intake openings with tape or a plastic cover.

Step 8 Place appropriate flags or markers on the equipment if needed for height and width restrictions.

3.0.0 Section Review

1. Fatalities related to skid steers most often _____.
   a. occur during a rollover
   b. involve the operators themselves
   c. involve workers in the immediate area
   d. occur when changing attachments

2. When do the instrument lights typically light up for a brief test?
   a. When the key is first turned to the On position
   b. When the key is first turned to the Start position
   c. When one of the instruments requires maintenance
   d. When the key is turned to the Off position for shutdown

3. Skid steers can be stopped by _____.
   a. using the service brake pedal
   b. using the service brake lever
   c. using the trigger switch on a joystick
   d. letting go of the joystick that controls machine movement

4. What would the rated operating capacity of a skid steer be if it has a tipping load of 2,480 pounds (1,125 kg)?
   a. 1,240 pounds (562.5 kg)
   b. 1,637 pounds (742.5 kg)
   c. 2,180 pounds (988.8 kg)
   d. 3,720 pounds (1,687.4 kg)

Module 22212 Review Questions

1. One feature that makes the skid steer more compact is the absence of _____.
   a. a radiator
   b. a transmission
   c. hydraulic pumps
   d. a hydraulic fluid reservoir

2. True or False: A skid steer engine has *no* direct relationship to the drive wheels.
   a. True
   b. False

3. The hydraulic pumps of a skid steer are typically located _____.
   a. at each drive wheel
   b. under the operator's seat
   c. in the engine compartment
   d. under the chassis, with one on each side

4. ROPS systems must be able to withstand an impact of _____.
   a. the weight of the machine
   b. twice the weight of the machine
   c. three times the weight of the machine
   d. four times the weight of the machine

5. With a fixed displacement hydraulic pump, _____.
   a. the flow rate of fluid is dependent on the engine speed
   b. a relief valve to allow continuous fluid flow is not required
   c. the flow rate of the fluid is controlled by valves inside the pump
   d. the amount of energy transferred through the fluid is greatly reduced

6. When the H-pattern of control is selected in a skid steer, the _____.
   a. left joystick controls the machine movement
   b. right joystick controls the machine movement
   c. joysticks only control the coupler and attachments while the skid steer sits still
   d. left joystick controls the left wheels while the right joystick controls the right wheels

7. True or False: It is common for skid steers to have both a foot- and hand-operated throttle.
   a. True
   b. False

8. Compared to pallet forks, bale forks are _____.
   a. more slender
   b. much heavier
   c. wider and longer
   d. vertical instead of horizontal

9. How would cold weather affect a skid steer lubrication schedule?
   a. Points requiring grease must be serviced half as often.
   b. Points requiring grease must be serviced twice as often.
   c. It would be performed in the morning, when the grease is coldest and thickest.
   d. It would be performed at the end of the day, when the grease and grease points are warmer.

10. Assuming standard five-day workweeks with normal working hours, what is the interval for a 500-hour inspection?
    a. Monthly
    b. Bimonthly
    c. Quarterly
    d. Semiannually

11. Which of the following items can be done to the ROPS by an operator?
    a. Drill a small hole and mount a cup holder.
    b. Weld a small bracket on it to mount something.
    c. Cut a small opening in one spot to improve visibility.
    d. Attach a copy of the inspection checklist using a spring clamp.

12. At shutdown, the skid steer should be allowed to continue operating to cool the fluids for approximately _____.
    a. 1 minute
    b. 2 minutes
    c. 5 minutes
    d. 15 minutes

13. True or False: When the skid steer is running, it will coast down almost any incline unless the parking brake is used.
    a. True
    b. False

14. When a single joystick is used to control skid steer travel, pushing the joystick forward and to the right will result in _____.
    a. the machine pivoting in place to the left
    b. the machine pivoting in place to the right
    c. a gradual turn to the left while moving forward
    d. a gradual turn to the right while moving forward

15. The operator can help the contents of the bucket move out when dumping by _____.
    a. bumping the tilt control back and forth
    b. bumping the machine forward and backward
    c. gently nudging the side of the truck tires a few times
    d. running the engine at idle speed when the bucket is tilted forward

Answers to odd-numbered review questions are found in the Module Review Answer Key at the back of this book.

Section Review Answer Key

Answer	Section	Objective
Section One		
1. b	1.1.1	1a
2. c	1.2.0	1b
3. b	1.3.0	1c
4. d	1.4.0	1d
Section Two		
1. b	2.1.1	2a
2. c	2.2.0	2b
Section Three		
1. c	3.1.2	3a
2. a	3.2.2	3b
3. d	3.3.0	3c
4. a	3.4.1	3d

APPENDIX 22102A

SDS for Solvent Cement

XXXXX Regular-Bleach₁ **Revision Date** June 12, 2023

7. HANDLING AND STORAGE

Precautions for safe handling

Handling Handle in accordance with good industrial hygiene and safety practice. Avoid contact with skin, eyes, and clothing. Do not eat, drink, or smoke when using this product.

Conditions for safe storage, including any incompatibilities

Storage Store away from children. Reclose cap tightly after each use. Store this product upright in a cool, dry area, away from direct sunlight and heat to avoid deterioration. Do not contaminate food or feed by storage of this product.

Incompatible Products Toilet bowl cleaners, rust removers, acids, and products containing ammonia.

8. EXPOSURE CONTROLS/PERSONAL PROTECTION

Control parameters

Exposure Guidelines

Chemical Name	ACGIH TLV	OSHA PEL	NIOSH IDLH
Sodium hypochlorite 7681-52-9	None	None	None

ACGIH TLV: American Conference of Governmental Industrial Hygienists - Threshold Limit Value. OSHA PEL: Occupational Safety and Health Administration - Permissible Exposure Limits. NIOSH IDLH: Immediately Dangerous to Life or Health.

Appropriate engineering controls

Engineering Measures
Showers
Eyewash stations
Ventilation systems

Individual protection measures, such as personal protective equipment

Eye/Face Protection If splashes are likely to occur: Wear safety glasses with side shields (or goggles) or face shield.

Skin and Body Protection Wear rubber or neoprene gloves and protective clothing such as long-sleeved shirt.

Respiratory Protection If irritation is experienced, NIOSH/MSHA approved respiratory protection should be worn. Positive-pressure supplied air respirators may be required for high airborne contaminant concentrations. Respiratory protection must be provided in accordance with current local regulations.

Hygiene Measures Handle in accordance with good industrial hygiene and safety practice. Wash hands after direct contact. Do not wear product-contaminated clothing for prolonged periods. Remove and wash contaminated clothing before re-use. Do not eat, drink, or smoke when using this product.

Source: U.S. Department of Labor/OSHA

Module Review Answer Key

MODULE 01 (22101)

Answer	Section
1. b	1.0.0
3. a	1.1.0
5. d	1.2.0
7. b	1.3.3
9. c	2.1.7

MODULE 02 (22103)

Answer	Section
1. b	1.1.1
3. d	1.1.3
5. a	1.1.5
7. b	1.1.7
9. d	1.2.1
11. a	2.2.2

MODULE 03 (22102)

Answer	Section
1. b	1.1.1
3. c	1.1.4
5. c	1.2.1
7. a	1.2.1
9. d	2.1.0
11. c	2.1.8
13. b	3.2.4
15. c	4.2.1

MODULE 04 (22108)

Answer	Section
1. c	1.4.1
3. a	2.1.2
5. b	2.2.1
7. d	3.1.0
9. a	4.1.0

MODULE 05 (22201)

Answer	Section
1. c	1.1.0
3. a	1.2.0
5. c	1.5.3
7. a	1.6.2
9. a	1.7.2
11. c	2.0.0; *Table 1*
13. b	2.1.8
15. c	2.2.2
17. d	2.5.1
19. b	3.2.0

MODULE 06 (22109)

Answer	Section
1. b	1.2.1
3. a	1.2.2
5. b	1.2.4
7. b	3.1.0

MODULE 07 (22105)

Answer	Section
1. c	1.1.1
3. d	1.2.1
5. c	1.3.2
7. b	1.3.10
9. d	1.4.4
11. a	2.1.2
13. d	2.3.2
15. b	2.4.2

MODULE 08 (22212)

Answer	Section
1. b	1.1.0
3. b	1.1.3
5. a	1.1.4
7. a	1.2.1
9. d	2.1.1
11. d	3.1.2
13. b	3.3.0
15. a	3.4.2

GLOSSARY

Aggregate: Manufactured or natural rock or sand that has a specific size characteristic. Used in foundations and subgrade work.

Articulated: Two parts connected by a joint so as to move independently.

Backhauling: The return trip of a piece of equipment after it has completed dumping its load.

Backslope: The area between the bottom of a ditch adjacent to a roadbed and the top of the cut into the existing natural grade.

Bead: The rigid inner ring of an inflatable tire that fits against the rim and creates an air-tight seal when the tire is inflated.

Bedrock: The solid layer of rock under Earth's surface. Solid rock, as distinguished from boulders.

Benchmarks: Permanent points of known or assumed elevation from which surveyors can establish grades.

Bucket: A U-shaped, closed-end scoop attached to a front-end loader or backhoe.

Centerline: The line that marks the center of a roadway or other construction project. This is marked on the plans by a line and on the ground by stakes.

Center of gravity: The point where all of an object's weight is evenly distributed.

Channeling devices: Cones, barrels, or markers that are used to move traffic from its normal lane to a desired lane.

Civil information modeling (CIM): The process of using data-rich objects for design of not only buildings but also civil structure and site components.

Cohesive: The property of a material that sticks or clumps to itself.

Consolidation: To become firm by compacting soil particles so they are closer together.

Core samples: Samples of earth taken from a cylindrical test boring.

Cross slope: The pitch or grade of a road's surface perpendicular to its centerline that helps drain water from the road as well as improve vehicle stability on curves.

Crystalline: Containing a mixture of closely adjoining mineral crystals. If the rock does not contain visible crystals, then it is non-crystalline.

Cycle time: The time it takes for a piece of equipment to complete a repetitive operation. This normally would include loading, hauling, dumping, and then returning to the starting point.

Demolition: The destruction and removal of a structure.

Dewatering: Removing standing water from a low area using a drain or pump.

Diesel exhaust fluid (DEF): Urea fluid that promotes the conversion of oxides of nitrogen into harmless gases.

Diesel particulate filter (DPF): A filter that catches particulate matter and burns it up before it exits the engine into the atmosphere.

Direct current (DC): An electrical current that flows in only one direction.

Dogged: Held or secured from moving with a mechanical device.

Draft: In the heavy construction equipment trade, refers to the load-pulling capacity of a vehicle.

Draglines: Crane-like excavating machines configured to drop a bucket from a boom and drag it toward the machine by a cable.

Electrically powered equipment (EPE): Types of equipment that are solely battery powered.

Electrolyte: A solution of sulfuric acid and water, which is highly poisonous and can cause severe chemical burns.

Embankment: Material piled in a uniform manner so as to build up the elevation of an area. Usually, the material is placed in long narrow strips.

Expansive soil: A clay-like soil that swells with an increase in moisture content and shrinks as it dries out.

Fixed displacement pump: A pump that cannot be adjusted to deliver more or less fluid volume in a pumping cycle.

Flaggers: Workers who are specially trained to direct traffic through and around a work zone.

Float mode: Placing the bucket or other attachment into the float mode allows the attachment to automatically follow the contour of the ground, remaining at the same elevation above it.

Foreslope: The area of roadway between the shoulder/hinge point and the bottom of the adjacent ditch or natural grade.

Geospatial information system (GIS): Computerized tools related to the physical mapping of data within a visual reference.

GLONASS: A Russian-owned Global Navigation Satellite System operated by the Ministry of Defense of the Russian Federation.

Glow plugs: Heating elements used to preheat the combustion chamber and aid in igniting fuel in a cold engine.

GNSS: An acronym for Global Navigation Satellite System. GNSS is the generic term used to describe a locating or grade control system that uses signals from GPS, GLONASS, BeiDou, or Japanese QZSS satellites.

Gradation: The classification of soils into different particle sizes.

Grade: The surface of a road, channel, or natural ground area. In civil engineering, the term usually refers to the surface level required by the plans or specifications.

Grading: Establishing the proper level or slope of the ground.

Groundwater: Water that resides and moves beneath the surface of the ground.

Haul road: A compacted dirt road used to move material and equipment on and off the site.

Hazard communication (HazCom): An OSHA standard that requires contractors to educate employees about working with or near hazardous chemicals.

Hybrid electric equipment (HEE): A type of application that has an internal combustion engine coupled with a power generation unit.

Hydraulic actuator: A device that converts the fluid energy of the hydraulic system into mechanical energy using a hydraulic cylinder or motor.

Hydrostatic drive: A system that relies on the flow and pressure of a fluid to transfer energy. In skid steers, the transfer of energy is from the engine, driven by diesel fuel, to the four drive wheels through hydraulic motors.

Hydrostatic: Relating to liquids at rest, or to the pressures they transmit, as in hydraulically powered controls or transmissions that stay in a neutral condition until the hydraulic fluid in them is put into motion or pressurized.

Igneous: Identifies earth materials that were formed from formerly molten (melted) rocks.

Impermeable: The property of a material that does not allow movement of a fluid through it; for example, an impermeable soil will not allow liquid water to pass through it.

Infrastructure: A system of public works such as highways, bridges, tunnels, and dams.

Inorganic: When considering earth materials, any substance that does not come from living organisms.

Integrity: Soundness of character or commitment to moral and ethical principles.

Lidar: A type of sensor (LIght Detection And Ranging) that sends out laser beam signals and then measures when they bounce back to the sensor. Lidar can generate a location image of those objects.

Lifts: When referring to soil compaction during construction, layers of soil that are placed in a fill area or on an embankment before compaction.

Loams: Soils containing clay, sand, and organic matter.

Lost-time injuries: Injuries sustained by employees at work that result in absenteeism or delays in the employees' normal workload.

Mine Safety and Health Administration (MSHA): A federal government agency established to ensure the safety of mining operations.

Occupational Safety and Health Administration (OSHA): The federal government agency established to ensure a safe and healthy environment in the workplace.

On-the-job learning (OJL): Job-related learning acquired while working.

Organic: When considering earth materials, any substance found in living organisms (plants, animals, or microorganisms) or produced by them.

Pascal's law: A scientific law stating that pressure at any one point in a closed static fluid system will be applied to all points in the system in every direction and will exert equal forces on equal areas. It is the basis of hydraulic theory and was developed by Blaise Pascal (1623–1662), a French philosopher and mathematician.

Pay item: A defined piece of material or work that the contractor is paid for. Pay items are usually expressed as unit costs.

Pinch points: The area in which two moving equipment parts come together.

Pit: An open excavation of some depth that usually does not require vertical shoring or bracing.

Plasticity index: The range of percent water content in which a soil displays plastic or pliable properties. Soils with high plasticity indexes are usually clays; soils with lower indexes are usually silt, sand, or gravel.

Pneumatic: Inflated with compressed air.

Policies: General statements establishing guidelines for a specific activity.

Polymers: Natural and synthetic materials made of long molecules containing many repeating chemical subunits. Familiar polymers include plastics and natural rubber. Soil stabilization polymers include acrylic materials.

Power takeoff (PTO): Rotating mechanical output device on an engine or transmission to which cables, belts, or shafts can be connected to drive attachments or tools.

Procedures: Instructions on how to carry out a policy.

Productivity: The rate at which work is accomplished.

Radar: A type of sensor (RAdio Detection And Ranging) that sends out radio waves and then measures how long it takes them to bounce back to the receiver to determine the distance of the objects in its viewpoint.

Rated operating capacity: The amount of weight that a skid steer is projected to handle safely through common maneuvers. The rated operating capacity is typically derived from the tipping load test and is equal to 50 percent of the tipping load value.

Retroreflective: A device or surface that reflects light with very little scattering.

Ripper: A towed attachment with teeth used on dozers, motor graders, and other machines to loosen heavily compacted soil and soft rock.

Riprap: Broken stone, in pieces weighing from 15 to 150 lb (6.8 to 68 kilograms) each, placed on the ground for protection against the action of water.

Riprap: Loose pieces of rock or manufactured materials placed on the slope of an embankment in order to stabilize the soil from the effects of erosional processes.

Robotic total station: A machine control system that uses infrared signals reflected by a target mounted on the grading equipment. This provides data to the machine's onboard computer about how to adjust the blade.

Rockshaft unit: A device that raises and lowers equipment mounted on a three-point hitch.

Rollover protective structure (ROPS): A metal structure attached to the equipment's cab designed to protect the operator should the equipment roll over.

Scarifier: An attachment with teeth used on motor graders to loosen soil.

Seat: To cause two surfaces to fit firmly together.

Sedimentary: Identifies rock materials that were originally deposited by wind or water.

Select material: Earth or manufactured material that meets a predetermined specification as to some physical property such as size, shape, or hardness.

Selective catalytic reduction (SCR) system: A system that allows the introduction of urea fluid into the exhaust stream to react with the NO_x present in the exhaust emissions.

Shear strength: A property of solid materials that measures its resistance to deforming along planes, called shearing, like spreading a deck of cards. In soil mechanics, it is measured by the angle of internal friction, which relates to the maximum steepness of a soil embankment before it slides due to gravity.

Shielding: The use of structures that are inserted into a trench to protect workers.

Shoring: A support system designed to prevent the trench walls from collapsing.

Silt: Soil deposited by running water.

Slope stakes: Surveyor's stakes placed at the intersection of a cut or fill slope with the natural ground. Instructions include the slope ratio to meet the hinge point of a roadway or other reference point. Also given the terms *cut* or *fill stakes*.

Slope: Ground that forms a natural or artificial incline upward or downward.

Sloping: The angling of trench walls in order to reduce the risk of trench collapse.

Soil testing: Any mechanical or electronic test used to determine the density, moisture content, and other properties of the soil that determines the amount of compaction required.

Sonic tracer system: Equipment that uses ultrasonic signals to maintain a set distance or elevation from an object, a design surface, or the ground.

Splines: Parallel grooves running lengthwise on a shaft that engage matching grooves on a part that slides onto the shaft.

Spoils: Excavated materials reserved for future use or unsuitable for construction that have been stockpiled.

Stakes: Markers that are used to mark cut, fill, or grade information; underground utilities; or other survey points.

Stormwater: Water directly resulting from rain or snow flowing over the ground's surface, or in natural or man-made water courses.

Temporary traffic control (TTC): Device and plan used to safely divert traffic when the normal use of the road is disrupted due to construction.

Test boring: A drilled or excavated hole created to take samples of the earth materials from different layers beneath the surface.

Test pit: Similar in purpose to a test boring but is a larger opening in the earth that provides a better understanding of the underlying ground structure and soil characteristics.

Tipping load: The maximum load that a skid steer can hold in the air, with the coupler outstretched, before the rear wheels begin to leave the ground.

Torque converter: A device that uses fluid, usually oil, to transmit torque from one shaft to another. Also known as a *fluid coupler*.

Toxic: The degree to which a substance can harm humans or animals.

Trench: A long, narrow, temporary excavation that workers fill in when work is completed.

Turbocharger: Exhaust-driven centrifugal air compressor used to increase air flow to an engine, which improves combustion efficiency and increases power.

Ultrasonic sensors: A type of sensor that projects sound waves, then measures the time it takes for them to bounce off the target and return to determine the location of the target. Ultrasonic sensors work well over short distances (1'–5').

Universal total station (UTS): A combination of a digital theodolite and an electronic distance measuring device that can integrate with a GNSS module to provide a point of reference for 3D site modeling.

Variable displacement pump: A pump that can deliver a variable fluid volume in a pumping cycle to support the changing needs of a hydraulically operated device.

Virtual reference station (VRS): An imaginary reference station that is established by a computer based on data from numerous GPS receivers and the grading equipment that serves as the source of data for grading control.

Water table: The depth below the ground's surface at which the soil is saturated with groundwater.

Wheel slip: The moving of wheels without transferring the power to the ground and propelling the equipment forward.

Windrows: Long, straight piles of placed material for the purpose of mixing or scraping up.

REFERENCES

Module 1 22101 Introduction to Heavy Equipment

Caterpillar: Modern Earthmoving Marvels, 2015. Frank Raczon and Keith Haddock. Beverly, MA: Motorbooks.

The Earthmover Encyclopedia, 2007. Keith Haddock. Beverly, MA: Motorbooks.

Heavy Equipment Maintenance Log Book, 2021. Samuel Akang. Independent publisher.

Heavy Equipment Operator Word Search Book: A Heavy Equipment Themed Puzzle Book, 2023. Rongh Studio. Independent publisher.

Module 2 22103 Classification of Heavy Equipment

Caterpillar: Modern Earthmoving Marvels, 2015. Frank Raczon and Keith Haddock. Beverly, MA: Motorbooks.

The Earthmover Encyclopedia, 2007. Keith Haddock. Beverly, MA: Motorbooks.

Fundamentals of Mobile Heavy Equipment, 2017. Gus Wright, Owen C. Duffy, and Scott A. Heard. Burlington, MA: Jones & Bartlett Learning.

Module 3 22102 Heavy Equipment Operations Safety

Handbook of OSHA Construction Safety and Health. Second Edition. Charles D. Reece and James V. Eidson. Boca Raton, FL: CRC Press.

Manual on Uniform Traffic Control Devices for Streets and Highways. Washington, DC: US Department of Transportation, Federal Highway Administration.

MSHA excavation guidelines. *www.msha.gov*.

The National Institute for Occupational Safety and Health (NIOSH) Centers for Disease Control and Prevention. Atlanta, GA. Available online at *www.cdc.gov*.

OSHA Standards 29 CFR, Part 1926 Safety and Health Regulations for the Construction Industry. Washington, DC: Occupational Safety and Health Administration (OSHA), US Department of Labor, US Government Printing Office.

United States Department of Labor, Occupational Safety and Health Administration Home Page. *www.osha.gov*.

Module 4 22108 Equipment Operation, Inspection, and Readiness

Caterpillar: Modern Earthmoving Marvels, 2015. Frank Raczon. Beverly, MA: Motorbooks.

The Earthmover Encyclopedia, 2007. Keith Haddock. Beverly, MA: Motorbooks.

Troubleshooting and Repairing Diesel Engines. Fifth Edition, 2018. Paul Dempsey. New York: McGraw Hill.

Module 5 22201 Introduction to Earthmoving

Caterpillar: Modern Earthmoving Marvels, 2015. Frank Raczon. Beverly, MA: Motorbooks.

The Earthmover Encyclopedia, 2007. Keith Haddock. Beverly, MA: Motorbooks.

Troubleshooting and Repairing Diesel Engines. Fifth Edition, 2018. Paul Dempsey. New York: McGraw Hill.

Module 6 22109 Introduction to GPS/GNSS

Arresting Fugitive Dust, October 1, 2021. Occupational Health and Safety website. *https://ohsonline.com/Articles/2021/10/01/Arresting-Fugitive-Dusts.aspx.*

Caterpillar Performance Handbook, Latest Edition. A CAT® Publication. Peoria, IL: Caterpillar, Inc.

National Asphalt Pavement Association (NAPA) website *www.asphaltpavement.org*.

Understanding GPS/GNSS Principles and Applications, 2017. Elliott D. Kaplan and Christopher J Hegarty. Boston, MA: Artech House Publishers.

Using Lime for Soil Stabilization and Modification, 2001. National Lime Association. www.lime.org.